COLLISIONAL LINE BROADENING AND SHIFTING OF ATMOSPHERIC GASES

A Practical Guide for Line Shape Modelling by Current Semi-classical Approaches

COLLISIONAL LINE BROADENING AND SHIFTING OF ATMOSPHERIC GASES

A Practical Guide for Line Shape Modelling by Current Semi-classical Approaches

Jeanna Buldyreva

University of Franche-Comte, France

Nina Lavrentieva

Tomsk Institute of Atmospheric Optics, Russia

Vitaly Starikov

Tomsk State University of Control Systems and Radio-Electronics, Russia

World Scientific

NEW JERSEY • LONDON • SINGAPORE • BEIJING • SHANGHAI • HONG KONG • TAIPEI • CHENNAI

Published by

Imperial College Press
57 Shelton Street
Covent Garden
London WC2H 9HE

Distributed by

World Scientific Publishing Co. Pte. Ltd.
5 Toh Tuck Link, Singapore 596224
USA office: 27 Warren Street, Suite 401-402, Hackensack, NJ 07601
UK office: 57 Shelton Street, Covent Garden, London WC2H 9HE

British Library Cataloguing-in-Publication Data
A catalogue record for this book is available from the British Library.

COLLISIONAL LINE BROADENING AND SHIFTING OF ATMOSPHERIC GASES
A Practical Guide for Line Shape Modelling by Current Semi-classical Approaches

ISBN-13 978-1-84816-596-0
ISBN-10 1-84816-596-X

Printed in Singapore.

To our parents,
who saw the black sky of the war
and dreamt that it became blue forever

Preface

The main goal of the present book is to review current semi-classical approaches for the calculation of line broadening and line shifting coefficients due to the pressure of atmospheric gases. These line shape parameters, together with the line positions and intensities, represent the basic characteristics of isolated spectral lines and are required for remote sensing of the Earth's and planetary atmospheres. They are determined by the individual properties of the absorbing active molecule as well as by its interaction with the surrounding medium (pressure, temperature, perturbers). The line broadening and line shifting coefficients have a fundamental character since they are directly related to the intermolecular interaction potential and in some cases are the single source of information about it.

However, no universal theoretical method exists to calculate these coefficients. Even within the semi-classical framework, different authors use various approximations for the scattering matrix, intermolecular interaction potential and relative trajectory of colliding particles. A similar situation takes place for the experimental determination of these coefficients. Their experimental values are deduced by fitting a theoretical (model) profile to the recorded shape of an isolated spectral line. Since there is no general profile able to cover a quite large pressure interval, various models are used, and the experimental data published for the same spectral line can differ significantly. A comprehensive review is therefore necessary not only for the theoretical but also for the experimental determination of line broadening and line shifting coefficients. It is made systematically for all considered molecular systems.

The second goal of the book is to provide the reader with exploitable theoretical expressions for the line broadening and line shifting coefficients. These formulae enable the writing of computer codes and calculations without consulting the original works. Moreover, for many particularly important vibrational bands, simple semi-empirical expressions for the line broadening coefficients are also given.

The first chapter contains fundamental definitions and a brief summary of main line shape models currently used in the spectroscopic literature. It gives the basis required for understanding the following chapters.

The second chapter reviews the modern semi-classical methods for calculating pressure broadening and shifting coefficients. Three trajectory models for the relative translational motion of colliding molecules are discussed: straight-line, parabolic and exact trajectories. For the last model, the expression for the line width is given for the most general case of asymmetric top molecules interacting by both long-range and short-range forces. For the long-range interactions, approximate analytical formulae are given for the real and imaginary parts of the resonance functions involved in the calculation of line widths and shifts.

The third chapter is devoted to the water vapour molecule. Experimental and theoretical studies of H_2O line broadening and shifting coefficients by pressure of various atmospheric gases are discussed, in particular, the semi-empirical method which ensures an accurate prediction of these coefficients based on a few adjustable parameters fitted on some experimental data. Analytical relations are given to calculate the self-, N_2-, O_2- and CO_2-broadening coefficients at high temperatures. The influence of the vibrotational coupling, vibrational excitation and accidental resonances on the line shape parameters is also analysed.

The fourth chapter gathers the results obtained for the line broadening and line shifting coefficients of other atmospheric molecules: asymmetric tops H_2S, SO_2, O_3, NO_2 of X_2Y type and asymmetric top C_2H_4, symmetric tops NH_3, CH_3Cl, CH_3F, CH_3D, spherical top CH_4 as well as linear CO_2, N_2O, C_2H_2 and diatomic CO, NO, OH, HF, HCl molecules. In the cases where analytical expressions for these coefficients are not available, their numerical values are given in the tables.

The appendices provide theoretical expressions and parameters required for line width and line shift computations. A table is also given for checking resonance functions obtained with exact trajectories.

The book is especially valuable as a concise practical guide for any specialist in the fields of atmospheric spectroscopy, physical chemistry and molecular physics. It can also be very useful for undergraduate, graduate and PhD students interested in spectral line shape theory.

The authors acknowledge the scientific cooperation and valuable discussions with Drs A.D. Bykov, O.V. Naumenko, S.N. Michailenko, A.E. Protasevich, L. Nguyen and Prof. L.N. Sinitsa.

J. Buldyreva, N. Lavrentieva, V. Starikov

Contents

Chapter 1

Basic definitions

1.1. Shape and parameters of a spectral line

A weak monochromatic radiation of wave number ν (cm^{-1}) and incident intensity I_0 (cm^{-2}) having propagated in a homogeneous medium through a distance L (cm) changes intensity I according to the Beer–Lambert law [1]:

$$I(\nu)=I_0(\nu)\,\mathrm{e}^{-K(\nu)L}. \tag{1.1.1}$$

The product $K(\nu)L$ in the exponent is called the spectral absorbance and $K(\nu)$ is referred to as the spectral absorption coefficient. This latter can be expressed as

$$K(\nu)=\frac{1}{L}\ln\left[\frac{I_0(\nu)}{I(\nu)}\right] \tag{1.1.2}$$

and easily determined experimentally.

When monochromatic radiation passes through a gas sample and the wave number ν is close to the wave number ν_0 corresponding to the transition between two (vib)rotational molecular levels i and f, this radiation is strongly absorbed by the gas molecules which change their quantum state from i to f. The high-resolution molecular absorption spectra therefore represent a set of lines associated with all possible (vib)rotational transitions in the absorbing (optically active) molecule.

For an isolated spectral line the absorption coefficient $K(\nu)$ can be written in terms of the normalised to unity line shape function $F(\nu)$ centred at ν_0:

$$K(\nu)=AF(\nu-\nu_0), \tag{1.1.3}$$

where

$$\int_{-\infty}^{\infty} F(\nu-\nu_0)\,d\nu=1. \tag{1.1.4}$$

It follows from Eq. (1.1.4) that the normalisation factor A in Eq. (1.1.3) represents the integral line absorption:

$$A = \int_{-\infty}^{\infty} K(\nu) d\nu . \qquad (1.1.5)$$

In practice, the integral absorption refers to the number of molecules per unit volume: $A_N = A / N$ (cm molecule^{-1}) or to the pressure: $A_P = A / P$ (cm^{-2} atm^{-1}) which are related to each other by

$$A_P = 2.6868 \cdot 10^{19} \frac{T_0}{T} A_N \qquad (1.1.6)$$

with the reference temperature $T_0 = 273.15$ K [2].

For any line shape $F(\nu - \nu_0)$ two main characteristics are the half-width at half-maximum (HWHM) noted typically γ and the shift of the line centre from ν_0 marked δ (Fig. 1.1). Sometimes, however, it is the full width at half-maximum (FWHM) which is physically meaningful, for example in the case of non-symmetric line shape.

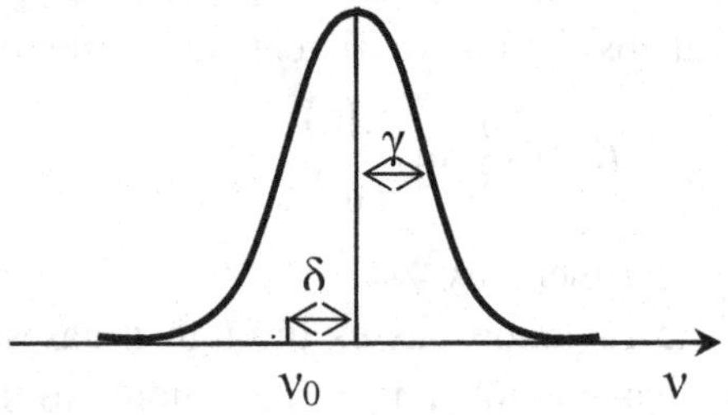

Fig. 1.1. Main characteristics of a spectral line: half-width on a half of maximum γ and shift δ.

1.2. Principle line-broadening mechanisms and model profiles

The evolution of the line shape with changing pressure and temperature is defined by the considered pressure range. The general behaviour of the line width with increasing pressure can be seen from Fig. 1.2.

Even at zero pressure the line width has a finite value γ_D due to thermal motion of the absorbing molecule (Doppler effect). While the gas pressure stays so weak that the mean free path l between two collisions is much greater than the radiation wavelength λ: $2\pi l >> \lambda$, one speaks about *Doppler regime*. With further pressure increasing molecular collisions become quite frequent and induce velocity changes that reduce the apparent molecular velocity and, as a consequence, the Doppler contribution in the line width. This process is widely known in the literature as molecular diffusion or the Dicke effect [3] leading to line narrowing. This intermediate pressure regime is the most complex to model since both principle mechanisms of line broadening take place simultaneously. If the pressure continues to increase, the line width becomes completely dominated by the collisional broadening and varies as a linear function of the pressure (in the binary collision approximation). This region of linear dependence is named *collisional regime*.

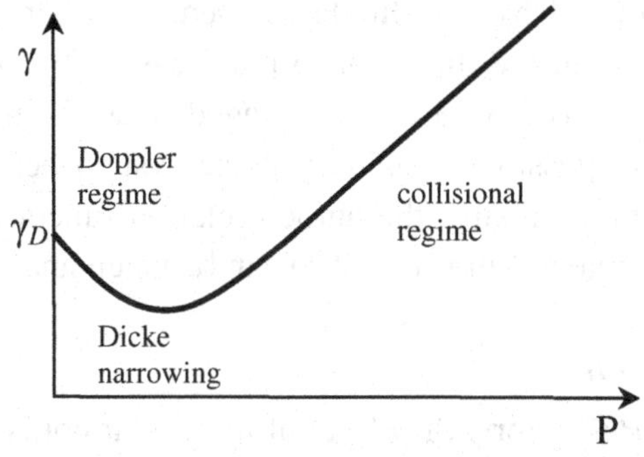

Fig. 1.2. General line-width dependence on pressure and corresponding pressure regimes.

Besides the Doppler effect and collisional broadening the line width is subjected to other mechanisms, such as natural broadening, saturation broadening or broadening by collisions with the cell walls. The always present natural line width is due to the finite lifetime (Heisenberg uncertainty principle) of the excited molecular states but has a negligible value for (vib)rotational molecular transitions situated in the infrared (IR) frequency region. The broadening by saturation takes place when the intensity of the incident radiation is too high so that the gas molecules have no time to evacuate the energy absorbed between two collisions. This phenomenon (more important for molecules with a high value of dipole moment, as HCl for example) is characterised by a distortion of the absorption profile [4] and can be eliminated by using sufficiently weak radiation intensities. The broadening by collisions with cell walls can be significant at small pressures if the cell dimensions are comparable to the mean free path. It can be easily excluded using a quite big cell.

The line-broadening mechanisms have been discussed by many authors (see Refs [5–11]), but it is unfortunately impossible to establish a universal line shape model which could be suitable for any pressure range. Different theoretical profiles are therefore used for different regimes.

In general, any line shape $F(\omega)$ in the frequency domain can be seen as a Fourier transform of some autocorrelation function $\Phi(t)$ in the time domain [5, 6, 12]:

$$F(\omega)=\frac{1}{\pi}\mathrm{Re}\int_0^\infty \Phi(t)\exp(i\omega t)dt, \tag{1.2.1}$$

where the angular frequency ω is related to the wave number ν by $\omega = 2\pi c\nu$ and $\Phi(t)=\langle\vec{\mu}(0)\cdot\vec{\mu}(t)\rangle$ is defined by the dipole moment operator $\vec{\mu}(t)$ coupling the molecule to the radiation in the dipole absorption case. The notation $\langle...\rangle$ means that the scalar product of two vectors is averaged over all initial conditions at $t = 0$ as well as over all possible evolution of the system between 0 and t. For some line-broadening mechanisms, the autocorrelation function $\Phi(t)$ has a quite simple form so that the integral in Eq. (1.2.1) can be taken analytically.

1.2.1. Doppler broadening

At very low pressure ($P \le 1$ Torr) the effect of molecular collisions is completely negligible and the line shape is determined by the kinetic motion of the absorbing molecule in the laboratory fixed frame. This motion modifies the absorbed frequency ω_0 by the Doppler effect so that the observed frequency ω depends additionally on the radiation wave vector $\vec{k}$ and on the molecule velocity $\vec{v}_a$: $\omega = \omega_0 + \vec{k}\cdot\vec{v}_a$. If the electromagnetic wave propagates along the laboratory z-axis, the frequency shift equals kv_{az} and induces during the free molecular motion the phase $kv_{az}t$ in the autocorrelation function [13]:

$$\Phi_D(t)=\langle\exp(-i\omega_0 t+ik\,v_{az}t)\rangle. \tag{1.2.2}$$

Since the molecular velocities are distributed according to the Maxwell–Boltzmann statistics

$$f(\vec{v}_a)d\vec{v}_a=\left(\frac{m_a}{2\pi k_B T}\right)^{3/2}\exp\left(-\frac{m_a v^2}{2k_B T}\right)d\vec{v}_a, \tag{1.2.3}$$

where m_a is the mass of the active molecule, k_B is the Boltzmann constant and T is the absolute temperature, the averaging procedure leads immediately to

$$\Phi_D(t)=\exp\left[-i\omega_0 t-\left(\frac{k\,v_{a0}\,t}{2}\right)^2\right] \tag{1.2.4}$$

with v_{a0} denoting the most probable velocity: $v_{a0} = (2k_BT/m_a)^{1/2}$.

The corresponding line profile calculated with Eq. (1.2.2) reads

$$F_D(\nu) = \sqrt{\frac{\ln 2}{\pi}} \frac{1}{\gamma_D} \exp\left[-\ln 2 \left(\frac{\nu - \nu_0}{\gamma_D} \right)^2 \right], \tag{1.2.5}$$

where the (half-)width γ_D is given by

$$\gamma_D = \sqrt{\frac{2 \ln 2 k_B T}{m_a c^2}} \nu_0 = 3.581 \cdot 10^{-7} \sqrt{\frac{T}{m_a (\mathrm{amu})}} \nu_0 \tag{1.2.6}$$

and in the infrared frequency domain takes typical values of about 10^{-3} cm^{-1} (the natural line width is about 10^{-8} cm^{-1}). The Doppler profile (1.2.5) represents a Gaussian function characterised by a rapid decreasing of intensity in the wings (Fig. 1.3) and in the Doppler regime agrees with the experimental line shapes typically within 1%. Since the velocity varies from one molecule to another this profile is inhomogeneous.

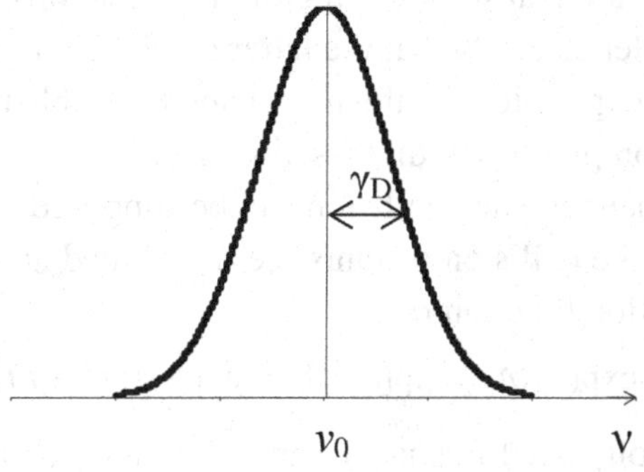

Fig. 1.3. Doppler broadening profile given by a Gaussian function.

Sometimes it is, however, mathematically convenient to work with dimensionless variables and the $1/e$ Doppler half-width denoted by $\Delta\nu'_D$ (cm^{-1}) or $\Delta\omega'_D$ (rad s^{-1}) [14] is commonly used for this purpose; the HWHM and the $1/e$ Doppler width are related by

$$\gamma_D = \sqrt{\ln 2}\Delta\nu'_D = \sqrt{\ln 2}\Delta\omega'_D / (2\pi c). \tag{1.2.7}$$

For any line shape the normalised frequency separation x from the line centre can be introduced as

$$x = (\nu - \nu_0) / \Delta\nu'_D = (\omega - \omega_0) / \Delta\omega'_D, \tag{1.2.8}$$

the standardised line width y as

$$y = \gamma / \Delta \nu'_D = \Gamma / \Delta \omega'_D \tag{1.2.9}$$

and the standardised line-shift s as

$$s = \delta / \Delta \nu'_D = \Delta / \Delta \omega'_D , \tag{1.2.10}$$

where Γ and Δ (collisions $s^{-1} \equiv$ rad s^{-1}) mean respectively the effective frequencies of line-broadening and line-shifting collisions.

In the dimensionless variables, the Doppler profile normalised according to Eq. (1.1.4) reads

$$F_D(\nu) = \sqrt{\frac{\ln 2}{\pi}} \frac{1}{\gamma_D} \exp(-x^2) . \tag{1.2.11}$$

1.2.2. Collisional broadening

At pressures superior to 1 Torr in the millimetre absorption domain or to 100 Torr in the infrared absorption region the line broadening by collisions dominates the Doppler effect so that the latter can be completely neglected (collisional regime). The expression for the line shape is established on the basis of the impact approximation [6, 15, 16] discussed in Sec. 2.2.

In a simplified manner, each collision can be supposed to interrupt completely the radiation. Since the collision instants are distributed according to the Poisson law, the autocorrelation function reads

$$\Phi_c(t) = \exp(-i\omega_0 t) \langle \exp[i\eta(t)] \rangle = \exp[-i\omega_0 t - \mathit{\Gamma} t], \tag{1.2.12}$$

where $\eta(t)$ is the collisional dephasing and $\mathit{\Gamma}$ is the complex-valued matrix of relaxation rates. This matrix contains the effects of averaging over all collision parameters:

$$\mathit{\Gamma} = n_b < \mathrm{v} \sigma > \tag{1.2.13}$$

and is defined by the number density of perturbing particles n_b, the relative velocity of the colliding molecular pair v and the differential scattering cross-section $\sigma = \sigma' + i\sigma''$ for the considered transition of the frequency ω_0. If σ is supposed to be independent from v (relative velocity is replaced by the mean thermal velocity $\bar{\mathrm{v}} = [8k_B T / (\pi m^\circ)]^{1/2}$ where m° is the reduced mass of the molecular pair), the relaxation matrix is given by

$$\mathit{\Gamma} = n_b \bar{\mathrm{v}} \sigma' + i n_b \bar{\mathrm{v}} \sigma'' \equiv \Gamma + i\Delta . \tag{1.2.14}$$

The corresponding line profile

$$F_c(\nu)=\frac{1}{\pi}\frac{\gamma_c}{(\nu-\nu_0-\delta_c)^2+\gamma_c^2} \tag{1.2.15}$$

depends on the collisional half-width $\gamma_c = n_b \overline{\mathrm{v}}\sigma'/(2\pi c)$ and the collisional shift $\delta_c = n_b \overline{\mathrm{v}}\sigma''/(2\pi c)$. It has a Lorentzian form (Fig. 1.4) characterised by a slow decreasing of intensity in the wings. The neglected dependence on the molecular velocity v means that this profile is homogeneous.

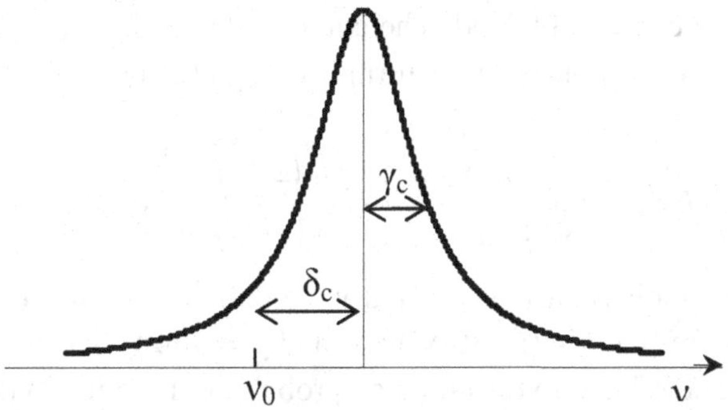

Fig. 1.4. Collisional broadening profile given by a Lorentzian function.

The collisional broadening profile in the dimensionless variables reads

$$F_c(x',y)=\frac{1}{\pi}\frac{y}{x'^2+y^2}, \tag{1.2.16}$$

where $x' \equiv x - s$, $s = (\ln 2)^{1/2}\delta_c/\gamma_D$ is introduced in Eq. (1.2.10) and the normalisation is made with dx' element.

1.2.3. Statistical independence of Doppler and collisional broadening

At intermediate pressures both principle mechanisms of line broadening – Doppler effect and molecular collisions – are present simultaneously, and various line shape models existing in the literature are based on different schemes of their coupling. As the first approximation, these two mechanisms can be considered as statistically independent. From a mathematical point of view it means that the resulting autocorrelation function is simply given by a product of autocorrelation functions of each process at zero frequency; the term $\exp(-i\omega_0 t)$ is introduced further to account for the real position of the line centre:

$$\Phi_V(t) = \exp(-i\omega_0 t)\langle\exp(ikv_{az}t)\rangle\langle\exp(i\eta(t))\rangle$$
$$= \exp\left[-i\omega_0 t - \Gamma t - \left(\frac{kv_{a0}t}{2}\right)^2\right]. \tag{1.2.17}$$

The resulting line profile represents a convolution of the line shapes associated with each mechanism; for example, a convolution of a Gaussian function $F_D(\nu)$ of Doppler half-width γ_D and a Lorentzian function $F_c(\nu)$ of collisional half-width γ_c. For this kind of profile named in the literature Voigt profile no analytical expression can be obtained. The autocorrelation function (1.2.17) enables its computation as a Fourier transform by Eq. (1.2.1). Moreover, it can be expressed as

$$F_V(\nu) = \sqrt{\frac{\ln 2}{\pi}} \frac{1}{\gamma_D} \frac{y}{\pi} \int_{-\infty}^{\infty} \frac{\exp(-\xi^2)}{y^2 + (x-\xi)^2} d\xi, \tag{1.2.18}$$

where the dimensionless variables x and y are defined by Eqs (1.2.8) and (1.2.9) with $\gamma = \gamma_c$, as in $x = (\ln 2)^{1/2}(\nu - \nu_0) / \gamma_D$ and $y = (\ln 2)^{1/2}\gamma_c / \gamma_D$. The integral in Eq. (1.2.18) can be related to the complex probability function $W(x,y)$ [14]:

$$W(x,y) = \frac{i}{\pi} \int_{-\infty}^{\infty} \frac{\exp(-\xi^2)}{x + iy - \xi} d\xi \tag{1.2.19}$$

which is computed by special numerical algorithms [17, 18]. The Voigt profile taking into account the line-shift ($x \rightarrow x'$) and normalised with the element dx' can be seen as its real part:

$$F_V(x', y) = \frac{y}{\pi} \int_{-\infty}^{\infty} \frac{\exp(-\xi^2)}{y^2 + (x' - \xi)^2} d\xi = \mathrm{Re}\, W(x', y). \tag{1.2.20}$$

The imaginary part of the complex probability function is not generally accounted for in spectroscopic studies since it represents a dispersion term. It becomes, however, very useful in fitting of asymmetric line shapes [19]. In the low-pressure limit ($\gamma_D >> \gamma_c$), the Voigt profile (1.2.20) tends to the Doppler profile (1.2.5) and in the high-pressure limit ($\gamma_c >> \gamma_D$) it reduces to the collisional broadening profile (1.2.15).

To fit the experimental line shapes, the Doppler component γ_D is typically fixed to its theoretical value (1.2.6) whereas the collisional component γ_c is considered as an adjustable parameter. In some studies [20, 21], however, a Voigt profile with floating Doppler width was used, which allows to analyse the evolution of the Doppler and collisional components with pressure increasing. A qualitative picture of this evolution is given in Fig. 1.5.

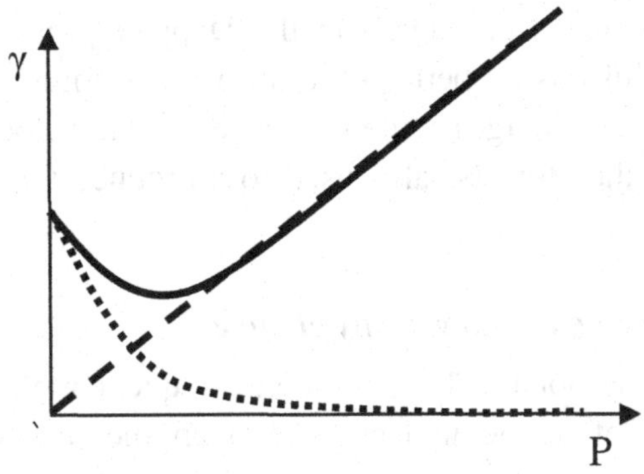

Fig. 1.5. Qualitative pressure dependence of both adjustable Doppler (dotted) and collisional (dashed) components of the line half-width (solid line) extracted from Voigt profile model.

The existence of the line-width minimum (minimum Dicke) and its value are defined by the competition between γ_D decreasing and γ_c increasing rates. A well-pronounced Dicke narrowing is observed in the case of relatively small collisional line widths; that is (typically), for the lines with high values of rotational quantum number J (numerous examples are given in Chapter 4). For diatomic molecules with rotational constant B, it occurs at $J \geq k_B T/(2Bh)$.

Despite the commodity of Voigt profile for atmospheric applications and its popularity, recent developments in high-resolution spectroscopy techniques revealed its departures from experimentally observed line shapes, namely for high values (>1000) of the signal-to-noise ratio, in both the microwave [22–25] and infrared [26–30] regions. The recorded line shapes were higher and narrower than those obtained with the Voigt model. This collisional narrowing of isolated spectral lines (Dicke effect [3]) can be modelled via theoretical line shapes accounting for the coupling between the Doppler effect and the collisional relaxation through molecular velocity. A quite exhaustive review of the literature pertaining to the observation and modelling of Dicke narrowing prior to 1996 can be found in Ref. [31]. The case of isolated line narrowing must be distinguished from the narrowing of the overlapped lines observed in the collisional regime at very high pressures when the collisions mix the internal states of the active molecule.

Two different approaches can be however adopted, based on different physical mechanisms of this coupling. On the one hand, the collisions induce velocity changes which reduce the apparent molecular velocity and, as a consequence, the Doppler contribution in the line width. The Dicke effect can therefore be taken into account by introducing a correction to the Doppler profile. On the other hand, the ability of collisions to perturb the absorption phase depends on the active molecule velocity. A Voigt profile corrected for the velocity dependence of relaxation rates can therefore be also used to reproduce the collisional line narrowing.

1.2.4. Collisional narrowing due to velocity changes

When the gas pressure is not too low, the quite frequent molecular collisions change the trajectories of active molecules through the diffusion effect and reduce their mean thermal velocity. They attenuate in consequence the Doppler effect and lead to a narrower, with respect to Voigt model, line profile. Since the mean free path becomes smaller and smaller, each molecule can be considered as localised in a small volume during some time, and line narrowing due to molecular collisions is often referred to in the literature as 'collisional confinement'.

The first diffusional model of Wittke and Dicke [32] did not consider the nature of molecular collisions and reproduced correctly the line profile only at high pressures (transition to the collisional regime). Two principle models corresponding to two limit kinds of shocks appeared in the 1960s: the model of 'soft' collisions by Galatry [7] and the model of 'hard' collisions by Rautian and Sobel'man [8].

The soft collision model of Galatry [7] assumes that a great number of collisions is needed to change the molecular velocity in a significant manner. It corresponds therefore to the case of a heavy active molecule in the atmosphere of a light buffer gas. The molecular displacement is modelled by the diffusion formula of Chandrasekhar [33] and the autocorrelation function reads

$$\Phi_G(t) = \exp\left[-i\omega_0 t - \Gamma t + \frac{1}{2}\left(\frac{kv_{a0}}{\beta}\right)^2 \left(1 - \beta t - e^{-\beta t}\right)\right], \tag{1.2.21}$$

where β (rad s^{-1}) is the narrowing rate (effective frequency of velocity-changing collisions). Sometimes it is identified with the dynamic friction coefficient $\beta_{diff} \equiv k_B T / (m_a D)$ defined by the mass diffusion constant D (cm^2 s^{-1}) [34], but this equivalence is more and more disputed by recent experimental studies (see Refs [35–38]). Its inverse gives the time necessary for the complete loss of initial velocity by the active molecule. Both parameters Γ and β are supposed to be linearly proportional to the buffer gas pressure. It is easily seen that Eq. (1.2.21) does represent the autocorrelation function of the Voigt profile $\Phi_V(t)$ with the Doppler component corrected for velocity changes induced by collisions. The corresponding line shape known in the literature as the Galatry profile is obtained by a numerical Fourier transform of $\Phi_G(t)$ and in the dimensionless variables (normalisation with dx' element) is written as [14]

$$F_G(x',y,z)=\frac{1}{\pi}\mathrm{Re}\left(\int_0^\infty \exp\left\{ix'\xi-y\xi+\frac{1}{2z^2}[1-z\xi-\exp(-z\xi)]\right\}d\xi\right)$$

$$=\frac{1}{\pi}\mathrm{Re}\left[\frac{1}{\dfrac{1}{2z}+y-ix'}M\left(1;1+\frac{1}{2z^2}+\frac{y-ix'}{z};\frac{1}{2z^2}\right)\right]. \tag{1.2.22}$$

In this equation, the new standardised narrowing parameter z is related to the narrowing rate β via

$$z\equiv\beta/\Delta\omega_D'=\beta/(2\pi c\Delta\nu_D'), \tag{1.2.23}$$

(i.e. $z=(\ln 2)^{1/2}\beta/\gamma_D$) and $M(\ldots)$ is the confluent hypergeometric function [39]. In the high-pressure limit ($z\to\infty$), the line profile (1.2.22) tends to the Lorentzian function (1.2.16).

The hard collision model of Rautian and Sobel'man [8] reposes on the hypothesis that each impact is so violent that the active molecule completely loses the memory of initial velocity and after the time period $1/\beta$ follows simply the Maxwell–Boltzmann distribution (no special assumption about the mass ratio are although made). Such a situation corresponds to a light molecule perturbed by a heavy buffer gas. The development starts from the kinetic equation and results in the line profile

$$F_R(x',y,z)=\frac{1}{\sqrt{\pi}}\mathrm{Re}\left[\frac{W(x',y+z)}{1-\sqrt{\pi}zW(x',y+z)}\right] \tag{1.2.24}$$

for uncorrelated collisional perturbations (where the collision is supposed to provoke either a dephasing of the absorbed radiation or a velocity change of the active molecule but nether the two effects simultaneously). If no change occurs in the molecular velocity after collision ($\beta=0$ and $z=0$), Eq. (1.2.24) reduces to the Voigt profile (1.2.20). In the opposite case of high pressures ($z \to \infty$), it tends to the Lorentzian shape of purely collisional line broadening (1.2.16).

In practice, when fitting the experimental line shapes to model profiles accounting for the Doppler effect, γ_D is usually fixed to its theoretical value (1.2.7) whereas γ_c and β are kept free. Their linear dependence on pressure provides an important criterion for validation of the chosen profile model.

Numerous studies have shown [40] that even the character of collisions is strictly opposite, both models of soft and hard collisions lead to quite similar line shapes [14, 41]. As a result, the collisional broadening line widths deduced from these two models are almost identical while the narrowing rates are typically greater for Galatry profile at low pressures but become equal with pressure increasing [31]. An unavoidable result of fitting with either a simple soft or hard collision model profile is an overestimation of the dynamic friction coefficient and consequently an underestimation of the mass diffusion constant which can be deduced from it.

1.2.5. Collisional narrowing due to speed dependence of relaxation rates

The fact that the real line shape for collisional broadening is given by a weighted sum of Lorentzian functions corresponding to different absorber speeds was accounted for in the quite general theoretical approach of Berman [42] and in its further development by Pickett [43] (Berman–Pickett theory). These authors introduced a speed-dependent Voigt (SDV) profile which needs a model describing the dependence of the relaxation rates Γ on the absorber speed v_a.

Among different approximations proposed for $\Gamma(v_a)$ [44–48] it is worthy to mention the semi-empirical quadratic form of Rohart *et al.* [47]

$$\Gamma(v_a) = \Gamma_0 + \Gamma_2\left[\left(\frac{v_a}{v_{a0}}\right)^2 - \frac{3}{2}\right] \tag{1.2.25}$$

which leads to an analytical correlation function

$$\Phi_{SDV}(t) = \frac{\exp[-i\omega_0 t - (\Gamma_0 - 3\Gamma_2/2)t]}{(1+\Gamma_2 t)^{3/2}} \exp\left[-\frac{(kv_{a0}t)^2}{4(1+\Gamma_2 t)}\right]. \tag{1.2.26}$$

In these equations, the phenomenological parameters $\Gamma_0 = \langle\Gamma(v_a)\rangle$ (the mean relaxation rate) and Γ_2 (characterising the speed dependence of the relaxation rate) are supposed to be linear functions of pressure. The corresponding line profile is computed numerically by Eq. (1.2.1). It was shown additionally [40] that for low pressure $3\Gamma_2 = \beta$, as it should be for this regime where the SDV and Galatry profiles are numerically indistinguishable.

The collisional relaxation becomes more efficient for high molecular speeds. It leads also to line narrowing but, in contrast with the velocity-changing mechanism described in Sec. 1.2.4, influences the line shape at any pressure.

1.2.6. Correlation between velocity changes and speed dependence of relaxation rates

In practice, both processes of velocity changes (molecular diffusion) and speed dependence of relaxation rates occur simultaneously. To describe them in a unified way, Ciuryło and Szudy [49] employed the Anderson–Talman [50, 51] classical phase-shift theory for collisional broadening and the Galatry diffusion model for the thermal motion of absorbing particles. They obtained a formula for the correlation function in the impact limit:

$$\begin{aligned}\Phi_{SDG}(t) &= \exp\left[-i\omega_0 t - \left(\frac{k\mathrm{v}_a}{2\beta}\right)^2\left(2\beta t - 3 + 4e^{-\beta t} - e^{-2\beta t}\right)\right] \\ &\times \frac{4}{\sqrt{\pi}}\int_0^\infty d\mathrm{v}^\circ\, \mathrm{v}^{\circ 2}\exp\left(-\mathrm{v}^{\circ 2}\right) Sinc\left[\left(1-e^{-\beta t}\right)\frac{k\mathrm{v}_a^2}{\beta \mathrm{v}_{a0}}\right] \\ &\times \exp\left[i\Delta\left(\mathrm{v}^\circ \mathrm{v}_{a0}\right)t - \Gamma\left(\mathrm{v}^\circ \mathrm{v}_{a0}\right)t\right]\end{aligned} \qquad (1.2.27)$$

with $\mathrm{v}^\circ \equiv \mathrm{v}_a/\mathrm{v}_{a0}$ and $Sinc(\xi) \equiv Sin(\xi)/\xi$, which can be regarded as a generalisation of the Galatry, Rautian and Sobel'man and speed-dependent Voigt models. When velocity-changing collisions are neglected ($\beta = 0$) but the Doppler-collision correlations are taken into account, it yields the speed-dependent Voigt model (1.2.26). In contrast, when the Doppler-collision correlations are omitted ($\Gamma(\mathrm{v}_a) = \Gamma$, $\Delta(\mathrm{v}_a) = \Delta$) but the velocity-changing collisions are included, it becomes identical to the Galatry formula (1.2.21). In the literature, this model of Ciuryło and Szudy is referred to as the speed-dependent Galatry profile.

In the dimensionless variables, it can be written as [14]

$$F_{SDG}(x, y, z, \zeta, s) = F_G(x', y, Z), \qquad (1.2.28)$$

where $Z \equiv z(1 - y/\zeta) - isz/\zeta$ and the new reduced variable ζ refers to the total collision frequency Ω (rad s^{-1}): $\zeta = \Omega/\Delta\omega_D'$, with the restriction y, s, $z \leq \zeta$. However, a strong correlation of two parameters (impossible to decouple without additional hypothesis) makes this model quite delicate for practical use.

An alternative approach to simultaneous including of velocity-changing collisions and speed dependence of relaxation rates was proposed by Duggan *et al.* [35] who simply performed a convolution of a Galatry profile and a weighted sum of Lorentzian profiles. In order to avoid confusion with the SDG profile of Eqs (1.2.27) and (1.2.28), their profile model is called S* ('soft' convolution) by some authors [31, 52]. However, this model fails to reproduce the SDV profile for $\beta \to 0$ and can lead to erroneous β-values [53].

The speed dependence of relaxation rates can be equally integrated into the hard collision model [8] leading to the so-called speed-dependent Rautian and Sobel'man profile (in dimensionless variables) [14]:

$$F_{SDR}(x, y, \zeta, s) = \frac{1}{\sqrt{\pi}} \mathrm{Re}\left[\frac{W(x,\zeta)}{1 - \sqrt{\pi}(\zeta - y - is)W(x,\zeta)}\right]. \qquad (1.2.29)$$

This profile is asymmetric and is useful only when the perturber is much heavier than the active molecule. (A later variant, named by the authors 'uncorrelated general Rautian and Sobel'man profile' was proposed by Lance *et al.* [54]; though analytical, it sometimes leads to an unconverging fitting procedure and needs generally too much computation time.)

By analogy with the S* profile, an H* profile can be defined as the convolution of a hard collision profile and a weighted sum of Lorentzian profile [31], which suffers, however, from the same problem of an incorrect limit for $\beta \to 0$ and gives approximately the same false values of β [53].

Different profile models result very often in quite distant values of the same line shape parameters. That is why the actual atmospheric applications require a unified profile model which could explain, for a given molecular pair, all the manifold of line shapes recorded under various experimental conditions (pressure, density, temperature). After the principle models mentioned above, a generalised speed-dependent dispersive Rautian–Galatry profile was proposed by Pine and Ciuryło [36], to explain, in particular, the line shape asymmetries due to the effects of correlation, hardness and collision duration (for other studies of asymmetric line shapes see Refs [55, 56] and works cited in Ref. [36]). Their line shape function is not analytical and contains many parameters to be adjusted simultaneously. A multi-spectrum fit procedure is therefore necessary to reduce the number of constraints and the uncertainties of the fitting parameters.

1.3. Broadening, shifting and narrowing coefficients

Fitting procedure with a profile model is carried out in practice by special computer codes which minimise the residua between the experimental and theoretical shapes for a given line at given pressure and temperature. The adjustable parameters, besides those proper to the model (narrowing rate, relaxation rates, and so on) and those of common use (width, line centre position, amplitude), sometimes include additional ones to correct experimental artefacts, in particular, the line asymmetry provoked by baseline distortion and/or by gas refractive index through residual standing waves [57]. Each parameter is then put as a function of pressure.

As an example, Fig. 1.6 shows the experimental positions of a water vapour line (ν_0 = 1187.022 cm^{-1}) shifted by nitrogen pressure [58]. Another example for the relaxation rates deduced from various profile models is given in Fig. 1.7 for the 602 GHz (J = 24 ←23) line of nitrous oxide perturbed by nitrogen at 296 K as a function of total gas pressure [24].

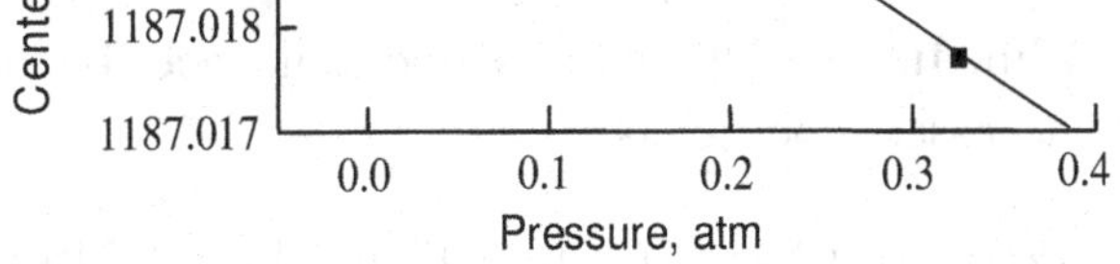

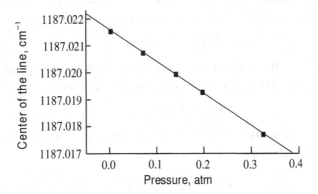

Fig. 1.6. Experimental positions of the water vapour line with ν_0 = 1187.022 cm^{-1} perturbed by nitrogen pressure [58].

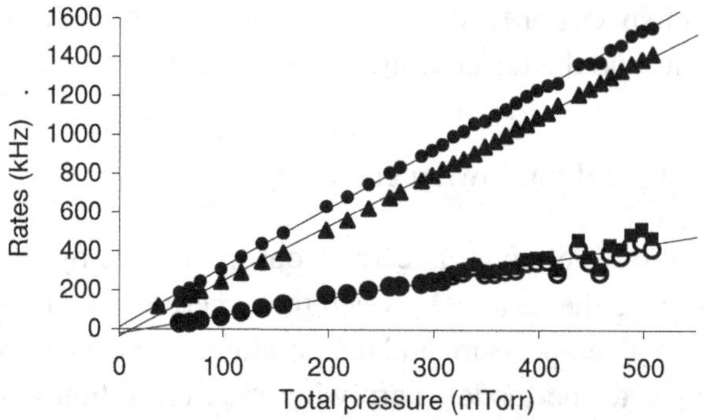

Fig. 1.7. Pressure dependence of the relaxation rates for N_2O–N_2 (8 mTorr of N_2O) $J = 24 \leftarrow 23$ line: Voigt profile (Γ_0 – ▲); speed-dependent Voigt profile (Γ_0 – ●; $3\Gamma_2$ – ○); Galatry profile (β – ■). Straight lines result from linear least squares fits of Γ_0 and Γ_2 for Voigt and SDV models [24]. Reproduced with permission.

Since all abovementioned profile models are obtained using the approximation of binary collisions, any adjustable parameter must be a linear function of buffer gas pressure P. For the collisional line width and shift we have therefore

$$\gamma_c(P) = \gamma_0 P, \quad \delta_c(P) = \delta_0 P, \tag{1.3.1}$$

where γ_0 and δ_0 are respectively the broadening and shifting coefficients expressed typically in cm^{-1} atm^{-1} for the infrared and in MHz $Torr^{-1}$ for the microwave absorption regions (see Table 1.1 for the conversion factors). The pressure dependence of the narrowing rate β can be specified as

$$\beta(P) = \beta_0 P \tag{1.3.2}$$

with β_0 denoting the narrowing coefficient. Again, the speed dependence parameter Γ_2 of the SDV profile as function of pressure reads

$$\Gamma_2(P) = \gamma_2 P, \tag{1.3.3}$$

and so on. Straight-line fitting of these pressure dependences by the least-squares method yields the necessary coefficients.

Table 1.1. Frequency conversion factors (1 atm = 760 Torr, 1 Torr = 133 Pa = 133.3 N m^{-2}).

Unit	cm^{-1}	MHz	μm
cm^{-1}	1	$2.99792458 \cdot 10^4$	10^{-4}
MHz	$3.33564 \cdot 10^{-5}$	1	$3.33564 \cdot 10^{-9}$
μm	10^4	$2.99792458 \cdot 10^8$	1

For a gaseous mixture with independent partial pressure contributions Eq. (1.3.1) can be generalised as

$$\gamma_c(P) = \sum_n \gamma_{0\,n} P_n, \quad \delta_c(P) = \sum_n \delta_{0\,n} P_n, \tag{1.3.4}$$

where P_n is the n-th gas pressure and $\gamma_{0\,n}$ and $\delta_{0\,n}$ are the broadening and shifting coefficients induced by this gas.

For binary mixtures of atmospheric molecules the collisional broadening coefficient ranges typically from 10^{-3} to 10^{-1} cm^{-1} atm^{-1} and the magnitude of the shifting coefficient depends on the molecular pair (for example, for the ν_2 band of self-broadened H_2O lines $\delta_c = -0.05 \div 0.05$ cm^{-1} atm^{-1} [26]). Table 1.2 gives examples of room-temperature collisional broadening (averaged over different experiments) and narrowing coefficients for H_2O ($\nu_0 = 7185.596$ cm^{-1}) [27] and NO ($\omega_0 / (2\pi) = 551.53$ GHz) [23] lines with various partners.

Table 1.2. Broadening and narrowing coefficients extracted from Galatry and Rautian and Sobel'man profile models for H_2O (in cm^{-1} atm^{-1}) [27] and NO (in MHz Torr^{-1}) [23] molecules.

System	β_0^G	β_0^R	γ_0	System	β_0^G	γ_0
H_2O–N_2	0.024(2)	0.032(3)	0.0469(5)	NO–N_2	0.50(15)	2.358(23)
H_2O–O_2	0.028(3)	0.034(5)	0.0258(8)	NO–O_2	0.42(7)	2.064(20)
H_2O–Ar	0.030(2)	0.040(5)	0.0192(4)	NO–Ar	0.70(9)	1.820(36)

1.4. Line interference

The notion of an isolated spectral line considered in the previous sections represents a theoretical abstraction and can be applied to the analysis of real line shapes only to some extent. Indeed, each (vib)rotational line is neighboured by other lines of the same band, and the integral absorption of Eq. (1.1.5) refers to all lines of this band (accidental resonances with other vibrational bands are disregarded here). Only if these lines are well spaced the absorption coefficient $K(\nu)$ can be expressed as a simple sum of individual lines:

$$K(\nu) = P_a \sum_{k=1}^{N} A_{P\,k} F_k(\nu - \nu_k), \tag{1.4.1}$$

where P_a (in atm) is the partial pressure of absorbing molecules and the lines (numbered by the index k) with the intensities $A_{P\,k}$ (cm^{-2} atm^{-1}) and centres at ν_k have the line shapes $F(\nu - \nu_k)$.

For the overlapping lines the intensity transfer between them takes place, so that the intensities recorded in a vibrational (or pure rotational) absorption band differ from Eq. (1.4.1). This phenomenon is known in the literature as 'line interference', 'line coupling' or 'line mixing'. In the case of strong overlapping resulting in a loss of rotational structure, the band shape can even collapse with further increasing pressure ('collisional narrowing'). Predicted by Baranger [5, 59] and Kolb and Griem [60] in 1958, this effect is being thoroughly studied in the infrared-absorption spectra during the last decades only. Since the line interference can strongly influence the values of molecular parameters obtained by fitting processes with assumed line shapes, the multi-line fitting procedure must take into account not only the widths and shifts of isolated spectral lines (diagonal elements Γ_{kk} of the relaxation matrix) but also the intensity transfer between different lines (off-diagonal elements Γ_{kl}, $l \neq k$). The description of integral vibrational band profile is, however, beyond the scope of the present book (see Ref. [61] for more detailed information on this subject), so that only two principle models of line interference are mentioned here.

For weakly overlapped lines the relaxation matrix Γ can be put in a diagonal form using the perturbation method of Rosenkranz [62], so that the absorption coefficient $K(\nu)$ satisfies Eq. (1.4.1). The corresponding line shapes F_k represent Lorentzian functions corrected for the line interference:

$$F_k(\nu - \nu_k) = \frac{1}{\pi} \frac{\gamma_k + (\nu - \nu_k) Y_k}{\gamma_k^2 + (\nu - \nu_k + \delta_k)^2}. \tag{1.4.2}$$

The individual line widths γ_k and line shifts δ_k as well as the correction factors

$$Y_k = 2 \sum_{l \neq k} \frac{d_l}{d_k} \frac{\Gamma_{lk} / (2\pi c)}{(\nu_k - \nu_l)} \tag{1.4.3}$$

(reduced dipole transition moments d_k are defined in Appendix A) are considered as adjustable parameters.

Another approach based on the profile model of Rautian and Sobel'man was used by Pine [29] who introduced the line mixing through the adjustable parameters in the absorption coefficient expression. For slightly overlapped lines these mixing parameters can be related to the off-diagonal elements of the relaxation matrix, and a direct computation of these elements in some cases allows one to calculate $K(\nu)$ without any fitting procedure.

Bibliography

1. Smith M.A.H., Rinsland C.P., Fridovich B. and Rao K.N. (1985). Intensities and Collision Broadening Parameters from Infrared Spectra. *Mol Spectrosc: Mod Res* 3: 111–228, Academic Press, New York.
2. Flaud J.M. and Camy-Peyret C. (1975). Vibration-rotation intensities in H_2O-type molecules. Application to the $2\nu_2$, ν_1, and ν_3 bands of $H_2{}^{16}O$. *J Mol Spectrosc* 55: 278–310.
3. Dicke R.H. (1953). The effect of collisions upon the Doppler width of spectral lines. *Phys Rev* 89: 472–473.
4. Townes C.H. and Schawlow A.L. (1955). *Microwave spectroscopy*. McGraw Hill, New York.
5. Baranger M. (1958). Problem of Overlapping Lines in the Theory of Pressure Broadening. *Phys Rev* 111: 494–504.
6. Fano U. (1963). Pressure broadening as a prototype of relaxation. *Phys Rev* 131: 259–268.
7. Galatry L. (1961). Simultaneous effect of Doppler and foreign gas broadening on spectral lines. *Phys Rev* 122: 1218–1223.
8. Rautian S.G. and Sobel'man I.I. (1966). Influence of collisions on Doppler's width of spectral lines. *Sov Phys Usp* 90: 209–236.
9. Smith E.W., Cooper J., Chapell W.R. and Dillon T. (1971). An Impact Theory for Doppler and Pressure Broadening. I. General Theory. *J Quant Spectrosc Radiat Trans* 11: 1547–1567.
10. Smith E.W., Cooper J., Chapell W.R. and Dillon T. (1971). An Impact Theory for Doppler and Pressure Broadening. II. Atomic and Molecular Systems. *J Quant Spectrosc Radiat Trans* 11: 1567–1576.
11. Sobelman I.I., Vainshtein L.A. and Yukov E.A. (1981). *Excitation of Atoms and Broadening of Spectral Lines*. Springer, Berlin.
12. Foster D. (1975). Hydrodynamic fluctuation, broken symmetry, and correlation functions: Advanced book program reading. W.A. Benjamin, Inc., Massachusetts.
13. D'Eu J.-F. (2001). PhD thesis. University of Sciences and Technologies of Lille, France.
14. Varghese P.L. and Hanson R.K. (1984). Collisional narrowing effects on spectral line shapes measured at high resolution. *Appl Opt* 23: 2376–2385.
15. Ben-Reuven A. (1966). Impact broadening of microwave spectra. *Phys Rev* 145: 7–22.
16. Mori H. (1965). Transport, Collective Motion, and Brownian Motion. *Prog Theor Phys* (Kyoto), 33: 423–455.
17. Humblicek J. (1982). Optimized computation of the Voigt and complex probability functions. *J Quant Spectrosc Radiat Trans* 27: 437–441.

18. Kuntz M. (1997). A new implementation of the Humblicek algorithm for the calculation of the Voigt profile function. *J Quant Spectrosc Radiat Trans* 57: 819–824.
19. Priem D. (1998). PhD thesis. University of Sciences and Technologies of Lille, France.
20. Giesen T., Schieder R., Winnewisser G. and Yamada K.M.T. (1992). Precise measurements of pressure broadening and shift for several H_2O lines in the ν_2 band by argon, nitrogen, oxygen and air. *J Mol Spectrosc* 153: 406–418.
21. Schmücker N., Trojan, Ch., Giesen T., Schieder R., Yamada K.M.T. and Winnewisser G. (1977). Pressure broadening and shift of some H_2O lines in the ν_2 band: revisited. *J Mol Spectrosc* 184: 250–256.
22. Fraser G.T. and Coy S.L. (1985). Absorber speed dependence of the coherence relaxation rate of the J = 0–1 transition of N_2O. *J Chem Phys* 83: 5687–5689.
23. Colmont J.-M., D'Eu J.F., Rohart F., Wlodarczak G. and Buldyreva J. (2001). N_2 and O_2 broadening and line shape of the 551.53 GHz line of ^{14}NO. *J Mol Spectrosc* 208: 197–208.
24. Nguyen L., Buldyreva J., Colmont J.-M., Rohart F., Wlodarczak G. and Alekseev E.A. (2006). Detailed profile analysis of millimeter 502 and 602 GHz N_2O–$N_2(O_2)$ lines at room temperature for collisional line-width determination. *Mol Phys* 104: 2701–2710.
25. Rohart F., Nguyen L., Buldyreva J., Colmont J.-M. and Wlodarczak G. (2007). Line shapes of the 172 and 602 GHz rotational transitions of $HC^{15}N$. *J Mol Spectrosc* 246: 213–227.
26. Toth R.A., Brown L.R. and Plymate C. (1998). Self-broadened widths and frequency shifts of water vapor lines between 590 and 2400 cm^{-1}. *J Quant Spectrosc Radiat Trans* 59: 529–562.
27. Lepère M., Henry A., Valentin A. and Camy-Peyret C. (2001). Diode-Laser Spectroscopy: Line Profile of H_2O in the Region of 1.39 μm. *J Mol Spectrosc* 208: 25–31.
28. Kissel A., Kronfeldt H.-D., Sumpf B., Ponomarev Yu.N., Ptashnik I.V. and Tikhomirov B.A. (1999). Investigation of line profiles in the ν_2 band of H_2S. *Spectrochim Acta Part A* 55: 2007–2013.
29. Pine A.S. (1997). N_2 and Ar broadening and line-mixing in the P and R branches of the ν_3 band of CH_4. *J Quant Spectrosc Radiat Trans* 57: 157–176.
30. Predoi-Cross A., Luo C., Sinclair P.M., Drummond J.R. and May A.D. (1999). Line Broadening and Temperature Exponent of the Fundamental Band in CO–N_2 Mixtures. *J Mol Spectrosc* 198: 291–303.
31. Henry A., Hurtmans D., Margottin-Maclou M. and Valentin A. (1966). Confinement narrowing and absorber speed-dependent broadening effects on CO lines in the fundamental band perturbed by Xe, Ar, Ne, He and N_2. *J Quant Spectrosc Radiat Trans* 56: 647–671.
32. Wittke J.P. and Dicke R.H. (1956). Redetermination of the Hyperfine Splitting in the Ground State of Atomic Hydrogen. *Phys Rev* 103: 620–631.
33. Chandrasekhar S. (1943). Stochastic problems in physics and astronomy. *Rev Mod Phys* 15: 1–89.
34. Hirschfelder J.O., Curtiss C.F. and Bird R.B. (1964). *Molecular Theory of Gases and Liquids.* Wiley, New York.
35. Duggan P., Sinclair P.M., Berman R., May A.D. and Drummond J.R. (1997). Testing Line Shape Models: Measurements for v = 1–0 CO Broadened by He and Ar. *J Mol Spectrosc* 186: 90–98.
36. Pine A.S. and Ciuryło R. (2001). Multispectrum Fits of Ar-broadened HF with a Generalized Asymmetric Line Shape: Effects of Correlation, Hardness, Speed Dependence, and Collision Duration. *J Mol Spectrosc* 208: 180–187; Ciurylo R., Pine A.S. and Szudy J. (2001). A generalized speed-dependent line profile combining soft and hard partially correlated Dicke-narrowing collisions. *J Quant Spectrosc Radiat Trans* 68: 257–271.

37. Wehr R., Vitcu A., Ciuryło R., Thibault F., Drummond J.R. and May A.D. (2002). Spectral line shape of the P(2) transition in CO–Ar: Uncorrelated *ab initio* calculation. *Phys Rev A* 66: 062502.
38. Brault J.W., Brown L.R., Chackerian C. Jr., Freedman R., Predoi-Cross A. and Pine A.S. (2003). Self-broadened CO line shapes in the v = 2–0 band. *J Mol Spectrosc* 222: 220–239.
39. Abramowitz M. and Stegun I.A. (1964). *Handbook of Mathematical Functions*. Dover Publications, New York.
40. D'Eu J.-F., Lemoine B. and Rohart F. (2002). Infrared HCN Line Shapes as a Test of Galatry and Speed-dependent Voigt Profiles. *J Mol Spectrosc* 212: 96–110.
41. Lance B., Ponsar S., Walrand J., Lepère M., Blanquet G. and Bouanich J.-P. (1998). Correlated and Non-correlated Line Shape Models under Small Line-shift Condition: Analysis of Self-perturbed CH_3D. *J Mol Spectrosc* 189: 124–134.
42. Berman P.R. (1972). Speed-dependent Collisional Width and Shift. Parameters in Spectral Profiles. *J Quant Spectrosc Radiat Trans* 12: 1331–1342.
43. Pickett H.M. (1980). Effects of velocity averaging on the shapes of absorption lines. *J Chem Phys* 73: 6090–6094.
44. Coy S.L. (1980). Speed dependence of microwave rotational relaxation rates. *J Chem Phys* 73: 5531–5555.
45. Nicolaisen H.W. and Mäder H. (1991). Rotational relaxation rates for the J = 0–1 transition of N_2O by self-collisions and foreign gas collisions. *Mol Phys* 73: 349–358.
46. Haekel J. and Mäder H. (1991). Speed-dependent T_2-relaxation rates of microwave emission signals. *J Quant Spectrosc Radiat Trans* 46: 21–30.
47. Rohart F., Mäder H. and Nicolaisen H.W. (1994). Speed dependence of rotational relaxation induced by foreign gas collisions: studies on CH_3F by millimeter wave coherent transients. *J Chem Phys* 101: 6475–6486.
48. Köhler T. and Mäder H. (1995). Measurement of speed dependent rotational relaxation rates using a microwave spectrometer with a circular waveguide. Studies on nitrous oxide. *Mol Phys* 86: 287–300.
49. Ciuryło R. and Szudy J. (1997). Speed-dependent pressure broadening and shift in the soft collision approximation. *J Quant Spectrosc Radiat Trans* 57: 411–423.
50. Anderson P.W. and Talman J.D. (1955). Conference on Broadening of Spectral Lines, University of Pittsburgh, unpublished *Bell Teleph Syst Tech Publ No 3117*.
51. Anderson P.W. (1952). A Method of Synthesis of the Statistical and Impact Theories of Pressure Broadening. *Phys Rev* 86: 809–809.
52. Mantz A.W., Thibault F., Cacheiro J.L., Fernandez B., Pedersen T.B., Koch H., Valentin A., Claveau C., Henry A. and Hurtmans D. (2003). Argon broadening of the ^{13}CO R(0) and R(7) transitions in the fundamental band at temperatures between 80 and 297 K: comparison between experiment and theory. *J Mol Spectrosc* 222: 131–141.
53. Wehr R., Ciuryło R., Vitcu A., Thibault F., Drummond J.R. and May A.D. (2006). Dicke-narrowed spectral line shapes of CO in Ar: Experimental results and a revised interpretation. *J Mol Spectrosc* 235: 54–68.
54. Lance B., Blanquet G., Walrand J. and Bouanich J.-P. (1997). On the Speed-dependent Hard Collision Line Shape Models: Application to C_2H_2 Perturbed by Xe. *J Mol Spectrosc* 185: 262–271.
55. Szudy J. and Baylis W.E. (1996). Profiles of line wings and rainbow satellites associated with optical and radiative collisions. *Phys Rep* 266: 127–227.

56. Boulet C., Flaud J.-M. and Hartmann J.-.M. (2004). Infrared line collisional parameters of HCl in argon, beyond the impact approximation: measurements and classical path calculations. *J Chem Phys* 120: 11053–11061.
57. Dore L. (2003). Using Fast Fourier Transform to compute the line shape of frequency-modulated spectral profiles. *J Mol Spectrosc* 221: 93–98.
58. Lavrentieva N.N. and Solodov A.M. (1999). Shift of water vapor lines of the $\nu_1 + \nu_2$, and $\nu_2 + \nu_3$ bands by oxygen and argon pressure. *Atmos Oceanic Opt* 4: 959–967.
59. Baranger M. (1958). General Impact Theory of Pressure Broadening. *Phys Rev* 112: 855–865.
60. Kolb A.C. and Griem H. (1958). Theory of Line Broadening in Multiplet Spectra. *Phys Rev* 111: 514–521.
61. Tonkov M.V., Filippov N.N., Timofeev Yu.M. and Polyakov A.V. (1996). A simple model of the line-mixing effect for atmospheric applications: theoretical background and comparison with experimental profiles. *J Quant Spectrosc Radiat Trans* 56: 783–795.
62. Rosenkranz P.W. (1975). Shape of the 5 mm oxygen band in the atmosphere. *IEEE Trans Antennas Propag* 23: 498–506.

Chapter 2

Semi-classical calculation of pressure-broadened line widths and pressure-induced line shifts

Among all the mechanisms of the spectral line broadening, it is the broadening by collisions which characterises the interaction of the active molecule with the buffer gas and which contains information on the latter's concentration and temperature. The collisional broadening and shifting coefficients constitute therefore the principal objective of all theoretical approaches describing the collisional regime at moderate pressures. This problem of line shape parameters has been studied by many authors, and an exhaustive review of all theories is hard to give. Only methods based on the fundamental Anderson–Tsao–Curnutte (ATC) [1, 2] and Fano [3, 4] theories are briefly discussed in this chapter.

2.1. Quantum system of two interacting molecules

The active molecule colliding with a buffer particle is additionally subjugated to the proton absorption process, so that the molecular pair taken alone is not conservative and its description in terms of the active-molecule variables constitutes a quite delicate problem. According to Ben-Reuven [5], 'By collisions with other molecules the photon absorbed is eventually dissipated to other degrees of freedom of the gas ... with a resulting broadening and shifting of the single molecule's resonance frequencies' which 'is expressed by replacing each spectral frequency by modified frequency with additional a non-Hermitian (or complex) perturbation to the resonance frequencies of the system'.

In the framework of the ATC theory [1, 2], the full Hamiltonian H_F of two interacting molecules in the radiation field is written as

$$H_F = H_a + H_b + V + H_{ar} + H_r, \tag{2.1.1}$$

where H_a and H_b are respectively the Hamiltonians of the active and perturbing molecules, V accounts for their time-dependent collisional interaction, and H_{ar} describes the interaction between the active molecule and the radiation field of Hamiltonian H_r (interaction of the perturbing molecule with the radiation is neglected). The unperturbed part of the full Hamiltonian $H_0 = H_a + H_b + H_r$ determines a complete set of orthonormalized time-independent eigenfunctions written as $\varphi_a \varphi_b \psi_{n_k}$, where φ_a and φ_b are the eigenfunctions of the isolated molecules and the functions ψ_{n_k} with occupation numbers n_k characterise the state of the radiation field (hereafter the product $\varphi_a \varphi_b$ will be noted φ).

The probability of a transition from the state $\Psi_i(0) = \varphi_i(0)\psi_{n_k}$ at time $t = 0$ to the state $\Psi_f(t) = \varphi_f(t)\psi_{n_k - 1}$ at time t is defined by the matrix element

$$P_{fi} = \left| \langle \Psi_f(t) | U_F | \Psi_i(0) \rangle \right|^2, \tag{2.1.2}$$

where U_F is the time evolution operator corresponding to the full Hamiltonian of Eq. (2.1.1). This transition probability defines the average energy w_{fi} absorbed per second and per unit photon intensity in the time interval from 0 to t [2]:

$$w_{fi} = \frac{\kappa}{t} \left| \int_0^t d\tau \, e^{-i\omega_k \tau} \langle \varphi_f(0) | \mu(\tau) | \varphi_i(0) \rangle \right|^2, \tag{2.1.3}$$

where $\kappa = 2\pi\omega_k/(\hbar c)$ is defined by the angular frequency of absorbed photon ω_k and $\mu(\tau) = U^{-1} \mu \, U$ is the dipole moment of the active molecule written in the Heisenberg representation. (The direction of photon polarization is taken along the laboratory z-axis but the index z of μ_z is omitted for brevity.) The evolution operator U corresponds to the Hamiltonian

$$H = H_a + H_b + V, \tag{2.1.4}$$

so that $\mu(\tau)$ satisfies the equation

$$\frac{d\mu(\tau)}{d\tau} = i\hbar^{-1} [H, \mu(\tau)]. \tag{2.1.5}$$

The averaged energy of Eq. (2.1.3) should be further summed over all final and averaged over all initial molecular states. For molecular transitions the initial state i and the final state f are degenerate over the magnetic quantum numbers m_i and m_f of the angular momentum operators $\vec{J}_i$ and $\vec{J}_f$ of the active molecule, so that additional averaging should be made over these magnetic quantum numbers (and those associated with the virtual states of the perturbing molecule). This removal of degeneracy on the magnetic quantum numbers constitutes an intrinsic part of the collisional line-broadening mechanism. The summation over the final states can be made using the Hermitian properties of the operator $\mu(\tau)$ and the normalisation relation $\sum_f |\varphi_f(0)\rangle\langle\varphi_f(0)| = 1$:

$$\sum_f w_{fi} = \frac{\kappa}{t}\int_0^t d\tau \int_0^t d\tau' e^{i\omega_k(\tau-\tau')}\langle\varphi_i(0)|\mu(\tau)\mu(\tau')|\varphi_i(0)\rangle, \tag{2.1.6}$$

the matrix element appearing now as the expectation value of the operator $\mu(\tau)\mu(\tau')$ for the initial state $\varphi_i(0)$. The averaging over the initial states can be approximately made with the density matrix ρ which depends solely on the internal molecular coordinates. We thus obtain one of the fundamental expressions [2]:

$$w = \frac{\kappa}{t}\int_0^t d\tau \int_0^t d\tau' e^{i\omega_k(\tau-\tau')} Tr[\rho\mu(\tau)\mu(\tau')], \tag{2.1.7}$$

where $Tr[\rho\,\mu(\tau)\mu(\tau')] = \sum_i\langle\varphi_i(0)|\mu(\tau)\mu(\tau')\varphi_i(0)\rangle$. The double integral in the above equation can be further simplified by introducing the classical path assumption which neglects the influence of the translational motion on the internal molecular states. The density matrix ρ thus becomes stationary and diagonal in the internal coordinates of both molecules. If, in addition, the natural line width is supposed to be much smaller than the line width due to collisions, the double integral takes a form proportional to t and w can be viewed as a constant absorption rate which determines the spectral shape [2]:

$$w = 2\kappa \,\mathrm{Re}\int_0^\infty dt e^{-i\omega_k t} Tr[\rho\mu U^{-1}(t)\mu U(t)]_{AV}. \tag{2.1.8}$$

In this equation, the subscript AV means the average over all possible paths for a given time interval $t = \tau - \tau'$ and the exponent in the integrand accounts for the phase shift. Formally, Eq. (2.1.8) reduces to the spectral distribution function which is convenient for further analysis of line broadening and shifting.

2.2. Spectral function

In the case of thermal equilibrium, the Hamiltonian of Eq. (2.1.4) has the eigenstates $|n\rangle$ which are time-independent and determined by the stationary Schrödinger equation

$$H|n\rangle = E_n|n\rangle, \tag{2.2.1}$$

where E_n are the corresponding eigenvalues. The spectral function of Eq. (2.1.8), by identification with the general expression of Eq. (1.2.1), leads to the correlation function

$$\Phi(t) = Tr[\rho\mu(0)\mu(t)] = \sum_n \langle n|\rho\mu(0)\mu(t)|n\rangle. \tag{2.2.2}$$

Further development is based on the concept of the Liouville operator L which allows finding a formal solution for the spectral distribution function without detailed information on the system and the bath. A general description of this formalism can be found in the original work of Fano [3] as well as in the review by Srivastava and Zaidi [4]. Only the main properties of this operator are briefly recalled here.

The operator L acts in the Hilbert space of ordinary quantum-mechanical operators X:

$$LX = \hbar^{-1}\,[H, X], \tag{2.2.3}$$

in which L is a superoperator. If $|m\rangle$ and $|n\rangle$ are eigenvectors of H, the operator $|m\rangle\langle n|$ is a basis vector for L:

$$L|m\rangle\langle n| = \hbar^{-1}\,(E_m - E_n)|m\rangle\langle n| = \omega_{mn}|m\rangle\langle n|, \tag{2.2.4}$$

and the corresponding eigenvalues ω_{mn} represent the frequencies of transitions between the states $|n\rangle$ and $|m\rangle$. The set of operators $|m\rangle\langle n|$ form a basis in the Hilbert space of superoperators which is called the Liouville space. The matrix elements of Eq. (2.2.3) are determined thus by

$$(LX)_{mn} = \sum_{m'n'} L_{mn,m'n'} X_{m'n'}, \tag{2.2.5}$$

$$L_{mn,m'n'} = \hbar^{-1}\left(H_{mm'}\delta_{nn'} - H_{nn'}\delta_{mm'}\right), \tag{2.2.6}$$

and the eigenvalues of L appear as diagonal matrix elements

$$L_{mn,mn} = \hbar^{-1}\,(H_{mm} - H_{nn}) = \hbar^{-1}\,(E_m - E_n) = \omega_{mn}. \tag{2.2.7}$$

The definition of the Liouville operator by Eq. (2.2.3) allows rewriting Eq. (2.1.5) as

$$\frac{d\mu(t)}{dt} = iL\mu(t) \tag{2.2.8}$$

which solution, due to the Hermitian property $\mu(t)^+ = \mu(t)$, is

$$\mu(t) = \exp(iLt)\mu(0) = \mu(0)\exp(-iLt). \tag{2.2.9}$$

Substitution of Eq. (2.2.9) in Eq. (2.2.2) and then in Eq. (1.2.1) gives

$$F(\omega) = -\frac{1}{\pi}\,\mathrm{Im}\,Tr\left[\mu(0)(\omega - L)^{-1}\rho\mu(0)\right]. \tag{2.2.10}$$

To further develop the right side of this equation, some specific approximations for the considered molecular system are needed. If the buffer gas is composed of foreign molecules, the Liouville operator L can be decomposed in accordance with Eq. (2.1.4) as

$$L = L_0 + L_c = L_0^a + L_0^b + L_c, \tag{2.2.11}$$

where

$$L_0^a X = \hbar^{-1}\,[H_a, X],\ L_0^b X = \hbar^{-1}\,[H_b, X],\ L_c X = \hbar^{-1}\,[V, X]. \tag{2.2.12}$$

The operators L_0^a and L_0^b are respectively diagonal in the eigenstate bases of H_a and H_b, so that

$$(L_0^a)_{if,i'f'} = \hbar^{-1}\,(E_i - E_f)\delta_{ii'}\delta_{ff'} = \omega_{if}\,\delta_{ii'}\delta_{ff'}, \tag{2.2.13}$$

$$(L_0^b)_{\alpha\beta,\alpha'\beta'} = \hbar^{-1}\,(E_\alpha - E_\beta)\delta_{\alpha\alpha'}\delta_{\beta\beta'} = \omega_{\alpha\beta}\delta_{\alpha\alpha'}\delta_{\beta\beta'}. \tag{2.2.14}$$

Here the indices i, i', f, f' correspond to the eigenstates of the Hamiltonian H_a and the indices $\alpha, \alpha', \beta, \beta'$ correspond to the eigenstates of the Hamiltonian H_b.

Using Eqs (2.2.13)–(2.2.14) the resolvent operator $(\omega\text{-}L)^{-1}$ can be written as

$$(\omega - L)^{-1} = (\omega - L_0)^{-1}[1 + M(\omega)(\omega - L_0)^{-1}]. \tag{2.2.15}$$

This equation can be considered as a formal relation for the determination of $M(\omega)$ and leads to

$$M(\omega) = L_c[1 - (\omega - L_0)^{-1}L_c]^{-1}. \tag{2.2.16}$$

The next approximation to be made concerns the density matrix ρ. If the latter does not include significant correlations between the active molecule and the bath ('initial chaos' approximation), it can be written as a product of the active-molecule and bath density matrices:

$$\rho = \rho^a \rho^b \tag{2.2.17}$$

and allows to separate the active-molecule and bath variables. Moreover, the abovementioned assumption of a stationary distribution of bath particles over the eigenstates of the Hamiltonian H_b leads to

$$\rho_{\alpha\alpha'}^b = \rho_{\alpha\alpha}^b \delta_{\alpha\alpha'}, \tag{2.2.18}$$

where the symbol $\delta_{\alpha\alpha'}$ makes vanishing the contribution of L_b^0 to L.

Taking into account Eqs (2.2.15) and (2.2.17)–(2.2.18), the trace operation over the bath variables transforms Eq. (2.2.10) into [3]

$$F(\omega) = -\frac{1}{\pi}\mathrm{Im}\,Tr_a\{\mu(0)(\omega - L_0^a)^{-1} \times [1 + \langle M(\omega)\rangle(\omega - L_0^a)^{-1}]\rho^a\mu(0)\}, \quad (2.2.19)$$

where Tr_a denotes that trace operation on the active-molecule variables and $\langle M(\omega)\rangle = Tr_b\,[M(\omega)\rho^b]$. If we write that

$$(\omega - L_0^a)^{-1}[1 + \langle M(\omega)\rangle(\omega - L_0^a)^{-1}] = [\omega - L_0^a - \langle M_c(\omega)\rangle]^{-1}, \quad (2.2.20)$$

this equation can be considered as a formal equation for the determination of $\langle M_c(\omega)\rangle$:

$$\langle M_c(\omega)\rangle = [1 + \langle M(\omega)\rangle(\omega - L_0^a)^{-1}]^{-1}\langle M(\omega)\rangle. \quad (2.2.21)$$

The spectral distribution function $F(\omega)$ is then expressed as

$$F(\omega) = -\frac{1}{\pi}\mathrm{Im}\,Tr_a\{\mu(0)[\omega - L_0^a - \langle M_c(\omega)\rangle]^{-1}\rho^a\mu(0)\}. \quad (2.2.22)$$

The operator $\langle M_c(\omega)\rangle$ is called the relaxation operator and describes the influence of the buffer gas on the optically active molecule (the unperturbed frequencies of L_0^a are shifted). The frequency dependence of the relaxation operator assumes the existence of 'memory' effects accounting for previous collisions (due to the perturbation expansion into powers of the interaction described by L_c).

In the particular case of $\langle M_c(\omega)\rangle = 0$ (no interaction with the surrounding molecules), we obtain the unperturbed line spectrum of the active molecule:

$$F(\omega) = -\frac{1}{\pi}\mathrm{Im}\sum_{if}\mu_{if}{}^*(\omega - \omega_{if})^{-1}(\rho^a\mu)_{if} \rightarrow \sum_{if}\mu_{if}{}^*\delta(\omega - \omega_{if})(\rho^a\mu)_{if}, \quad (2.2.23)$$

where $\mu_{if} = \langle i|\,\mu(0)|\,f\rangle$, $\mu \equiv \mu(0)$ and the asterisk means the complex conjugation.

A line spectrum also arises when using a transformation operator T the operator $L_0^a + \langle M_c(\omega)\rangle$ is put into a form diagonal in the Liouville space. In this case, the eigenvalues of the transformed operator $T\,[\,L_0^a + \langle M_c(\omega)\rangle]T^{-1}$ are generally complex since $\langle M_c(\omega)\rangle$ is non-Hermitian.

Further simplification of Eq. (2.2.22) is achieved owing to the so-called 'impact approximation' which is based on two assumptions:

i) the collisions are binary, i.e. the mean collision duration τ_c is much smaller then the mean time interval between collisions;
ii) the collisions are instantaneous, i.e. τ_c is small in comparison with a typical duration of the interaction process.

The first assumption means that the relaxation operator is proportional to the density of the bath molecules n_b. The second one allows us to neglect the memory effects and consider $\langle M_c(\omega)\rangle$ as frequency-independent: $\langle M_c(\omega)\rangle \cong \langle M_c(0)\rangle$. In the impact approximation, the relaxation operator therefore can be written as [3, 5]

$$\langle M_c(\omega)\rangle \equiv \Lambda(\omega) \cong \Lambda(0) \equiv \Lambda = n_b\, Tr_b[\rho^b\, \mathrm{m}(0)], \tag{2.2.24}$$

where $\mathrm{m}(\omega)$ is a binary-collision transition operator in the Liouville space. The spectral function of Eq. (2.2.22) after applying the trace operation over the active-molecule variables consequently takes the form

$$F(\omega) = -\frac{1}{\pi}\,\mathrm{Im}\left\{\sum_{if,i'f'}\langle i'|\mu(0)|f'\rangle^* [\omega - L_0^a - \Lambda]^{-1}_{i'f',if}\, \rho_i^a \langle i|\mu(0)|f\rangle\right\}, \tag{2.2.25}$$

where ρ_i^a is the population of the i-th energy level of the active molecule. This equation defines the absorption coefficient $K(\omega)$ which, for the case of weakly interfering lines and in the first order of the perturbation theory on Λ, can be reduced to the form of Eq. (1.4.2).

In the general case of a relaxation process related to the 2^l-pole moment of the active molecule, the electric dipole moment operator μ must be replaced by the corresponding moment $X^{(l)}$ and Eq. (2.2.25) reads

$$F(\omega) = -\frac{1}{\pi}\,\mathrm{Im}\left\{\sum_{if,i'f'}\langle i'|X^{(l)}|f'\rangle^* [\omega - L_0^a - \Lambda^{(l)}]^{-1}_{i'f',if}\, \rho_i^a \langle i|X^{(l)}|f\rangle\right\}. \tag{2.2.26}$$

Equations (2.2.25) and (2.2.26) are valid for isolated as well as for overlapping lines. The non-Hermitian operator Λ can be expressed in terms of two Hermitian operators Δ and Γ in the Liouville space:

$$\Lambda = \Delta - i\Gamma = -i\Gamma, \tag{2.2.27}$$

where both Δ and Γ are frequency independent and proportional to the buffer gas density. As mentioned in Sec. 1.2, the operator Δ defines the line shifts whereas the operator Γ is responsible for the line widths.

For isolated lines, Λ has only diagonal matrix elements

$$\Lambda_{if,\,if} = \Delta_{if} - i\Gamma_{if}, \tag{2.2.28}$$

which real and imaginary parts give the relaxation rates Δ_{if} and Γ_{if} respectively for the pressure-induced line shift and the pressure-broadened line width corresponding to the transition $i \to f$. With these notations the spectral function of Eq. (2.2.25) takes the form

$$F(\omega) = -\frac{1}{\pi}\mathrm{Im}\left[\sum_{if}\rho_i^a\left|\mu_{if}\right|^2\left(\omega-\omega_{if}-\Lambda_{if}\right)^{-1}\right] = \frac{1}{\pi}\sum_{if}\rho_i^a\left|\mu_{if}\right|^2\frac{\Gamma_{if}}{\left(\omega-\omega_{if}-\Delta_{if}\right)^2+\Gamma_{if}^2}, \tag{2.2.29}$$

which represents a superposition of Lorentzian functions.

In the general case of overlapping lines, the relaxation matrix $\Lambda^{(l)}$ contains non-zero off-diagonal matrix elements which are responsible for the line interference. As a result, the spectral function deviates from a simple sum of Lorentzians.

2.3. General expressions for line half-width and shift calculation

According to Eq. (2.2.28) the half-width γ_{if} and the shift δ_{if} of the line corresponding to the optical transition $i \to f$ are defined respectively by the minus imaginary and the real parts the of the diagonal matrix element $\Lambda_{if} \equiv \Lambda_{if,\,if}$ of the relaxation matrix Λ:

$$\gamma_{if} = -(2\pi c)^{-1}\,\mathrm{Im}\,\Lambda_{if}, \tag{2.3.1}$$

$$\delta_{if} = (2\pi c)^{-1}\,\mathrm{Re}\,\Lambda_{if} \tag{2.3.2}$$

(in cm^{-1}). For the general case of the 2^l-pole moment relaxation the matrix element $\Lambda^{(l)}_{if,i'f'}$ was derived by Ben Reuven [5] and Cherkasov [6] neglecting the Doppler effect and collisional narrowing. This derivation is briefly presented below, following Ref. [6].

Let $\vec{J}$ and $\vec{l}$ be the total angular momenta of the active and bath molecules characterised by the quantum numbers J and l respectively. Let m and m_2 be the corresponding magnetic quantum numbers describing the projections on the space-fixed z-axis (other quantum numbers will be omitted hereafter for brevity). The angular momentum $\vec{l}$ includes the angular momentum $\vec{J}_b$ of the rotation of the bath molecule in its molecule-fixed frame and the orbital angular momentum $\vec{l}_b$ of its translational motion relative to the active molecule: $\vec{l} = \vec{J}_b + \vec{l}_b$. The vectors $|Jm\rangle$ and $|lm_2\rangle$ are eigenvectors of the Hamiltonians H_a and H_b respectively.

The vectors $|Jm\ lm_2\rangle$ ($\equiv |Jm\rangle|lm_2\rangle$ in the classical-trajectory approximation) generate the basis $|Jm\ lm_2\rangle\langle J'm'\ l'm_2'|$ which could be employed for the calculation of the relaxation matrix elements.

Instead of these vectors it is, however, much more convenient to use the rotationally invariant Liouville vectors [5, 6] which are eigenvectors of the angular momentum $\vec{K} = \vec{J} + \vec{l}$ having the projection M on the space-fixed z-axis. In the vector coupling scheme (see, e.g., Ref. [7]), the vectors $\vec{j} = \vec{j}_i - \vec{j}_f$ and the vectors $\vec{l} = \vec{l}_i - \vec{l}_f$ are set to zero for all bath molecules (the vectors $\vec{l}_b$ are separated and treated classically), so that $\vec{K} = \vec{J}_i - \vec{J}_f$, where $\vec{J}_i$ and $\vec{J}_f$ are the angular momenta for the initial and the final states of the active molecule. The selection rules for the projection M depend on the operator $X^{(l)}$: for the electric dipole transitions $l = 1$ and $\Delta M = 0, \pm 1$.

To compute the matrix elements $\Lambda^{(l)}_{if,i'f'}$ the binary-collision operator m(0) from Eq. (2.2.24) is presented in the form

$$\mathrm{m}(0) = -i(1 - \mathrm{S}^{-1}\mathrm{S}),$$

where S is the semi-classical scattering operator acting on the final states whereas S^{-1} acts on the initial states. In this case, the matrix element $\Lambda^{(l)}_{if,i'f'}$ reduces to the matrix elements of S and S^{-1} in the basis of invariant Liouville vectors:

$$\Lambda^{(l)}_{if,i'f'} = -in_b \sum_j \rho_j$$

$$\times \sum_{\substack{m_i m_f m_2 M \\ m_i' m_f' m_2' jj'}} (-1)^{m_i + m_f + m_i' + m_f'} C^{J_i m_i}_{J_f m_f KM} C^{J_i' m_i'}_{J_f' m_f' KM} (2j+1)^{-1} \left[(2J_i + 1)(2J_i' + 1)\right]^{-1/2} \quad (2.3.3)$$

$$\times \left\{\delta - \langle J_i m_i\, j m_2 | \mathrm{S}^{-1} | J_i' m_i'\, j' m_2' \rangle \langle J_f' m_f'\, j' m_2' | \mathrm{S} | J_f m_f\, j m_2 \rangle \right\},$$

where

$$\delta = \delta_{J_i J_i'} \delta_{J_f J_f'} \delta_{m_i m_i'} \delta_{m_f m_f'} \delta_{jj'} \delta_{m_2 m_2'} \quad (2.3.4)$$

and $C^{J_i m_i}_{J_f m_f KM}$ denotes the Clebsch–Gordan coefficients [7]. Introducing the distribution function for the translational motion we can write that

$$\sum_j \rho_j = \sum_{J_2} \rho_{J_2} \int_0^\infty \mathrm{v}\, d\,\mathrm{v}\, f(\mathrm{v}) \int_0^\infty db 2\pi b, \quad (2.3.5)$$

where J_2 are the rotational quantum numbers for the bath molecule ρ_{J_2} are the populations of the corresponding energy levels, $f(\mathrm{v})$ is the Maxwell–Boltzmann velocity distribution (normalised to unity) and b is the impact parameter. The matrix element of Eq. (2.3.3) thus becomes [6]

$$\Lambda_{if,i'f'}^{(l)} = -in_b \sum_{J_2} \rho_{J_2} \int_0^{\infty} \mathrm{v} d\,\mathrm{v}\, F(\mathrm{v}) \int_0^{\infty} db 2\pi b S(if,i'f'|b), \tag{2.3.6}$$

where

$$\begin{aligned} S(if,i'f'|b) \equiv \sum_{MJ_2'} \sum_{\substack{m_i m_f m_2 \\ m_i' m_f' m_2'}} C_{J_f m_f KM}^{J_i m_i} C_{J_f' m_f' KM}^{J_i' m_i'} (2J_2+1)^{-1} [(2J_i+1)(2J_i'+1)]^{-1/2} \\ \times \{\delta - \langle J_i m_i J_2 m_2 | \mathrm{S}^{-1} | J_i' m_i' J_2' m_2' \rangle \langle J_f' m_f' J_2' m_2' | \mathrm{S} | J_f m_f J_2 m_2 \rangle \} \end{aligned} \tag{2.3.7}$$

is called the 'interruption function'. The properties of the Clebsch–Gordan coefficients [7] allow expressing the function $S(b) \equiv S(if, if \mid b)$ as [2]

$$\begin{aligned} S(b) = 1 - \sum_{\substack{m_i m_f m_2 \\ m_i' m_f' m_2'}} \sum_{MJ_2'} C_{J_f m_f KM}^{J_i m_i} C_{J_f' m_f' KM}^{J_i' m_i'} (2J_i+1)^{-1} (2J_2+1)^{-1} \\ \times \langle J_i m_i J_2 m_2 | \mathrm{S}^{-1}(b) | J_i m_i' J_2' m_2' \rangle \langle J_f m_f' J_2' m_2' | \mathrm{S}(b) | J_f m_f J_2 m_2 \rangle . \end{aligned} \tag{2.3.8}$$

Here $\mathrm{S} = (U_0)^{-1} U$ where $U_0 = e^{-i(H_a+H_b)t/\hbar}$ is the time evolution operator for the Hamiltonian $H_a + H_b$ and U is the time evolution operator for the Hamiltonian H of Eq. (2.1.4). For each individual collision, S is solution of the equation

$$i\hbar \dot{\mathrm{S}} = (U_0^{-1} V U_0) \mathrm{S} \tag{2.3.9}$$

(with the initial condition $\mathrm{S}(\infty) = 1$) which, neglecting the non-commutative character of $V(t)$ and $V(t')$, gives

$$\mathrm{S}(b) = \theta \exp[\frac{1}{i\hbar} \int_{-\infty}^{\infty} dt\, \tilde{V}(t)] \tag{2.3.10}$$

with θ denoting the time ordering operator and $\tilde{V}(t)$ standing for the interaction potential in the Heisenberg representation:

$$\tilde{V}(t) = e^{i(H_a+H_b)t/\hbar} V e^{-i(H_a+H_b)t/\hbar} = (U_0)^{-1} V U_0 . \tag{2.3.11}$$

The scattering matrix S can be computed classically or quantum-mechanically. In the purely quantum approach, its computation is quite tedious and is limited in practice to simple molecular systems. Some quantum results for HF and HCl in Ar are discussed in Sec. 4.5.

The interruption function of Eq. (2.3.8) integrated over the impact parameter values and averaged over the relative molecular velocities yields the (complex-valued) optical cross-section:

$$\sigma_{J_2} = \int_0^{\infty} \mathrm{v} d\mathrm{v} f(\mathrm{v}) \int_0^{\infty} 2\pi b db S(b), \tag{2.3.12}$$

which leads to a compact expression for the line-shift δ_{if} and half-width γ_{if} (see also Eq. (1.2.14)):

$$\gamma_{if} + i\delta_{if} = (2\pi c)^{-1} n_b \sum_{J_2} \rho_{J_2} \sigma_{J_2} \,. \tag{2.3.13}$$

2.4. Anderson–Tsao–Curnutte theory

The first suitable method for line-width and shift calculations was proposed by Anderson [1] in 1949. In 1962, it was systematised by Tsao and Curnutte [2] and extended to the case of Raman scattering by Fiutak and van Kranendonk [8, 9]. The improved version has become widely adopted under the name of the Anderson–Tsao–Curnutte (ATC) theory.

The method is based on the perturbation theory for the scattering matrix S. Supposing that the intermolecular interaction potential (perturbation) is small in comparison with the intramolecular energies, the matrix S(b) in Eq. (2.3.10) can be expanded into perturbation series

$$\mathrm{S} = \mathrm{S}_0 + \mathrm{S}_1 + \mathrm{S}_2 + \ldots, \tag{2.4.1}$$

where

$$\mathrm{S}_0 = 1, \tag{2.4.2}$$

$$\mathrm{S}_1 = \frac{1}{i\hbar} \int_{-\infty}^{\infty} U_0^{-1} V(t) U_0 \, dt \,, \tag{2.4.3}$$

$$\mathrm{S}_2 = -\frac{1}{\hbar^2} \int_{-\infty}^{\infty} \left[U_0^{-1} V(t') U_0\right] dt' \int_0^{t'} \left[U_0^{-1} V(t'') U_0\right] dt'' \,, \tag{2.4.4}$$

and the summation is usually stopped on the second-order term.

In accordance with Eq. (2.4.1), the interruption function of Eq. (2.3.8) reads

$$S(b) = S_0(b) + S_1(b) + S_2(b) + \ldots. \tag{2.4.5}$$

The zero-order term $S_0(b)$ is defined by $\mathrm{S}_0 = \mathrm{S}_0^{-1} = 1$ and, due to the properties of the Clebsch–Gordan coefficients [7], the sum in the right side of Eq. (2.3.8) is equal to one, so that $S_0(b) = 0$. The first-order term $S_1(b)$ depends on $\mathrm{S}_0^{-1}\,\mathrm{S}_1$ and $\mathrm{S}_0\mathrm{S}_1^{-1}$ and contains consequently two (purely imaginary) terms:

$$S_1(b) = i[\Delta_i - \Delta_f], \tag{2.4.6}$$

where

$$\Delta_i = \sum_{m_i m_2} \frac{\langle J_i m_i J_2 m_2 | P | J_i m_i J_2 m_2 \rangle}{(2J_i + 1)(2J_2 + 1)} \tag{2.4.7}$$

and $P \equiv i\mathrm{S}_1$ (Δ_f is obtained by simple replacing of i by f in Eq. (2.4.7)). The second-order contribution $S_2(b)$ is due to $\mathrm{S}_0^{-1}\mathrm{S}_2$, $\mathrm{S}_0\,\mathrm{S}_2^{-1}$ and $\mathrm{S}_1^{-1}\,\mathrm{S}_1$, so that three terms appear:

$$S_2(b) = S_{2,i}^{outer}(b) + S_{2,f}^{outer}(b) + S_2^{middle}(b) \tag{2.4.8}$$

with

$$S_{2,i}^{outer}(b) = \frac{1}{2} \sum_{m_i m_2} \frac{\langle J_i m_i J_2 m_2 | P^2 | J_i m_i J_2 m_2 \rangle}{(2J_i + 1)(2J_2 + 1)}, \tag{2.4.9}$$

$$S_2^{middle}(b) = -\sum_{\substack{m_i m_i' m_f m_f' \\ m_2 m_2' J_2' M}} C_{J_f m_f KM}^{J_i m_i} C_{J_f m_f' KM}^{J_i m_i'} (2J_i + 1)^{-1} (2J_2 + 1)^{-1}$$
$$\times \langle J_f m_f J_2 m_2 | P | J_f m_f' J_2' m_2' \rangle \langle J_i m_i' J_2' m_2' | P | J_i m_i J_2 m_2 \rangle \tag{2.4.10}$$

and $P^2 \equiv -2\mathrm{S}_2$ (the corresponding final state contribution $S_{2,f}^{\;outer}(b)$ is obtained from Eq. (2.4.9) by replacing the subscript i by the subscript f). However, the formulae (2.4.5)–(2.4.10) can not yet be used for practical calculations, since a preliminary summation over magnetic quantum numbers is needed. This operation is described in Sec. 2.6.

For small values of the impact parameter b the collision becomes so strong that it 'either interrupts the absorption completely or it results in the arbitrary phase-shift (which averages to zero) even if the molecule remains in the same non-degenerate state' [2]. As a result, the function $S(b)$ diverges at $b \to 0$. To overcome this problem Anderson introduced an artificial cut-off value b_0 such that for $b<b_0$ the interruption function equals to unity. The optical cross-section of Eq. (2.3.12) thus becomes

$$\sigma_{J_2} = \int_0^\infty \mathrm{v} d\mathrm{v} f(\mathrm{v}) \left[\pi b_0^2 + \int_{b_0}^\infty 2\pi b db S(b) \right], \tag{2.4.11}$$

$$S(b) = i(\Delta_i - \Delta_f) + S_{2,i}^{outer} + S_{2,f}^{outer\,*} + S_2^{middle}. \tag{2.4.12}$$

In the original version of the ATC theory, the principle objective was the line-width calculation, so that $S_2(b)$ was assumed to be real and the cut-off parameter b_0 was determined from the equation

$$S_2(b_0) = 1. \tag{2.4.13}$$

Later, in order to include the line shifts, the imaginary part of $S_2(b)$ was accounted for and the cut-off b_0 was properly modified [10]:

$$\mathrm{Re}\, S(b_0) + \mathrm{Im}\left| S(b_0) \right| = 1. \tag{2.4.14}$$

As the lower-limit estimation b_{min} for b_0, the 'kinetic' diameter r_0 [11] can be taken which represents the b value averaged over all collisions, so that b_{min} replaces b_0 in Eq. (2.4.11).

The optical cross-section of Eq. (2.4.11) is significantly simplified if the mean thermal velocity $\overline{\mathrm{v}}$ is used for the relative molecular motion:

$$\sigma_{J_2} = \overline{\mathrm{v}}[\pi b_0^2 + \int_{b_{sup}}^{\infty} 2\pi b db S(b)], \tag{2.4.15}$$

where $b_{sup} = \mathrm{Sup}\{b_0, b_{min}\}$.

2.5. Interaction potential and irreducible tensors formalism

In order to put the interruption function $S(b)$ in a suitable computation form, the intermolecular potential should be specified. For long-range interactions, it is generally written as an expansion into irreducible tensor operator components for 2^l-pole moments. The full charge of the molecule (scalar) is an irreducible tensor of rank $l = 0$, the dipole moment (vector) is an irreducible tensor of rank $l = 1$, the quadrupole moment has the rank $l = 2$, $l = 3$ corresponds to the octupole moment, and so on. Any irreducible tensor operator T^l is completely defined [7] by the set of linear operators T_m^l ($m = -l,\ldots, l$) which are self-images of the Hilbert ket-vectors space. Their space is rotationally irreducible, and during rotation all m-components of the operator T^l linearly transform one through another, according to the irreducible representation of the l-th rotational group. If A_{lk} are the components of an irreducible tensor operator in the molecular frame and T_m^l are its components in the space-fixed frame, then [12]

$$T_m^l = \sum_{k=-l}^{l} A_{lk} D_{km}^l, \tag{2.5.1}$$

where the rotational matrices $D_{km}^l(\varphi,\theta,\chi)$ are defined according to Ref. [13].

For example, if μ_x, μ_y, μ_z are the Cartesian components of dipole moment along the molecular axes Ox, Oy, Oz, then its standard spherical components A_{lk} (specifically noted q_{lk} for electrostatic interactions) in this frame are

$$q_{11} = -(\mu_x + i\mu_y)/\sqrt{2}, \quad q_{10} = \mu_z, \quad q_{1-1} = (\mu_x - i\mu_y)/\sqrt{2}. \tag{2.5.2}$$

In the same way, if M_X, M_Y, M_Z are the Cartesian components of the dipole moment along the space-fixed axes, the corresponding standard spherical components $T^1_m \equiv M^1_m$ in this space-fixed system are written as

$$M^1_1 = -(M_X + iM_Y)/\sqrt{2}, \quad M^1_0 = M_Z, \quad M^1_{-1} = (M_X - iM_Y)/\sqrt{2}. \tag{2.5.3}$$

Analogously, for the quadrupole moment with the Cartesian components

$$Q_{\alpha\beta} = e\sum_i Z_i\left(3r_{i\alpha}r_{i\beta} - \delta_{\alpha\beta}r_i^2\right)/2 \tag{2.5.4}$$

(Z_i denoting the charges in units of elementary charge e) in the molecular frame ($\alpha, \beta = x, y, z$), the associated spherical components in the same frame are

$$q_{20} = Q_{zz},\ q_{2\pm1} = -(2/3)^{1/2}\,(Q_{xz} \pm i\,Q_{yz}),\ q_{2\pm2} = (Q_{xx} - Q_{yy} \pm 2i\,Q_{xy})/6^{1/2}. \tag{2.5.5}$$

Moreover, since the quadrupole moment is a symmetric traceless tensor ($\Sigma_\alpha Q_{\alpha\alpha} = 0$), in the principle-axes molecular coordinate system it reduces to a diagonal form and only q_{20}, $q_{2\pm2}$ components are non-vanishing.

The intermolecular potential for two interacting molecules 1 and 2 in the form invariant with respect to the overall rotations of the system has been obtained by Leavitt [14] and Gray and Gubbins [15]. The long-range potential expression with the usual order of Euler angles (φ, θ, χ) in the rotational matrices reads [15]

$$V(1,2,\vec{r}) = \sum_{l_1l_2lk_1k_2} V^{k_1k_2}_{l_1l_2l}(r)\sum_{m_1m_2} C^{lm_1+m_2}_{l_1m_1l_2m_2} D^{l_1}_{m_1k_1}(1)^* D^{l_2}_{m_2k_2}(2)^* C_{lm_1+m_2}(\hat{r})^*, \tag{2.5.6}$$

where the radial potential components

$$V^{k_1k_2}_{l_1l_2l}(r) \equiv \sum_n B^{(n)}_{l_1l_2l}(r) A^{(n)}_{l_1k_1} A^{(n)}_{l_2k_2} \tag{2.5.7}$$

are introduced, the spherical harmonics $C_{lm}(\hat{r}) = [4\pi/(2l+1)]^{1/2} Y_{lm}(\theta, \varphi)$ characterise the orientation of the intermolecular vector $\vec{r}$ in the laboratory frame and the asterisk stands for the complex conjugation. The superscript (n) distinguishes different (electrostatic, induction and dispersion) interactions of the same tensor rank. These interactions are briefly reviewed below, following the notations of Leavitt [14].

2.5.1. Electrostatic interactions

For the electrostatic interactions the molecular parameters $A^{(elec)}_{lk} \equiv q_{lk}$ are the expectation values of the corresponding multipole moment operator Θ_{lk} in the electronic ground state:

$$q_{lk} = \langle 0|\,\Theta_{lk}\,|0\rangle. \tag{2.5.8}$$

The operator itself is defined as

$$\Theta_{lk} = e\sum_i Z_i r_i^l C_{lk}(\hat{r}_i) \tag{2.5.9}$$

with $\hat{r}_i$ standing for the angular orientation of the i-th atom in the molecular frame. For $l = 0$, Θ_{00} corresponds to the charge of the molecule. For $l = 1$, Θ_{1k} give the dipole moment components of Eq. (2.5.2). For $l = 2$, Θ_{2k} determine the spherical components (2.5.5) of the quadrupole moment. For a given molecular point group symmetry, the number of non-vanishing spherical components q_{lk} for $l = 1, 2, 3$ can be found in Table III of Ref. [14]. For the C_{2v} group, for example, there is one dipole moment component q_{10}, two quadrupole moment components q_{20}, $q_{22} = q_{2-2}$ and two octupole moment components.

The radial-dependence coefficients $B_{l_1 l_2 l}^{(n)}(r)$ are specified for the electrostatic interactions as [14]

$$B_{l_1 l_2 l}^{(elec)}(r) = (-1)^{l_2} \begin{pmatrix} 2l_1 + 2l_2 \\ 2l_1 \end{pmatrix}^{1/2} \delta_{l,l_1+l_2} / r^{l_1+l_2+1}, \tag{2.5.10}$$

$$\begin{pmatrix} a \\ b \end{pmatrix} = \frac{a!}{b!(a-b)!}.$$

If the interacting molecules are neutral, the leading term of Eq. (2.5.6) is given by the dipole–dipole interaction ($l_1 = l_2 = 1$) which varies as r^{-3}.

2.5.2. Induction and dispersion interactions

A neutral active molecule 1 with a permanent dipole moment is a source of an electric field which induces a dipole moment in the perturbing (polar or not) molecule 2. This induced dipole moment interacts with the permanent dipole moment of the first molecule, and the potential energy of this interaction depends on the polarizability (anisotropic and frequency-dependent) of the second molecule. Since the tensor ranks l_1 and l_2 for both molecules are determined by the vector sums $s_1 + s_1'$ and $s_2 + s_2'$ respectively, the molecular parameters due to the permanent dipole moment of the active molecule are defined by [14]

$$A_{l_1 k_1}^{(ind\,1)} = P_{l_1 k_1}^{(s_1 s_1')}(1), \qquad A_{l_2 k_2}^{(ind\,1)} = \alpha_{l_2 k_2}^{(s_2 s_2')}(2), \tag{2.5.11}$$

where

$$P_{l_1 k_1}^{(s_1 s_1')}(1) = \sum_{t_1} q_{s_1 t_1} q_{s_1' k_1 - t_1} C_{s_1 t_1 s_1' k_1 - t_1}^{l_1 k_1} \tag{2.5.12}$$

and the generalised polarizabilities

$$\alpha_{l_2k_2}^{(s_2s_2')}(2)=\sum_{t_2p_2\neq 0}\left(E_{p_2}^{(2)}-E_0^{(2)}\right)^{-1}\langle 0|\Theta_{s_2t_2}|p_2\rangle\langle p_2|\Theta_{s_2'k_2-t_2}|0\rangle C_{s_2t_2s_2'k_2-t_2}^{l_2k_2} \qquad (2.5.13)$$

are summed over all excited ($p_2\neq0$) states of the second molecule having the energies $E_{p_2}^{(2)}$. The spherical components $q_{s_1t_1}$ for the molecule 1 are defined by Eq. (2.5.8) and the multipole moment components $\Theta_{s_2t_2}$ of the molecule 2 should be taken from Eq. (2.5.9). For the induction energy due to the permanent moment of molecule 2, the indices 1 and 2 in Eqs (2.5.11)–(2.5.13) have to be exchanged.

Besides the induction terms, the interaction of instantaneous dipoles in both molecules yields an additional dispersion contribution to the potential energy. According to Ref. [14], the corresponding molecular parameters are given by

$$A_{l_1k_1}^{(disp)}=\alpha_{l_1k_1}^{(s_1s_1')}(1), \qquad A_{l_2k_2}^{(disp)}=\alpha_{l_2k_2}^{(s_2s_2')}(2), \qquad (2.5.14)$$

and the radial-dependence coefficients are related by

$$B_{l_1l_2l}^{(ind)}(r)=B_{l_1l_2l}^{(disp)}(r)/\bar{u}, \qquad (2.5.15)$$

where $\bar{u}=u_1u_2/(u_1+u_2)$ is defined by the ionization energies u_1 and u_2 of molecules 1 and 2. A general expression for $B_{l_1l_2l}^{(ind)}(r)$ coefficients and their particular values for the induction terms proportional to r^{-6} and r^{-7} can be found in Ref. [14].

Spherical components α_{lk} of the polarizability tensor with $l = 0$ (isotropic part) and with $l = 2$ (anisotropic part) are defined through the molecular-frame Cartesians components as

$$\begin{aligned}
\alpha_{00}&=-(\alpha_{xx}+\alpha_{yy}+\alpha_{zz})/3^{1/2}, \\
\alpha_{20}&=(2\alpha_{zz}-\alpha_{yy}-\alpha_{xx})/6^{1/2}, \\
\alpha_{2\pm1}&=\mp(\alpha_{xz}\pm i\alpha_{yz}), \\
\alpha_{2\pm2}&=[(\alpha_{xx}-\alpha_{yy})\pm 2i\alpha_{xy}]/2.
\end{aligned} \qquad (2.5.16)$$

In the space-fixed axes, the spherical components of the polarizability tensor are defined by the same Eq. (2.5.16) but with the corresponding laboratory-frame Cartesian components. It is also noteworthy that, in contrast with the quadrupole moment tensor, the polarizability tensor is not symmetric in the general case.

It follows from Eqs (2.5.6) and (2.5.7) that the first contribution to the isotropic part of induction and dispersion energies is given by the term with $l_1 = l_2 = 0$:

$$V_{000}^{(ind+disp)}=\left[P_{00}^{(11)}(1)\alpha_{00}^{(11)}(2)+P_{00}^{(11)}(2)\alpha_{00}^{(11)}(1)+\bar{u}\,\alpha_{00}^{(11)}(1)\alpha_{00}^{(11)}(2)\right]B_{000}^{(ind)}(r). \qquad (2.5.17)$$

Putting $B_{000}^{(ind)}(r) = -2/r^6$, $P_{00}^{(11)} = -\mu^2/3^{1/2}$ and $\alpha_{00}^{(11)} = -3^{1/2}\alpha/2$, where $\alpha = -\alpha_{00}/3^{1/2}$ is the mean polarizability, we obtain

$$V_{000}^{(ind+disp)} = -\left(\mu_1^2\alpha_2 + \mu_2^2\alpha_1 + \frac{3}{2}\bar{u}\alpha_1\alpha_2\right)r^{-6} \qquad (2.5.18)$$

(the indices 1 and 2 refer to the active and perturbing molecules, respectively).

As shown by Leavitt [16], the spherical components A_{lk} follow the relation

$$A_{lk} = (-1)^{\sigma-l+k} A_{l-k}^*, \qquad (2.5.19)$$

where $(-1)^{\sigma}$ is the parity of A_{lk} given by $(-1)^l$ for electrostatic interactions and by $(-1)^{s+s'}$ for induction and dispersion interactions.

2.5.3. Atom–atom interactions

For non-polar molecules or a polar molecule perturbed by an atom, the long-range part of the interaction potential is not sufficient to correctly reproduce the line widths and shifts. The short-range forces should be additionally accounted for. The most common approach to do this consists in adding pair atom–atom interactions to the long-range (electrostatic) terms. The potential energy of atom–atom interactions approximated by Lennard–Jones dependences reads [17]

$$V^{(aa)} = \sum_{i,j}\left(\frac{d_{ij}}{r_{1i,2j}^{12}} - \frac{e_{ij}}{r_{1i,2j}^{6}}\right), \qquad (2.5.20)$$

where d_{ij} and e_{ij} are the atomic pair energy parameters and $r_{1i,2j}$ is the distance between the i-th atom of the first molecule and the j-th atom of the second molecule. In order to put Eq. (2.5.20) into the form of Eq. (2.5.6), the two-centre expansion of the function $r_{1i,2j}^{-n}$ (for $r > r_{1i} + r_{2j}$) has to be made [18, 19]:

$$r_{1i,2j}^{-n} = \sum_{l_1l_2l} f_{l_1l_2l}^{n}(r_{1i}, r_{2j}, r) \sum_{m_1m_2m} C_{l_1m_1l_2m_2}^{lm} Y_{l_1m_1}(\hat{r}_{1i}) Y_{l_2m_2}(\hat{r}_{2j}) C_{lm}(\hat{r})^* \qquad (2.5.21)$$

with

$$f_{l_1l_2l}^{n}(r_{1i}, r_{2j}, r) \equiv \frac{(-1)^{l_2} 4\pi[(2l_1+1)(2l_2+1)]^{1/2}}{r^n} C_{l_10l_20}^{l0}$$
$$\times \sum_{p,q}\left(\frac{r_{1i}}{r}\right)^p \left(\frac{r_{2j}}{r}\right)^q \frac{(n+p+q-l-3)!!(n+p+q+l-2)!!}{(n-2)!(p-l_1)!!(p+l_1+1)!!(q-l_2)!!(q+l_2+1)!!} \qquad (2.5.22)$$
$$\times[1+\delta_{n1}(\delta_{pl_1}\delta_{ql_2}\delta_{p+q,l}-1)],$$

$p = l_1, l_1+2, l_1+4,...$ and $q = l_2, l_2+2, l_2+4,...$ Since for polyatomic molecules the atoms i,j may be located out of the molecular axes described by $(\varphi_1, \theta_1, \chi_1)$ and $(\varphi_2, \theta_2, \chi_2)$ (cf. Eq. (2.5.6)), the spherical harmonics $Y_{l_1m_1}(\hat{r}_{1i}) \equiv Y_{l_1m_1}(\theta_{1i}, \varphi_{1i})$ and $Y_{l_2m_2}(\hat{r}_{2j}) \equiv Y_{l_2m_2}(\theta_{2j}, \varphi_{2j})$ must be rewritten as functions of $(\varphi_1, \theta_1, \chi_1)$ and $(\varphi_2, \theta_2, \chi_2)$ [15]:

$$Y_{l_xm_x}(\hat{r}_{xi}) = \sum_{k_x} D^{l_x}_{m_xk_x}(\hat{r}_x)^* Y_{l_xk_x}(\hat{r}'_{xi}), \tag{2.5.23}$$

where $\hat{r}'_{xi}$ is the orientation of the i-th atom of the x-th molecule in the molecular frame.

The radial components of the atom–atom potential in form of Eq. (2.5.6) read therefore

$$V^{k_1k_2}_{l_1l_2l}(r) = \sum_{ij}[d_{ij} f^{12}_{l_1l_2l}(r_{1i}, r_{2j}, r) - e_{ij} f^{6}_{l_1l_2l}(r_{1i}, r_{2j}, r)] Y_{l_1k_1}(\hat{r}'_{1i}) Y_{l_2k_2}(\hat{r}'_{2j}) \cdot \tag{2.5.24}$$

Some models of pair potentials are discussed in Ref. [11]. Molecular parameters for $V^{(aa)}$ are given in Appendix B.

2.5.4. Matrix elements of the tensor operators

The formalism of irreducible tensor components T^l_μ for physical quantities appears to be very suitable since it allows us to separate the magnetic quantum numbers m, m' from the other quantum numbers J, {q} by applying the Wigner–Eckart theorem [7] to the matrix elements $\langle \{q\}Jm | T^l_\mu | \{q'\}J'm' \rangle$:

$$\langle \{q\}Jm | T^l_\mu | \{q'\}J'm' \rangle = (2J+1)^{-1/2} C^{Jm}_{J'm'l\mu} \langle \{q\}J \| T^l \| \{q'\}J' \rangle, \tag{2.5.25}$$

where $\langle \{q\}J \| T^l \| \{q'\}J' \rangle$ is the reduced matrix element characterising the tensor operator and the factor $(2J+1)^{-1/2}$ is introduced for convenience. Taking into account the properties of the Clebsch–Gordan coefficients [7, 12], Eq. (2.5.25) gives the selection rules for the rotational and magnetic quantum numbers:

$$\begin{aligned} &\mu = m - m', \\ &|J - J'| \le l \le J + J'. \end{aligned} \tag{2.5.26}$$

In the basis of symmetric top wave function $|JKm\rangle$ (K is the projection of the total momentum $\vec{J}$ on the molecular z-axis), the matrix element of Eq. (2.5.1) reads

$$\langle JKm | T^l_\mu | J'K'm' \rangle = \sum_k A_{lk} \langle JKm | D^l_{k\mu} | J'K'm' \rangle. \tag{2.5.27}$$

Since the rotational matrix $D^l_{k\mu}$ is proportional to the wave function $|lk\mu\rangle$, then

$$\left\langle JKm\middle|D^l_{k\mu}\middle|J'K'm'\right\rangle = [(2J+1)(2J'+1)]^{1/2}C^{Jm}_{J'm'l\mu}C^{JK}_{J'K'lk} = [(2J+1)(2J'+1)]^{1/2}C^{J'm'}_{Jml-\mu}C^{J'K'}_{JKl-k}\,, \tag{2.5.28}$$

and Eq. (2.5.27) becomes

$$\left\langle JKm\middle|T^l_\mu\middle|J'K'm'\right\rangle = (2J+1)^{-1/2}C^{Jm}_{J'm'l\mu}\sum_k A_{lk}\left\langle JK\middle\|D^l_{k\circ}\middle\|J'K'\right\rangle, \tag{2.5.29}$$

where the reduced matrix element $\langle JK\|\, D^l_{k\circ}\,\|J'K'\rangle$ is defined by

$$\left\langle JK\middle\|D^l_{k\circ}\middle\|J'K'\right\rangle = (2J'+1)^{1/2}C^{JK}_{J'K'lk} = (-1)^{J-J'-k}(2J+1)^{1/2}C^{J'K'}_{JKl-k} \tag{2.5.30}$$

and obeys the relation

$$\langle JK\|\, D^l_{k\circ}\,\|J'K'\rangle = (-1)^{J-J'-k}\langle J'K'\|\, D^l_{-k\circ}\,\|JK\rangle. \tag{2.5.31}$$

(In Eqs (2.5.28)–(2.5.30), the properties of the Clebsch–Gordan coefficients were used.) It is also worthy to note that because of various possible definitions of D-matrices in Eq. (2.5.1) (with usual or inversed order of arguments) it is necessary to chose the appropriate definition of the rotational wave functions $|JKm\rangle$. The Clebsch–Gordan coefficient in Eq. (2.5.30) yields the selection rule for the quantum numbers K and K':

$$k = K - K'. \tag{2.5.32}$$

The reduced matrix element of the operator T^l can be obtained from Eq. (2.5.29):

$$\left\langle JK\middle\|T^l\middle\|J'K'\right\rangle = \sum_k A_{lk}\left\langle JK\middle\|D^l_{k\circ}\middle\|J'K'\right\rangle. \tag{2.5.33}$$

In the general case of asymmetric top molecules, it is necessary to calculate the matrix elements of T^l in the basis $|J\tau m\rangle$, where the label $\tau\,(\tau = -J, \ldots, J)$ distinguishes the $2J$+1 different rotational states:

$$\left\langle J\tau m\middle|T^l_\mu\middle|J'\tau'm'\right\rangle = (2J+1)^{-1/2}C^{Jm}_{J'm'l\mu}A_l(J\tau;J'\tau')\,, \tag{2.5.34}$$

$$A_l(J\tau;J'\tau') = \sum_k A_{lk}\left\langle J\tau\middle\|D^l_{k\circ}\middle\|J'\tau'\right\rangle = (-1)^{\sigma-l+J-J'}A^*_l(J'\tau';J\tau)\,. \tag{2.5.35}$$

In any case, the calculation of the matrix element $A_l(J\tau;J'\tau')$ reduces to its evaluation in the basis of symmetric top wave functions $|JK\rangle$ by Eq. (2.5.30) (explicit formulae are listed in Appendix A).

For a particular case of two linear molecules the rotational matrices reduce to the usual spherical harmonics [12]:

$$D^l_{m0}(\varphi,\theta,\chi)^* = [4\pi/(2l+1)]^{1/2}Y_{lm}(\theta,\varphi), \tag{2.5.36}$$

and the general potential of Eq. (2.5.6) simplifies to

$$V(1,2,\vec{r})=\sum_{l_1l_2l}V_{l_1l_2l}(r)\sum_{m_1m_2}C_{l_1m_1l_2m_2}^{lm_1+m_2}Y_{l_1m_1}(1)Y_{l_2m_2}(2)C_{lm_1+m_2}(\hat{r})^*. \qquad (2.5.37)$$

For a linear molecule interacting with an atom ($l_2 = 0$), moreover, $Y_{00}(\theta, \varphi) = (4\pi)^{1/2}$, and the interaction potential is expanded into a series of Legendre polynomials $P_l(\cos\theta)$:

$$V(r,\theta) = \sum_l V_l(r)P_l(\cos\theta), \qquad (2.5.39)$$

where θ is the angle between the molecular axis and the intermolecular distance vector.

2.6. Interruption function for the long-range intermolecular potential

Let i and f respectively denote the sets of quantum numbers of the initial (lower) and final (upper) states of the absorbing molecule. The practical calculation of the line widths and shifts with Eqs (2.4.6)–(2.4.10) requires the summation over magnetic quantum numbers. The first-order contribution S_1 given by Eqs (2.4.6) and (2.4.7) is diagonal on m_i and m_f and comes therefore from the isotropic part of the interaction potential [14]:

$$\Delta_i = \frac{1}{\hbar}\sum_n \frac{A_0^{(n,1)}(i;i)A_0^{(n,2)}(2;2)F_{0000}^{(n)}(0)}{\sqrt{(2J_i+1)(2J_2+1)}}, \qquad (2.6.1)$$

where $A_{l_1}^{(n,1)}(i';i) \equiv A_{l_1}^{(n,1)}(J_i'\tau_i';J_i\tau_i)$, $A_{l_2}^{(n,2)}(2';2) \equiv A_{l_2}^{(n,2)}(J_2'\tau_2';J_2\tau_2)$ and $F_{l_1l_2lm}^{(n)}(\omega)$ are the Fourier transforms of the interaction potential coefficients:

$$F_{l_1l_2lm}^{(n)}(\omega) = \int_{-\infty}^{\infty} dt\, e^{i\omega t}C_{lm}(\hat{r}(t))B_{l_1l_2l}^{(n)}(r(t)) \qquad (2.6.2)$$

The substitution of the isotropic part (2.5.18) (induction and dispersion interactions) of the interaction potential in Eq. (2.6.2) and then in Eq. (2.6.1) gives

$$\Delta_i = -\frac{3\pi}{8\hbar v b^5}\left[\mu_1^2\alpha_2 + \mu_2^2\alpha_1 + \frac{3}{2}\bar{u}\alpha_1\alpha_2\right]_i. \qquad (2.6.3)$$

After the summation over magnetic quantum numbers the second-order contributions of Eqs (2.4.9) and (2.4.10) read [14]

$$S_{2,i}^{outer} = \frac{1}{2\hbar^2(2J_i+1)(2J_2+1)} \times \sum_{\substack{i'2' \\ l_1l_2l\,nn'}} \frac{A_{l_1}^{(n',1)}(i';i)A_{l_1}^{(n,1)}(i';i)^* A_{l_2}^{(n',2)}(2';2)A_{l_2}^{(n,2)}(2';2)^*}{(2l_1+1)(2l_2+1)} \sum_m \varphi_{l_1l_2lm}^{(nn')}, \tag{2.6.4}$$

$$S_2^{middle} = -\frac{1}{\hbar^2(2J_2+1)} \sum_{\substack{2' \\ l_1l_2l\,nn'}} \frac{A_{l_1}^{(n',1)}(i;i)A_{l_1}^{(n,1)}(f;f)^* A_{l_2}^{(n',2)}(2';2)A_{l_2}^{(n,2)}(2';2)^*}{(2l_1+1)(2l_2+1)} \times W(J_i pl_1 J_f; J_f J_i) \sum_m \mathrm{Re}\,\varphi_{l_1l_2lm}^{(nn')}, \tag{2.6.5}$$

where $W(J_i\ pl_1\ J_f;J_f\ J_i)$ is the Racah coefficient [7] and the quantities $\varphi_{l_1l_2lm}^{(nn')}$ are defined by the products of integrals (2.6.2).

For commodity, the second-order contributions are sometimes written as a series over l_1 and l_2 values:

$$S_2(b) = \sum_{l_1l_2} {}^{l_1l_2}S_2(b) = \sum_{l_1l_2} [{}^{l_1l_2}S_{2,i}^{outer}(b) + {}^{l_1l_2}S_{2,f}^{outer}(b) + {}^{l_1l_2}S_2^{middle}(b)]. \tag{2.6.6}$$

The sums over m in Eqs (2.6.4.) and (2.6.5) can be therefore expressed as

$$\sum_m \varphi_{l_1l_2lm}^{(nn')} \equiv M_{pp'}^{l} = {}^{(nn')}C_{pp'}^{l}[{}^{l_1l_2}f_p^{p'}(k) - iI\,{}^{l_1l_2}f_p^{p'}(k)]. \tag{2.6.7}$$

In this equation, the coefficients ${}^{(nn')}C_{pp'}^{l}$ are defined by the interaction type whereas ${}^{l_1l_2}f_p^{p'}(k)$ and $I\,{}^{l_1l_2}f_p^{p'}(k)$ are respectively so-called (real-valued) resonance functions and their Hilbert transforms (imaginary parts) of the parameter

$$k = b(\omega_{i'i} + \omega_{2'2})/v \tag{2.6.8}$$

(called adiabaticity parameter), where $\omega_{i'i} = (E_{i'} - E_i)/\hbar$; $\omega_{2'2} = (E_{2'} - E_2)/\hbar$ are the angular frequencies of the collision-induced transitions in the active and perturbing molecules. The detailed expressions for these resonance functions are discussed in the next section. The most complete list of the different contributions in $S_2(b)$ in form of Eq. (2.6.6) is given in Ref. [14] and is widely used for molecular systems with dominant long-range forces. In this reference, the contributions ${}^{l_1l_2}S_{2,i}$ corresponding to the induction and dispersion interactions are put together and depend on the quantity

$$C_{l_1l_2}^{(k_1k_1';k_2k_2')}(1'2';12) = P_{l_1}^{(k_1k_1')}(1';1)\alpha_{l_2}^{(k_2k_2')}(2';2) + \alpha_{l_1}^{(k_1k_1')}(1';1)P_{l_2}^{(k_2k_2')}(2';2) + \bar{u}\,\alpha_{l_1}^{(k_1k_1')}(1';1)\alpha_{l_2}^{(k_2k_2')}(2';2), \tag{2.6.9}$$

where $P_{l_1}^{(k_1k_1')}(1';1)$ etc are the matrix elements defined by Eq. (2.5.35). For asymmetric top molecules X_2Y the following relations for physical quantities A_{lk} should be used in this expression:

$$P_{11}^{(12)} = -P_{1-1}^{(12)} = -\frac{\mu}{\sqrt{2}}\left(q_{20}\frac{1}{\sqrt{10}} - q_{22}\sqrt{\frac{3}{5}}\right),$$

$$\alpha_{00}^{(11)} = -\frac{\sqrt{3}}{2}\alpha, \quad P_{20}^{(11)} = -\frac{\mu^2}{\sqrt{6}}, \tag{2.6.10}$$

$$P_{22}^{(11)} = P_{2-2}^{(11)} = -\frac{\mu^2}{2}, \quad P_{00}^{(11)} = -\frac{\mu^2}{\sqrt{3}},$$

where, as previously, α is the mean polarizability of the molecule, μ is its dipole moment (in the molecular frame such that the symmetry axis is the dipole moment axis), and q_{20}, $q_{22} = q_{2-2}$ are the spherical components of the quadrupole moment given by Eq. (2.5.5).

For linear molecules the eigenstates of the rotational Hamiltonian are labelled by the rotational quantum number J, i.e. these are vectors $|JK\rangle$ of the symmetric top with $K = 0$. In this case, for the matrix element,

$$A_l^{(n)}(J;J') = A_{l0}^{(n)}(2J'+1)^{1/2} C_{J'0l0}^{J0} \tag{2.6.11}$$

the value of $J + l + J'$ must be even since the terms S_2^{middle} vanish for l odd. We also have

$$P_{l0}^{(kk')} = q_{k0}q_{k'0}C_{k0k'0}^{l0}. \tag{2.6.12}$$

The molecular parameters for linear rotors can be written in the usual form:

$$q_{10} = \mu, \quad q_{20} = Q, \quad q_{30} = \Omega, \quad \alpha_{00}^{(11)} = -\frac{\sqrt{3}}{2}\alpha,$$

$$\alpha_{20}^{(11)} = \sqrt{\frac{3}{2}}\alpha\gamma, \quad \alpha_{10}^{(12)} = -\frac{1}{\sqrt{10}}(A_{\text{II}} + 2A_{\perp}), \tag{2.6.13}$$

$$\alpha_{30}^{(12)} = \frac{\sqrt{3}}{2\sqrt{5}}\left(A_{\text{II}} - \frac{4}{3}A_{\perp}\right).$$

Here μ is the dipole moment, Q is the quadrupole moment, Ω is the octupole moment, α is the polarizability, γ is the polarizability anisotropy, A_{II} and $A_{\perp}$ are higher polarizabilities.

One more detail is worthy to note for the calculation of contributions to $S_2(b)$. The denominators of these contributions contain the factors $(2J_i + 1)$, $(2J_2 + 1)$

but the squares of the reduced matrix elements are proportional to these factors too, so that finally the terms ${}^{0l_2}S_2$ do not depend on J_i and J_f as well as the terms ${}^{l_10}S_2$ do not depend on J_2. In some cases in computer programs, it is therefore more convenient to use the reduced matrix elements and the expressions for the second-order contributions without the abovementioned factors.

2.7. Resonance functions in the straight-line trajectory approximation

For the sake of commodity, the ${}^{l_1l_2}S_2$ terms are usually written in terms of the resonance functions ${}^{l_1l_2}f_p^{p'}$ (and ${}^{l_1l_2}g_p^{p'}$ coming from the polarization part of the interaction potential) normalised to unity at $k = 0$. Indeed, Eq. (2.6.7) can be rewritten as

$$M^{\,l}_{pp'} = {}^{(nn')}C^{\,l}_{pp'}\,{}^{l_1l_2}f_p^{p'} = \sum_{m=-l}^{l} I^{p\,*}_{lm} I^{p'}_{lm}\,, \tag{2.7.1}$$

where the integrals

$$I^{p}_{lm} = \int_{-\infty}^{\infty} dt \frac{e^{i\omega t} C_{lm}(\hat{r})}{r(t)^p} \tag{2.7.2}$$

represent the Fourier transforms of the potential coefficients and differ from Eq. (2.6.2) only by a constant factor. The spherical harmonics $C_{lm}(\theta,\varphi) = [4\pi/(2l+1)]^{1/2} Y_{lm}(\theta,\varphi)$ are tied to the orientation of the intermolecular distance vector $\vec{r}$ in the laboratory frame. If the collision takes place in the *YOZ*-plane ($\varphi = \pi/2$) and $\theta = \pi/2 - \Psi$ (i.e. $x(t) = 0$, $y(t) = b$, $z(t) = \mathrm{v}\,t$), solely the angle Ψ governs the angular collisional dynamics. Moreover, we have

$$\begin{aligned} r(t) &= (b^2 + \mathrm{v}^2 t^2)^{1/2}, \\ \cos\Psi &= b/r(t), \\ \sin\Psi &= \mathrm{v}t/r(t). \end{aligned} \tag{2.7.3}$$

If the spherical harmonics are expressed as

$$C_{lm}\left(\frac{\pi}{2} - \Psi, \frac{\pi}{2}\right) = (\cos\Psi)^m \sum_s a(l,m,s)(\sin\Psi)^s\,, \tag{2.7.4}$$

where the coefficients $a(l, m, s)$ are given by Table C1 of Appendix C, Eq. (2.7.2) becomes

$$I^{p}_{lm} = \sum_s a(l,m,s) \int_{-\infty}^{\infty} dt \frac{e^{i\omega t} (\cos\Psi)^m (\sin\Psi)^s}{r(t)^p}. \tag{2.7.5}$$

Introducing further the dimensionless variables $u = \mathrm{v}t/b$ and $k = \omega b/\mathrm{v}$ transforms this equation into

$$I_{lm}^{p} = \sum_{s} a(l,m,s) \frac{1}{\mathrm{v}b^{p-1}} \int_{-\infty}^{\infty} \frac{\mathrm{e}^{iku} u^{s} du}{\left(1+u^{2}\right)^{(p+m+s)/s}}, \tag{2.7.6}$$

where any I_{lm}^{p} is completely defined by the integrals

$$I_{n} = \int_{-\infty}^{\infty} dx \frac{e^{ikx}}{\left(1+x^{2}\right)^{n}} \tag{2.7.7}$$

since the integration by parts

$$\int_{-\infty}^{\infty} dx \frac{\mathrm{e}^{ikx} x}{\left(1+x^{2}\right)^{n+1/2}} = \left. \frac{\mathrm{e}^{ikx}}{(-2n+1)\left(1+x^{2}\right)^{n-1/2}} \right|_{-\infty}^{\infty} + \frac{ik}{2n-1} \int_{-\infty}^{\infty} dx \frac{\mathrm{e}^{ikx}}{\left(1+x^{2}\right)^{n-1/2}} \tag{2.7.8}$$

can be used. If $2n - 1 > 0$ the first term on the right-hand side of Eq. (2.7.8) vanishes. For the integer n values the integrals I_n from Eq. (2.7.7) reduce to sums of products of powers of k and modified Bessel functions of the second kind $K_n(k)$. Otherwise they appear as products of an exponential function and some polynomial of k.

If in Eq. (2.7.1) $l_1 = l_2 = 1$ (dipole–dipole interaction), then $l = l_1 + l_2 = 2$ and

$$M_{pp'}^{2} = C_{pp'}^{2}\, {}^{11}f_{p}^{p'}(k) = \left(I_{20}^{p} I_{20}^{p'*} + 2 I_{21}^{p} I_{21}^{p'*} + 2 I_{22}^{p} I_{22}^{p'*}\right), \tag{2.7.9}$$

where

$$\begin{aligned} I_{20}^{p} &= \frac{1}{\mathrm{v}b^{p-1}} \left(I_{\frac{p}{2}} - \frac{3}{2} I_{\frac{p+2}{2}} \right), \\ I_{21}^{p} &= \frac{1}{\mathrm{v}b^{p-1}} \sqrt{\frac{3}{2}} ik \frac{1}{p} I_{\frac{p}{2}}, \\ I_{22}^{p} &= \frac{1}{\mathrm{v}b^{p-1}} \frac{1}{2} \sqrt{\frac{3}{2}} I_{\frac{p+2}{2}}. \end{aligned} \tag{2.7.10}$$

Putting $p = 3$ and the integrals from Appendix C in Eq. (2.7.10) gives the normalised to unity resonance function (noted $f_1(k)$ in Refs [2, 14])

$${}^{11}f_{3}^{3}(k) = \frac{k^{4}}{4} \left[K_{2}(k)^{2} + 4K_{1}(k)^{2} + 3K_{0}(k)^{2} \right].$$

A quite complete list of resonance functions is given by Leavitt [14]. The most important ones correspond to the electrostatic and the leading induction and dispersion interactions:

$$^{12}f_4^4(k) \equiv f_2(k) = \frac{k^6}{64}\left[K_3(k)^2 + 6K_2(k)^2 + 15K_1(k)^2 + 10K_0(k)^2\right],$$

$$^{22}f_5^5(k) \equiv f_3(k) = \frac{k^8}{2304}[K_4(k)^2 + 8K_3(k)^2 + 28K_2(k)^2 + 56K1(k)^2 + 35K_0(k)^2],$$

$$\ldots \qquad (2.7.11)$$

$$^{20}f_6^6(k) \equiv g_1(k) = \frac{e^{-2k}}{63}[63 + 126k + 126k^2 + 84k^3 + 39k^4 + 12k^5 + 2k^6],$$

$$^{10}f_7^7(k) \equiv g_3(k) = \frac{e^{-2k}}{225}[225 + 450k + 414k^2 + 228k^3 + 81k^4 + 18k^5 + 2k^6].$$

For the short-range part of the interaction potential the necessary functions are

$$^{20}f_6^{12}(k) = (e^{-2k} / 45360)\ [24J_3J_6 - 3J_3J_7 - 6J_4J_6 + J_4J_7 + k^2J_3J_6],$$

$$^{20}f_{12}^{12}(k) = (e^{-2k} / 33\ 041\ 925)\ [48(J_6)^2 - 12J_6J_7 + (J_7)^2 + k^2(J_6)^2]$$

(the polynomials J_n can be found in Table C2).

The correspondence between the different interactions and the kind of resonance functions is presented in Table C3 for $k > 0$. For $k < 0$ $f_n(-k) = f_n(k)$, $g_n(-k) = g_n(k)$. The resonance functions $g_n(k)$ can be written in the form

$$g_n(k) = e^{-2k}\sum_{m=0}^{L_n} a_m^{(n)} k^m. \qquad (2.7.12)$$

The coefficients $a_m^{(n)}$ for $n = 1$–7 taken from Ref. [20] are listed in Table 2.1.

Table 2.1. Coefficients $a_m^{(n)}$ [20] ($a_0^{(n)} = 1$, $a_1^{(n)} = 2$). Reproduced with permission.

n/m	2	3	4	5	6	7	8	9	10
1	2	4/3	13/21	4/21	2/63	–	–	–	–
2	88/43	184/129	1292/1677	580/1677	668/5031	24/559	6/559	–	–
3	46/25	76/75	9/25	2/25	2/225	–	–	–	–
4	872/425	1832/1275	964/1275	392/1275	364/3825	16/765	2/765	–	–
5	101/50	103/75	209/300	41/150	37/450	4/225	1/450	–	–
6	2694/ 1325	5564/ 3975	26332/ 35775	11336/ 35775	4184/ 35775	4088/ 107325	1186/ 107325	4/ 1431	4/ 7155
7	68/35	128/105	58/105	4/21	16/315	16/1575	2/1575	–	–

The line-shift calculations need the imaginary parts of resonance functions. In the notations of Boulet *et al.* [21]), they are defined as

$$If_n(k) = \frac{1}{\pi} P.P. \int_{-\infty}^{\infty} dk' \frac{f_n(k')}{k - k'}, \qquad Ig_n(k) = \frac{1}{\pi} P.P. \int_{-\infty}^{\infty} dk' \frac{g_n(k')}{k - k'}, \qquad (2.7.13)$$

where the symbol *P.P.* denotes the Cauchy principal part. The resonance functions $If_1(k)$, $If_2(k)$, $If_3(k)$ are given in Ref. [21] (for $If_1(k = 3.0)$ the corrected value 0.7492 should be used). For $Ig_n(k)$ the formula of Ref. [20] can be applied:

$$Ig_n(k) = \frac{1}{\pi}\left[e^{-2k} Ei(2k) \sum_{m=0}^{L_n} a_m^{(n)} k^m - e^{2k} Ei(-2k) \sum_{m=0}^{L_n} a_m^{(n)} (-k)^m - \sum_{m=0}^{R_n} c_m^{(n)} k^{2m+1} \right] \quad (2.7.14)$$

with $Ei(k)$ denoting the integral exponent and the coefficients $a_m^{(n)}$, $c_m^{(n)}$ given in Tables 2.1–2.2; $R_n = 2$ for $n = 1, 3$; $R_n = 3$ for $n = 2, 4, 5, 7$; $R_n = 4$ for $n = 6$.

Table 2.2. Coefficients $c_m^{(n)}$ [20]. Reproduced with permission.

n/m	0	1	2	3	4
1	183/42	46/63	2/63	–	–
2	13031/3354	5323/5031	803/5031	6/559	–
3	13/5	91/225	2/225	–	–
4	617/170	3737/3825	409/3825	2/765	–
5	139/40	401/450	83/900	1/450	–
6	9491/2385	36307/35775	5168/35775	1366/107325	4/7155
7	31/10	43/63	89/1575	2/1575	–

2.8. Advanced semi-classical methods for line-broadening calculation

For many decades the semi-classical approach of Anderson, Tsao and Curnutte [1, 2] has been widely used for the calculation of pressure broadening of isolated lines. However, it only gave a good agreement with experimental data solely for the polar molecules with strong electrostatic interactions. Its model of straight-line trajectories for the relative molecular motion and especially the development of the scattering matrix by the perturbation theory up to the second order were very questionable, mainly due to the unphysical cut-off procedure needed to avoid the divergence for small values of the impact parameter. For molecular systems with significant short-range forces (perturbation by a non-polar linear molecule or by an atom) the ATC theory failed completely to reproduce the experimentally observed line broadening and shifting.

Many refinements of the ATC approach has therefore been proposed by different authors for the trajectory model, non-perturbative treatment of the scattering operator, higher orders including, interactions at small intermolecular distances, and so on. A short review of some methods can be found in Ref. [14]. We summarise here a slightly larger set of improvements the most frequently used for the line-width and shift calculations.

2.8.1. Murphy and Boggs method

The improvement proposed by these authors in 1967 [22] consisted in the including of some higher order terms via an exponential form of the scattering matrix and avoiding thus the cut-off procedure. Neglecting the correlations between the initial and final states i and f, they obtained the real part of the interruption function as

$$\mathrm{Re}\,S(b) = [1 - \exp(2\,\mathrm{Re}\,S_{2,i}^{outer'})]/2 + [1 - \exp(2\,\mathrm{Re}\,S_{2,f}^{outer'})]/2, \qquad (2.8.1)$$

where the prime on the second-order *outer* contributions means that they do not include the terms diagonal on the quantum numbers of the active molecule. Similarly, all other contributions with the quantum numbers of the active molecule unchanged after collision (e.g. ${}^{0l_2}S_2$) are omitted too. It means that the important effects of reorientation are not taken into account and the elastic broadening mechanism is completely disregarded. Despite this drawback and still unrealistic trajectory model, the authors obtained a quite satisfactory agreement with experiment for some molecular systems [22–24].

2.8.2. Cattani method

In 1972, Cattani applied the Murthy–Boggs idea of exponential collisional matrix element development to the basic $S(b)$ expression of the ATC theory [25]. Including both terms in the same exponent, his formula

$$S(b) = 1 - \exp(-S_{2,i}^{outer'} - S_{2,f}^{outer'*}) \qquad (2.8.2)$$

accounts partially for the correlation between the initial and final states but the reorientation contributions are still absent.

2.8.3. Cherkasov method

Cherkasov improved the result of Cattani by including in the treatment the vibrational dephasing contributions [26]. The interruption function in this approach has the form:

$$S(b) = 1 - \exp[-iS_1 + \mathrm{Re}(S_{2v} + S_{2r}) + i\,\mathrm{Im}(S_{2v} + S_{2r})], \qquad (2.8.3)$$

where S_{2v} and S_{2r} denote the vibrational and rotational second-order contributions.

2.8.4. Korff and Leavitt method

Published for the first time in 1975 [27, 28], this method uses the linked cluster theorem for degenerate states and leads to the interruption function

$$S(b) = 1 - (1 - S_2^{middle}) \exp(-i\Delta_i + i\Delta_f - S_{2,i}^{outer} - S_{2,f}^{outer*}). \qquad (2.8.4)$$

This equation includes the reorientation, inelastic collisions, phase shifts and agrees term by term with the expression of the ATC theory (2.4.12) when expanded in the Taylor series for large values of the impact parameter. The term ${}^{00}S_2$ is, however, not accounted for.

2.8.5. Herman and Jarecki method

In 1975–1976 [29, 30], Herman and Jarecki studied the case of HF vibrotational lines broadened and shifted by rare gases introducing a more realistic trajectory description for close collisions (modified cut-off) and including the vibrational and rotational phase shift effects up to infinite order. Their line-width calculations agreed with the measurements within 30% whereas the line shifts exhibited an overestimation of 30–100%.

2.8.6. Smith, Giraud and Cooper method

Smith, Giraud and Cooper [31] developed an infinite-order semi-classical approach (peaking approximation) with curved classical trajectories governed by the isotropic part of the interaction potential in the Lennard–Jones 12–6 form. Although sceptical on the validity of such a trajectory approximation, the authors stated a good agreement with experimental values and quantum-mechanical close-coupling calculations for some simple molecular systems.

2.8.7. Davis and Oli method

Although it can not be called a semi-classical method, this method proposed in 1978 provides γ_{if} and δ_{if} expressions analogous to those of the ATC theory [32]. The resonance functions are calculated taking into account the quantum character of the relative translational motion, so that this method is referred to as quantum Fourier transform (QFT) method.

2.8.8. Salesky and Korff method

Obtained in 1979 [33], this expression for the interruption function is similar to that of Cattani:

$$S(b) = 1 - \exp(-i\Delta_i + i\Delta_f - S_{2,i}^{outer'} - S_{2,f}^{outer'*}) \qquad (2.8.5)$$

but the primes here indicate that the contributions diagonal on the quantum numbers of either active or perturbing molecules are omitted. It means that not only the reorientation effects and ${}^{0l_2}S_2$ terms are neglected for the active molecule but also the atomic perturbers can not be considered since for them the single non-zero contributions are those of ${}^{l_1 0}S_2$ type. The term ${}^{00}S_2$ is omitted too.

2.8.9. Robert and Bonamy formalism

The merit and the great advantage of this approach proposed in 1979 [34] consist in putting together separate improvements proposed previously by different authors: the exponential form of the interruption function like Korff and Leavitt method, the curved trajectories governed by the isotropic Lennard–Jones part of the interaction potential like Smith–Giraud–Cooper model [31] (simplified to parabolic trajectories in Ref. [35]) as well as in an analytical development of close collision contributions via the atom–atom potential model. The $S(b)$ function is written in the form

$$S(b)=1-(1-S_2^{middle'})\exp(-i\Delta_i+i\Delta_f-S_{2,i}^{outer}-S_{2,f}^{outer*}-S_2^{middle''}), \qquad (2.8.6)$$

where $S_2^{middle''}$ is the part of S_2^{middle} diagonal in the quantum numbers of the perturbing molecule and $S_2^{middle'}=S_2^{middle}-S_2^{middle''}$ (the term ${}^{00}S_2$ is omitted). This method will be described in detail below. The fully complex implementation of the RB formalism (CRB) was proposed by Lynch and Gamache [36] in 1996 for simultaneous computation of collisional line widths and shifts. An invalid assumption of the RB formalism concerning the use of cumulant expansion has been recently evoked and corrected by Ma *et al.* [37] (Modified Robert–Bonamy formalism — MRB). No significant change for the line widths has been observed for the case of weakly interacting molecular systems; for the systems with strong interactions the MRB results for both line widths and line shifts had worse agreement with experimental data than the corresponding RB values.

2.8.10. Exact trajectory model

The idea to incorporate in the ATC theory exact classical trajectories issued from the classical equations of motion for a particle in an isotropic potential field [38] was put forward by Bykov *et al.* [39] who studied the vibrational (first-order) contributions to H_2O and CH_4 vibrotational line shifts in the visible region. An attempt to study the changes resulting from the use of exact trajectories in the pure vibrational dephasing mechanism in H_2–He collisions in the framework of the RB formalism was made by Joubert *et al.* [40]. The second-order contributions defined by the anisotropic potential were considered in the framework of the RB formalism with exact trajectories (RBE) for the case of linear molecules by Buldyreva *et al.* [41]; the resonance functions describing the line widths were obtained for the electrostatic and atom–atom interactions. This RBE approach was further extended to symmetric [42] and asymmetric [43] tops.

For diatomic active molecules with great values of the rotational constant the fully quantum close coupling (CC) method is often used leading to a very good agreement with experimental values (see Ref. [44]). However, these computations need refined potential energy surfaces, and the main current limitation for their application to more complex molecular systems is either the absence or the insufficient precision of the intermolecular interaction potential. Another difficulty is a too high CPU cost, so that much effort is constantly made to develop realistic approximations without a noticeable loss of precision.

The natural way to do this is the classical modelling of the relative molecular motion (classical path approximation). The most complete formalism of this kind was proposed by Neilsen and Gordon (NG) [45]. In their approach, the trajectory for given kinetic energy and impact parameter is computed from the isotropic potential using the equations of classical mechanics. Each trajectory is treated as a sequence of small time intervals, and each new space point is determined by numerical solving of the classical equations for the intermolecular distance and the phase angle. Then the evolution operator for the scattering matrix is calculated for each time interval. The results obtained with the CC and NG methods for HF and HCl lines are briefly discussed in Secs 2.14 and 4.5.

Among all the semi-classical computation methods mentioned above we discuss below in detail the original Robert and Bonamy formalism with parabolic trajectories [34] since it includes not only long-range but also the short-range part of the intermolecular potential. The electrostatic long-range forces are written separately whereas the induction and dispersion terms are accounted for via pair atom–atom Lennard–Jones type potentials of Eq. (2.5.20) which contain also the short-range part of the interaction:

$$V = V^{(elec)} + V^{(aa)}. \tag{2.8.7}$$

This potential is further developed in terms of spherical harmonics (or rotational matrices for symmetric and asymmetric tops) tied to the orientations of two molecular axes in particular ‘r-frames’ (z-axis of each molecular frame coincides with the intermolecular distance vector $\vec{r}$). For the atom–atom part the inversed interatomic distances $(r_{1i,2j})^{-1}$ are expanded into series over powers of the inversed intermolecular distance r. For the case of an asymmetric top active molecule X_2Y colliding with a diatom AB (in terms of the point symmetry groups, for the systems $C_{2v} - C_{\infty v}$), this expansion was originally obtained by Labani *et al.* [46], but later Neshyba and Gamache [47] corrected the signs (dependent on the choice of the molecular axes) in the reported potential of these authors, so that below we keep the expressions of Ref. [47] for the so called II^{R} molecule-fixed-frame representation. The inversed interatomic distance is written as [47]

$$(r_{1i,2j})^{-1} = r^{-1}\left\{1 + \sum_{l_1 l_2 m} f_m^{l_1 l_2}(r_{1i}, r_{2j}) D_{m0}^{l_1}(\Omega_{1i}) D_{m0}^{l_2 *}(\Omega_{2j}) / r^{l_1+l_2}\right\} \quad (2.8.8)$$

with $\Omega_{1i} = (\alpha_{1i}, \beta_{1i}, 0)$ and $\Omega_{2j} = (\alpha_{2j}, \beta_{2j}, 0)$ denoting the orientations of the i-th atom of molecule 1 and the j-th atom of molecule 2 respectively and

$$f_m^{l_1 l_2}(r_{1i}, r_{2j}) = (-1)^{l_2+m} \frac{(l_1 + l_2)!\, r_{1i}^{l_1} r_{2j}^{l_2}}{\sqrt{(l_1 - |m|)!(l_2 - |m|)!(l_1 + |m|)!(l_2 + |m|)!}}. \quad (2.8.9)$$

Equation (2.8.8) is further converted into a form containing explicitly the orientations of molecular frames Ω_1, Ω_2, and the atom–atom potential reads [47]

$$V^{(aa)} = \sum_{\substack{l_1 l_2 q \\ m_1 m_2 m}} \left\{ {}^{m_1 m_2}_{\;\;m}b_q^{l_1 l_2} \frac{{}^{m_1 m_2}D_{12+q}^{l_1 l_2}}{r^{12+q}} - {}^{m_1 m_2}_{\;\;m}c_q^{l_1 l_2} \frac{{}^{m_1 m_2}E_{6+q}^{l_1 l_2}}{r^{6+q}} \right\} D_{mm_1}^{l_1}(\Omega_1) D_{-m-m_2}^{l_2}(\Omega_2), \quad (2.8.10)$$

where ${}^{m_1 m_2}D_{12+q}^{l_1 l_2}$ and ${}^{m_1 m_2}E_{6+q}^{l_1 l_2}$ are analytical functions of the atom–atom parameters d_{ij} and e_{ij} (see Eq. (2.5.20)), of Euler angles α_{1i}, β_{1i}, α_{2j}, β_{2j} and of the intermolecular distances r_{1i}, r_{2j} whereas ${}^{m_1 m_2}_{\;\;m}b_q^{l_1 l_2}$ and ${}^{m_1 m_2}_{\;\;m}c_q^{l_1 l_2}$ are numerical coefficients. The principle difficulty in the calculation of this atom–atom potential up to very high orders on q is the reducing of the high-order terms with D-matrices products to the form of the first order. To overcome this difficulty Neshyba and Gamache used symbolic computation codes. To keep the coherence with the considered electrostatic contributions, the summation on q was truncated at the fourth order (in the framework of the CRB formalism [36] it is done at $q = 8$). The expressions of the analytical functions and numerical coefficients for Eq. (2.8.10) (reproduced with permission from Ref. [47]) are given in Appendix D. The case of asymmetric rotors $X_2Y - X_2Y$ is considered in Ref. [48].

The atom–atom potential of Eq. (2.8.10) contains the isotropic (angular-independent) and anisotropic parts:

$$V^{(aa)} = V_{iso} + V_{aniso}^{(aa)}, \quad (2.8.11)$$

$$V_{iso} = \sum_{i,j}\left[\frac{d_{ij}}{r^{12}} - \frac{e_{ij}}{r^6}\right]. \quad (2.8.12)$$

When only the long-range part of the isotropic potential (for example, for X_2Y-A systems) is estimated through the molecular parameters:

$$V_{iso}^{long-range} = -\left(\mu_1^2\alpha_2 + \frac{3}{2}\bar{u}\alpha_1\alpha_2\right)r^{-6}, \tag{2.8.13}$$

it corresponds approximately to one half of the complete (long-range plus short-range) V_{iso} of Eq. (2.8.12). However, for the first-order line-shift computations it is used instead of Eq. (2.8.12) since the vibrational dependence of the molecular parameters is given explicitly. For atomic perturbers the electrostatic contributions are absent, and the intermolecular potential is defined solely by the pair atom–atom interactions (for X_2Y-A systems, see Appendices D, E).

For a linear molecule interacting with an atom the first- and the second-order contributions are given in Ref. [34]. The model interaction potential is taken there as a series of Legendre polynomials $P_l(\cos\theta)$ up to the second order:

$$V(r,\theta) = V_0(r) + V_1(r,\theta) + V_2(r,\theta) =$$
$$= 4\varepsilon\left\{\left[\left(\frac{\sigma}{r}\right)^{12} - \left(\frac{\sigma}{r}\right)^6\right] + \left[R_1\left(\frac{\sigma}{r}\right)^{12} - A_1\left(\frac{\sigma}{r}\right)^6\right]P_1(\cos\theta) + \left[R_2\left(\frac{\sigma}{r}\right)^{12} - A_2\left(\frac{\sigma}{r}\right)^6\right]P_2(\cos\theta)\right\} \tag{2.8.14}$$

with the Lennard–Jones parameters ε = 202 K, σ = 3.37 Å and the adjustable coefficients $R_1 = 0.37$, $R_2 = 0.65$, $A_1 = 0.33$, $A_2 = 0.14$ for HCl–Ar [34], leading to

$$\begin{aligned} {}^{10}S_{2,i} = \left(\frac{4\varepsilon\sigma}{\hbar v_c'}\right)^2 \Bigg[& \frac{25\pi^2}{1536}A_1^2\left(\frac{\sigma}{r_c}\right)^{12}\sum_{J_i'}\left(C_{J_i 0 1 0}^{J_i' 0}\right)^2 {}^{10}f_7^7 \\ & - \frac{160\pi}{2079}A_1R_1\left(\frac{\sigma}{r_c}\right)^{17}\sum_{J_i'}\left(C_{J_i 0 1 0}^{J_i' 0}\right)^2 {}^{10}f_7^{12} \\ & + \frac{131\,072}{1\,440\,747}R_1^2\left(\frac{\sigma}{r_c}\right)^{22}\sum_{J_i'}\left(C_{J_i 0 1 0}^{J_i' 0}\right)^2 {}^{10}f_{12}^{12}\Bigg], \end{aligned} \tag{2.8.15}$$

$$\begin{aligned} {}^{20}S_{2,i} = \left(\frac{4\varepsilon\sigma}{\hbar v_c'}\right)^2 \Bigg[& \frac{21\pi^2}{2560}A_2^2\left(\frac{\sigma}{r_c}\right)^{10}\sum_{J_i'}\left(C_{J_i 0 2 0}^{J_i' 0}\right)^2 {}^{20}f_6^6 \\ & - \frac{63\pi^2}{5120}A_2R_2\left(\frac{\sigma}{r_c}\right)^{16}\sum_{J_i'}\left(C_{J_i 0 2 0}^{J_i' 0}\right)^2 {}^{20}f_6^{12} \\ & + \frac{48\,951\pi^2}{10\,485\,760}R_2^2\left(\frac{\sigma}{r_c}\right)^{22}\sum_{J_i'}\left(C_{J_i 0 2 0}^{J_i' 0}\right)^2 {}^{20}f_{12}^{12}\Bigg], \end{aligned} \tag{2.8.16}$$

$$
{}^{20}S_2^{middle} = -2(-1)^{J_i+J_f}(2J_i+1)(2J_f+1)C_{J_i0\,20}^{J'_i0}C_{J_f0\,20}^{J'_f0}\left(\frac{4\varepsilon\sigma}{\hbar v'_c}\right)^2 W(J_iJ_fJ_iJ_f;12)
$$

$$
\times\left[\frac{21\pi^2}{2560}A_2^2\left(\frac{\sigma}{r_c}\right)^{10}{}^{20}f_6^6(0)-\frac{63\pi^2}{5120}A_2R_2\left(\frac{\sigma}{r_c}\right)^{16}{}^{20}f_6^{12}(0)\right.
$$

$$
\left.+\frac{48\,951\pi^2}{10\,485\,760}R_2^2\left(\frac{\sigma}{r_c}\right)^{22}{}^{20}f_{12}^{12}(0)\right], \tag{2.8.17}
$$

where only the real parts of resonance functions are taken into account. The term ${}^{20}S_2^{middle}$ is corrected here by the factor ${}^{20}f_p^{p'}(0)$ which is not equal to unity at $k = 0$ in the general case. The quantities v'_c and r_c are relative to the trajectory model choice (parabolic trajectories) described in the next section.

2.9. Parabolic trajectory approximation

The original formalism of Robert and Bonamy [34] employs the parabolic trajectory model [35] which includes the influence of the isotropic intermolecular potential (in the ATC approach with straight-line trajectories $V_{iso} = 0$). The time dependence of the intermolecular distance vector is given by

$$
\vec{r}(t) = \vec{r}_c + \vec{v}_c t + \frac{1}{2}\vec{a}t^2, \tag{2.9.1}
$$

where the acceleration $\vec{a} = \vec{F}_c / m^\circ$ (m° is the reduced mass of the molecular pair) is due to the force derived from the isotropic potential at the distance of the closest approach $\vec{r}_c$ and $\vec{v}_c$ is the corresponding relative velocity (Fig. 2.1). The trajectory is a parabola with the minimum at $\vec{r}_c$ (taken as the origin of time $t = 0$).

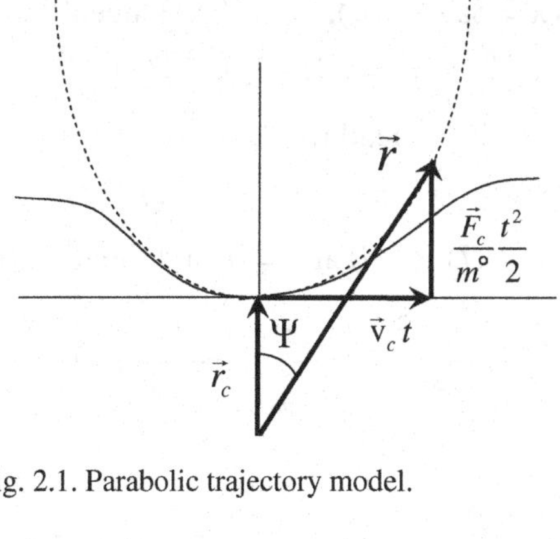

Fig. 2.1. Parabolic trajectory model.

The force $\vec{F}_c$ is defined by the relation

$$\vec{F}_c = -\left(\frac{dV_{iso}}{dr}\right)\Bigg|_{r=r_c} \frac{\vec{r}_c}{r_c} \tag{2.9.2}$$

and with the isotropic potential in the Lennard–Jones 12-6 form

$$V_{iso} = 4\varepsilon\left[\left(\frac{\sigma}{r}\right)^{12} - \left(\frac{\sigma}{r}\right)^{6}\right] \tag{2.9.3}$$

has an analytical expression

$$\vec{F}_c = \frac{24\varepsilon}{\sigma}\left[2\left(\frac{\sigma}{r_c}\right)^{13} - \left(\frac{\sigma}{r_c}\right)^{7}\right]\frac{\vec{r}_c}{r_c}. \tag{2.9.4}$$

The absolute value of the intermolecular distance can be therefore written as

$$r(t) = (r_c^2 + \mathrm{v}_c'^2 t^2)^{1/2}, \tag{2.9.5}$$

where the apparent relative velocity v_c' reads

$$\mathrm{v}_c'^2 = \mathrm{v}_c^2 + \frac{\vec{F}_c \cdot \vec{r}_c}{m^\circ} = \mathrm{v}_c^2 + \frac{F_c r_c}{m^\circ} \tag{2.9.6}$$

since $\vec{F}_c \| \vec{r}_c$ and $\vec{F}_c \perp \vec{\mathrm{v}}_c$ (Fig. 2.1). The conservation of the angular momentum

$$\mathrm{v}_c r_c = \mathrm{v}b \tag{2.9.7}$$

and of the energy

$$\frac{m^\circ \mathrm{v}^2}{2} = \frac{m^\circ \mathrm{v}_c^2}{2} + 4\varepsilon\left[\left(\frac{\sigma}{r}\right)^{12} - \left(\frac{\sigma}{r}\right)^{6}\right] \tag{2.9.8}$$

allows us to relate v_c' to the initial relative velocity v and the reduced Lennard–Jones parameters $\lambda = 8\varepsilon / (m^\circ \mathrm{v}^2)$, $\beta = \sigma / r_c$. Indeed, Eq. (2.9.8) gives

$$\mathrm{v}_c^2 = \mathrm{v}^2[1 - \lambda(\beta^{12} - \beta^6)], \tag{2.9.9}$$

and Eqs (2.9.4) and (2.9.6) lead to

$$\mathrm{v}_c'^2 = \mathrm{v}^2[1 + \lambda(5\beta^{12} - 2\beta^6)]. \tag{2.9.10}$$

According to Eq. (2.9.7) $\mathrm{v}_c = 0$ at $b = 0$, and the minimal r_c value is defined by

$$\frac{m^\circ \mathrm{v}^2}{8\varepsilon} = \left(\frac{\sigma}{r_{c\,min}}\right)^{12} - \left(\frac{\sigma}{r_{c\,min}}\right)^{6}, \tag{2.9.11}$$

so that

$$r_{c\,min} = \sigma\left[\frac{2}{1+(1+4/\lambda)^{1/2}}\right]^{1/6}. \qquad (2.9.12)$$

Putting this value in the expression for the parameter β and using Eq. (2.9.9) one obtains the apparent velocity at the minimal distance of the closest approach:

$$v'_{c\,min} = v\left\{6\left[1+\frac{\lambda}{4}\left(1+\sqrt{1+4/\lambda}\right)\right]\right\}^{1/2}. \qquad (2.9.13)$$

As can be seen from Fig. 2.1, the phase angle Ψ in the parabolic trajectory model is also influenced by the isotropic potential:

$$\sin\Psi = \frac{v_c t}{r(t)} = \frac{v_c t}{\sqrt{r_c^2 + v_c'^2 t^2}} \equiv \xi_1 \frac{r_c \zeta}{r}, \qquad (2.9.14)$$

where the notations $\xi_1 = (v_c / v'_c)$ and $\zeta = v'_c t / r_c$ are introduced. If only the terms up to the second order on t are retained, this equation leads to

$$\cos\Psi = \sqrt{1-\sin^2\Psi} \cong \frac{r_c}{r}\left(1+\xi_2\zeta^2\right) \qquad (2.9.15)$$

with $\xi_2 = \left(1-\xi_1^2\right)/2$. Taking $\xi_1 = 1$, $\xi_2 = 0$ corresponds to the straight-line trajectory.

Since the trajectory is now characterised by r_c and not more by b, the integration in the differential cross-section of Eq. (2.3.12) should be made in terms of r_c. The necessary change of variables $bdb = b(db/dr_c)dr_c$ is easily obtained from Eqs (2.9.7) and (2.9.10):

$$bdb = (v'_c / v)^2 r_c dr_c,$$

so that

$$\int_0^\infty bdb \to \int_{r_{c\,min}}^\infty r_c dr_c \left(\frac{v'_c}{v}\right)^2 \qquad (2.9.16)$$

with the ratio $(v'_c / v)^2$ defined by Eq. (2.9.10).

Since for the parabolic trajectories the intermolecular distance and the phase angle have the time dependences different from those of the straight-line trajectory model, the resonance functions introduced in the framework of the ATC theory should be redefined. For instance, according to the general definition,

$$^{10}f_p^{p'}(k) = N[I_{10}^p I_{10}^{p*} + 2I_{11}^p I_{11}^{p*}], \qquad (2.9.17)$$

where N is the normalisation factor. Using Eqs (2.9.14) and (2.9.15) we obtain

$$I_{10}^{p} = \frac{1}{v_c' r_c^{p-1}} \frac{ik}{p-1} \xi_1 I_{\frac{p-1}{2}}, \tag{2.9.18}$$

$$I_{11}^{p} = \frac{1}{v_c' r_c^{p-1}} \left(-\frac{1}{\sqrt{2}}\right) \left[I_{\frac{p+1}{2}} + \xi_2 \left(I_{\frac{p-1}{2}} - I_{\frac{p+1}{2}} \right) \right] \tag{2.9.19}$$

with the integrals given in Appendix C; for example, for $p = p' = 7$ we have

$$^{10}f_7^7(k) = \frac{e^{-2k}}{[J_4(0)]^2} \left\{ (k\xi_1 J_3)^2 + [J_4 + \xi_2(6J_3 - J_4)]^2 \right\}.$$

These resonance functions differ from the ATC functions by the factor ξ_1 and additional ξ_2-dependent terms. Their detailed expressions can be found in Refs [34, 49]. For the resonance functions $^{10}f_p^{p'}$, $^{20}f_p^{p'}$ with $p, p' = 7, 9, 13, 15$ their numerical values are given in Appendix E.

The main features of the RB approach can thus be summarised in four points:

i) the parabolic trajectories curved by the isotropic potential provide a more realistic description of the relative molecular motion;
ii) the short-range interactions are included which are indispensable for non-polar colliders;
iii) the long-range induction and dispersion contributions are accounted for through the pair atom–atom potential model;
iv) the non-perturbative exponential form of the scattering matrix allows to avoid the unphysical cut-off procedure at the small values of the impact parameter.

2.10. Resonance functions within the exact trajectory model

At the beginning of the 1990s, Bykov *et al.* [39, 50] proposed to employ the exact classical trajectories for the long-range interactions (for more details, see also [51, 52]). According to the classical equations of motion for a particle in an isotropic potential field, the time t and the phase Ψ are functions of the intermolecular distance r [38]:

$$t(r) = \int_{r_c}^{r} \frac{dx}{\sqrt{\frac{2}{m^\circ}[E - V_{iso}(x)] - \frac{M^2}{m^{\circ 2} x^2}}} + c_1, \tag{2.10.1}$$

$$\Psi(r) = \int_{r_c}^{r} \frac{\frac{M}{x^2}dx}{\sqrt{2m^{\circ}[E - V_{iso}(x)] - \frac{M^2}{x^2}}} + c_2 . \tag{2.10.2}$$

Here $E = \overset{\circ}{m}\mathrm{v}^2/2$ and $M = \overset{\circ}{m}b\mathrm{v}$ are the energy and the angular momentum of the colliding pair respectively. If the collision takes place in the *XOY* plane so that at $t = 0$ we have $r = r_c$ and $\Psi = 0$, the integration constants c_1 and c_2 vanish. From Eq. (2.10.1) the time dependence of the intermolecular distance is

$$\frac{dr(t)}{dt} = \sqrt{\frac{2}{m^{\circ}}[E - V_{iso}(r)] - \frac{M^2}{m^{\circ 2} r^2}} , \tag{2.10.3}$$

and the integration over t in Eq. (2.3.12) can be easily replaced by the integration over r.

For each value of the impact parameter b the corresponding distance of the closest approach r_c is obtained from Eqs (2.9.7) and (2.9.9):

$$b / r_c = \sqrt{1 - V^{\circ}_{iso}(r_c)} , \tag{2.10.4}$$

where the dimensionless isotropic potential $V^{\circ}_{iso}(r) = 2V_{iso}(r)/(m^{\circ}\mathrm{v}^2)$ can be taken, for example, in the Lennard–Jones form of Eq. (2.9.3) leading to $V^{\circ}_{iso}(r) = \lambda[\beta^{12}(r_c / r)^{12} - \beta^6 (r_c / r)^6]$.

Introducing the reduced intermolecular distance $z = r/r_c$, the variable $y = z^2$ and the notations

$$f(y) = y - 1 - yV^{\circ}_{iso}(r_c\sqrt{y}) + V^{\circ}_{iso}(r_c) , \tag{2.10.5}$$

$$A_n(y) = \frac{1}{2}\int_1^y \frac{dy'}{y'^{n/2}[f(y')]^{1/2}} , \tag{2.10.6}$$

Eqs (2.10.1) and (2.10.2) can be rewritten as

$$t(y) = \frac{r_c}{\mathrm{v}} A_0(y) , \tag{2.10.7}$$

$$\Psi(y) = \sqrt{1 - V^{\circ}_{iso}(r_c)} A_2(y) . \tag{2.10.8}$$

Moreover, for the isotropic Lennard–Jones potential Eq. (2.10.5) reads

$$f(y) = y - 1 + \lambda[\beta^{12}(1 - y^{-5}) - \beta^6 (1 - y^{-2})] . \tag{2.10.9}$$

The integrals of Eq. (2.7.5) take therefore the form

$$I^p_{lm}(k_c) = \frac{2}{\mathrm{v} r_c^{p-1}} \sum_s a(l,m,s) \int_0^{\infty} u du \frac{e^{ik_c A_0(u)} (\cos\Psi)^m (\sin\Psi)^s}{(1+u^2)^{p/2}[f(u)]^{1/2}} , \tag{2.10.10}$$

where $u^2 = y-1$ and the new resonance parameter corresponding to exact trajectory is defined by r_c (and not by b) value: $k_c = r_c\omega/\mathrm{v}$. The real parts of the resonance functions (2.7.1) are now defined as

$$^{l_1 l_2} f_{p'}^{p}(k_c) = N \sum_{m=-l}^{l} I_{lm}^{p^*}(k_c) I_{lm}^{p'}(k_c), \tag{2.10.11}$$

and after normalisation to unity at $k_c = 0$ they can be written as [50]

$$^{l_1 l_2} f(k_c) = \frac{2(2L)!N'}{(2l_1+1)!(2l_2+1)!} \sum_{L+m\ even} \frac{(L+m-1)!!(L-m-1)!!}{(L+m)!!(L-m)!!} \times \left\{ \int_1^{\infty} dz \frac{\cos[A_0(z)k_c + m\sqrt{1-V_{iso}^{\circ}(r_c)}A_2(z)]}{z^L \sqrt{z^2 - 1 - z^2 V_{iso}^{\circ}(zr_c) + V_{iso}^{\circ}(r_c)}} \right\}^2 \tag{2.10.12}$$

(p and p' are omitted for brevity), where N' is the normalisation factor and

$$A_n(z) = \int_1^z \frac{dz'}{z'^{n-1}[f(z')]^{1/2}}. \tag{2.10.13}$$

In some cases, the integrals in Eqs (2.10.6) and (2.10.13) can be calculated analytically. Indeed, Eq. (2.10.9) can be expanded in Taylor series in the vicinity of $y_0 = 1$:

$$f(y) = f(y_0) + \left.\frac{df}{dy}\right|_{y=y_0} (y-1) + \ldots \tag{2.10.14}$$

with $f(y_0 = 1) = 0$ and, for the Lennard–Jones 12-6 isotropic potential,

$$\left.\frac{df}{dy}\right|_{y=y_0} = f_y'(1) = 1 + \lambda\left[5\beta^{12} - 2\beta^6\right] = \left(\frac{\mathrm{v}_c'}{\mathrm{v}}\right)^2. \tag{2.10.15}$$

Limiting Eq. (2.10.14) to the terms linear on (y–1) one can integrate analytically Eq. (2.10.6), so that Eqs (2.10.7)–(2.10.8) become

$$t(y) = \frac{r_c}{\mathrm{v}} \frac{1}{2} \int_1^y \frac{dy'}{[f_{y'}'(1)]^{1/2}(y'-1)^{1/2}} = \frac{r_c}{\mathrm{v}_c'}(y-1)^{1/2} = \frac{(r^2 - r_c^2)^{1/2}}{\mathrm{v}_c'} \tag{2.10.16}$$

$$\Psi(y) = \frac{b}{r_c} A_2(y) = \frac{\mathrm{v}_c}{\mathrm{v}_c'} \operatorname{arctg}\sqrt{y-1} = \frac{\mathrm{v}_c}{\mathrm{v}_c'} \operatorname{arctg}\sqrt{\frac{r^2}{r_c^2} - 1}. \tag{2.10.17}$$

The first of these equations is equivalent to Eq. (2.9.5) of the parabolic trajectory model. The second one can be written as

$$\sin^2\left(\frac{v'_c}{v_c}\Psi\right) = \frac{r^2 - r_c^2}{r^2} = \frac{(v'_c t)^2}{r_c^2 + (v'_c t)^2} \qquad (2.10.18)$$

which slightly differs from Eq. (2.9.14). For $V_{iso} = 0$ ($r_c = b$, $v_c = v'_c = v$) Eqs (2.10.16)–(2.10.17) reduce to the straight-line formulae of Eq. (2.7.3).

For the model of exact trajectories the resonance functions are not analytical and should be calculated numerically [53, 54]. For example, the numerical values of the resonance function ${}^{10}f_7^7(k_c)$ obtained by Eq. (2.10.12) with the Lennard–Jones 12-6 isotropic potential can be found in Appendix F.

An alternative method for the calculation of these functions was proposed in Ref. [55]. First, the integrals $A_n(y, \lambda, \beta)$ of Eq. (2.10.6) are calculated for a given y and various λ and β (for $1 \le y \le 5$ with the step 0.2 and for $5 < y \le 9$ with the step 0.5 for $\lambda = 0.5, 1.0, 1.5, 2.0, 2.5, 3.0, 3.5$ and $\beta = 0.3, 0.5, 0.7, 0.8, 0.85, 0.9, 0.95, 1.0, 1.01, 1.02, 1.04, 1.08$). Then these integrals are approximated by

$$\tilde{A}_0(u,\lambda,\beta) = t(u)\frac{v\beta}{\sigma} = \frac{b_1(\lambda,\beta)u}{1 + b_2(\lambda,\beta)u}, \qquad (2.10.19)$$

$$\tilde{A}_2(u,\lambda,\beta) = b_3(\lambda,\beta)\mathrm{arctg}(u), \qquad (2.10.20)$$

where

$$b_n(\lambda,\beta) = \frac{1 - \delta_{n2} + b_{n1}\lambda\beta^6(\beta^6 - b_{n2})}{1 + b_{n3}\lambda\beta^6(\beta^6 - b_{n2})}, \qquad (2.10.21)$$

$u = (z^2 - 1)^{1/2}$, σ is the Lennard–Jones potential parameter, δ_{n2} is Kronecker delta symbol, and the parameters b_{nm} collected in Table 2.3 are obtained by least squares fitting to $A_n(y, \lambda, \beta)$ of Eq. (2.10.6). The analytical Eqs (2.10.19)–(2.10.21) are chosen from the conditions that in the straight-line trajectory approximation $A_0(y,\lambda=0) = \tilde{A}_0(y,\lambda=0) = u$ and $A_2(y,0) = \tilde{A}_2(y,0) = \mathrm{arctg}(u)$. The functions $\tilde{A}_0$ and $\tilde{A}_2$ can be used for the calculation of the integrals in Eq. (2.10.10) which define the resonance functions of Eq. (2.10.11) or even the resonance functions $f(k_c, k_c')$ which appear in the off-diagonal matrix elements of the scattering matrix $S(if, i'f' \mid b)$ of Eq. (2.3.7).

Table 2.3. Parameters b_{nm} for analytical representation of integrals $A_n(y, \lambda, \beta)$ [55]. Reproduced with permission.

Parameter	Value	Parameter	Value	Parameter	Value
b_{11}	0.2867	b_{21}	–0.3374	b_{31}	0.3513
b_{12}	0.5577	b_{22}	0.5173	b_{32}	0.5441
b_{13}	1.7291	b_{23}	1.4561	b_{33}	1.5409

2.11. Approximation for the real parts of exact-trajectory resonance functions

As mentioned above, in the particular case $V_{iso}(r) = 0$ the resonance functions of Eq. (2.10.12) reduce to the well known ATC functions given by Eq. (2.7.11). For the dipole–dipole, dipole–quadrupole and quadrupole-quadrupole interactions the tabulated ATC functions $f_1(k)$, $f_2(k)$ and $f_3(k)$ are shown on Fig. 2.2. These functions can be approximated by the truncated series

$$f^{model}(k) = a_0 + a_1\text{th}(z) + a_2\text{th}^2(z) + a_3\text{th}^3(z), \qquad (2.11.1)$$

where $z = \alpha(k - k_e)$ and the parameters a_0, a_1, a_2, a_3, α and k_e are obtained by a least-squares fit to the corresponding ATC function. Since the parameters a_2 and a_3 are correlated, one of them should be set to zero. Two equivalent representations thus become possible for the resonance functions:

$$f^{model}(k) = a_1[\text{th}(z) - 1] + a_2[\text{th}^2(z) - 1], \qquad (2.11.2)$$

$$f^{model}(k) = a_1[\text{th}(z) - 1] + a_3[\text{th}^3(z) - 1]. \qquad (2.11.3)$$

To ensure the correct asymptotic behaviour at large k values the parameter a_0 should be set to $-a_1-a_2$ or $-a_1-a_3$, respectively. These model resonance functions are in a very good agreement with the tabulated ones (Fig. 2.2).

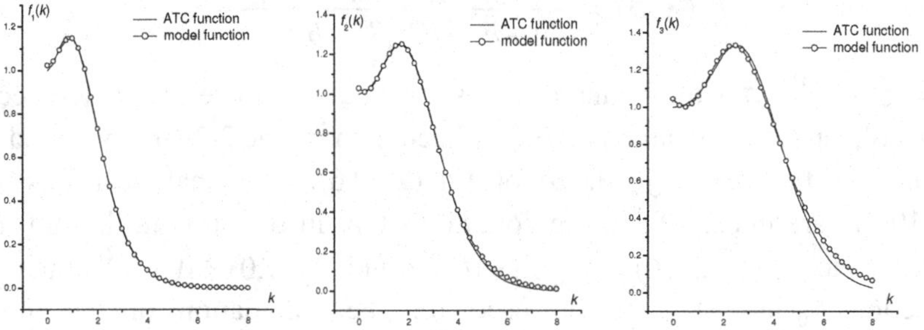

Fig. 2.2. Comparison of the ATC resonance functions from Eq. (2.7.11) and approximate resonance functions from Eq. (2.11.3) for dipole–dipole (a), dipole–quadrupole (b) and quadrupole-quadrupole (c) interactions [53]. Reproduced with permission.

The analytical functions of Eq. (2.11.3) can also be used to represent the exact-trajectory resonance functions for the isotropic potential in the Lennard–Jones form. The fitting parameters therefore become λ- and β-dependent, so that

$$f^{model}(k,\lambda,\beta) = a_1(\lambda,\beta)\{\text{th}[z(\lambda,\beta)] - 1\} + a_3(\lambda,\beta)\{\text{th}^3[z(\lambda,\beta)] - 1\}, \qquad (2.11.4)$$

with $z(\lambda, \beta) = \alpha(\lambda, \beta)\,[k - k_e(\lambda, \beta)]$ (the index c in k_c is omitted here for brevity). It should be kept in mind that the λ and β parameters in Eq. (2.11.4) are not constant. They are defined by the relative velocity v and the impact parameter b: $\lambda = \lambda(\mathrm{v}, b)$ and $\beta = \beta(\mathrm{v}, b)$. As previously, the coefficients $a_1(\lambda, \beta)$, $a_3(\lambda, \beta)$, $\alpha(\lambda, \beta)$ and $k_e(\lambda, \beta)$ are determined by the least-squares fitting of Eq. (2.11.4) to the calculated points $f(k_i, \lambda, \beta)$. As follows from the least-squares analysis, for the induction and dispersion resonance functions the parameter a_3 should be set to zero, and the corresponding approximation reads

$$g^{model}(k, \lambda, \beta) = a_1(\lambda, \beta)\{\mathrm{th}[z(\lambda, \beta)] - 1\}. \qquad (2.11.5)$$

The numerical values of the coefficients $a_1(\lambda, \beta)$, $a_3(\lambda, \beta)$, $\alpha(\lambda, \beta)$ and $k_e(\lambda, \beta)$ for the model functions $f_n^{model}(k, \lambda, \beta)$ ($n = 1, 2, 3$) with $\beta \le 1.0$ are given in Ref. [56]. Figure 2.3 shows the β-dependence of $a_1(\lambda, \beta)$ and $a_3(\lambda, \beta)$ coefficients for various λ in the case of dipole–quadrupole resonance function f_2^{model} (this dependence is typical for a_1, a_3, α and k_e parameters of all resonance functions).

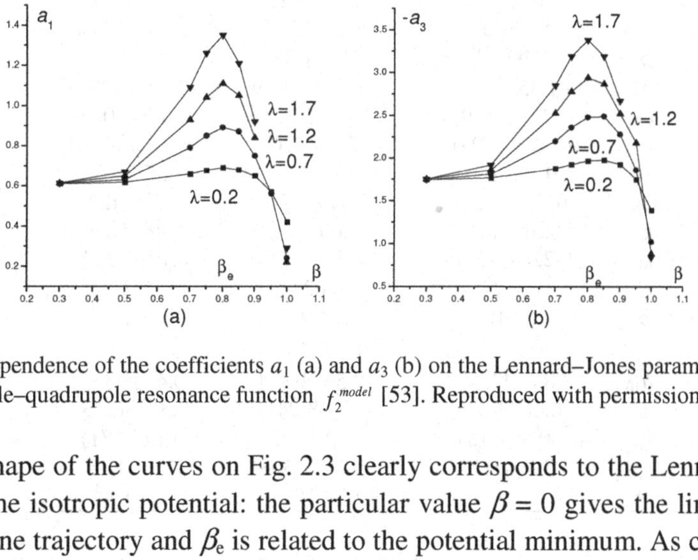

Fig. 2.3. Dependence of the coefficients a_1 (a) and a_3 (b) on the Lennard–Jones parameters λ and β for the dipole–quadrupole resonance function f_2^{model} [53]. Reproduced with permission.

The shape of the curves on Fig. 2.3 clearly corresponds to the Lennard–Jones form of the isotropic potential: the particular value $\beta = 0$ gives the limit case of straight-line trajectory and β_e is related to the potential minimum. As can be seen from this figure, the λ-dependence of the model resonance function coefficients changes drastically for various intervals of the β value. Analytical formulae for these coefficients were proposed in Refs [53, 56]:

$$a_n(\lambda, \beta) = a_{n0} + a_{n1}\{\mathrm{th}[a_{n\beta}(\beta - \beta_{ne})] + (\mathrm{th}[a_{n\beta}(\beta - \beta_{ne})])^2\},\ n = 1, 3;$$
$$\alpha(\lambda, \beta) = \alpha_0 + \alpha_{\lambda\beta\beta}\lambda\beta^2; \qquad (2.11.6)$$
$$k_e(\lambda, \beta) = k_{e0} + k_{e\lambda\beta\beta}\lambda\beta^2.$$

The adjustable parameters a_{n0}, …, $k_{e\lambda\beta\beta}$ obtained by fitting the model functions of Eq. (2.11.4) to the calculated points $f(k_i,\lambda,\beta)$ of the exact functions of Eq. (2.10.12) for all pairs (λ, β) simultaneously are given in Tables 2.4–2.6 with

$$\sigma = \sqrt{\sum_{i=1}^{N} \frac{\left[f(k_i,\lambda,\beta) - f^{model}(k_i,\lambda,\beta)\right]^2}{N-L}},$$

where N is the number of points $f(k_i,\lambda,\beta)$ and L is the number of fitted parameters. It is not possible to find a single set of parameters for all β values because the behaviour of resonance functions is different for $\beta<1$ and for $\beta>1$.

Table 2.4. Parameters for the real-part model functions $f_n^{model}(k, \lambda, \beta)$ ($\lambda \le 3.2$) [55]. Reproduced with permission.

	$f_1(k)$	$f_2(k)$	$f_3(k)$
		$0.3 \le \beta \le 1.0$	
a_{10}	0.3723(264)	0.6313(276)	0.7834(271)
a_{11}	−1.0514(1210)	−0.8643(1127)	−0.8119(1029)
$a_{1\beta}$	4.726(278)	5.890(377)	6.260(461)
β_{1e}	0.940(7)	0.911(7)	0.901(7)
a_{30}	−1.4518(261)	−1.7721(310)	−1.9567(323)
a_{31}	1.1893(1093)	1.1612 (0.1069)	1.2176(1069)
$a_{3\beta}$	4.7883(2709)	5.7028(3365)	5.8228(3536)
β_{3e}	0.897(6)	0.877(6)	0.8735(6)
α_0	0.6722(73)	0.4794 (51)	0.3862(37)
k_{e0}	0.512(19)	1.002(22)	1.4685(25)
$k_{e\lambda\beta\beta}$	0.2264(129)	0.1864(155)	0.1913(204)
N	1624	1624	1512
σ	$2.2\cdot10^{-3}$	$2.8\cdot10^{-3}$	$3.1\cdot10^{-3}$
		$1.0 < \beta \le 1.13$	
a_{10}	−0.6649(462)	−0.5674(394)	−0.4739(186)
α_0	0.4813(236)	0.3648(241)	0.3585(213)
k_{e0}	0.898(15)	1.920(237)	3.175(181)
$k_{e\lambda\beta\beta}$	−0.3041(134)	−0.4531(308)	−0.6068(383)
σ	$2.7\cdot10^{-3}$	$4.4\cdot10^{-3}$	$4.3\cdot10^{-3}$

Hereafter, the functions of Eq. (2.11.4) are referred to as the Model resonance functions obtained in the Exact Trajectory approach (MET). For some particular values of λ and β parameters they differ significantly from the ATC or parabolic-trajectory (PT) [34] resonance functions, as it can be seen in Fig. 2.4 for ${}^{20}f_6^{\,6} = g_1$ and ${}^{10}f_7^{\,7} = g_3$ at (λ,β) = (2.0, 0.9) and (λ,β) = (2.0, 1.05). This difference is especially pronounced for $\beta>1$ (close collisions region).

Table 2.5. Parameters for the real-part model functions ${}^{10}f_p^{p'}(k, \lambda, \beta)$, ($\lambda \leq 3.5$).

	${}^{10}f_7^7$	${}^{10}f_9^9$	${}^{10}f_{12}^{12}$	${}^{10}f_{13}^{13}$	${}^{10}f_{15}^{15}$
			$\beta \leq 1.0$		
a_{10}	−0.6593(72)	−0.6839(136)	−0.7749(130)	−0.7472(168)	−0.7714(187)
a_{11}	0.2965(60)	0.3371(156)	0.3598(57)	0.3726(162)	0.3854(164)
$a_{1\beta}$	8.395(298)	6.911(632)	9.054(250)	7.178(667)	7.277(681)
β_{1e}	0.816(168)	0.805(3)	0.810(17)	0.800(4)	0.800(4)
α_0	0.3957(171)	0.3263 (218)	0.2737(72)	0.2392(171)	0.2132(157)
$\alpha_{\lambda\beta\beta}$	0.2773(233)	0.3099(369)	0.1305(133)	0.2642(308)	0.2464 (286)
k_{e0}	1.688(41)	1.886(73)	1.572(33)	1.946(101)	1.957(112)
$k_{e\lambda\beta\beta}$	0.3546(244)	0.2469(364)	0.7195(353)	0.3740(470)	0.4392(515)
N	2010	868	2013	868	868
σ	$1.7 \cdot 10^{-3}$	$4.6 \cdot 10^{-3}$	$3.2 \cdot 10^{-3}$	$5.2 \cdot 10^{-3}$	$5.5 \cdot 10^{-3}$
			$\beta > 1.0$ ($k_{e0} = 0$)		
a_{10}	−0.5781(123)	−0.5166(133)	−0.5093(410)	−0.4708(128)	−0.4550(125)
α_0	0.2460(62)	0.1933(63)	0.1647 (69)	0.1519(53)	0.1333(49)
$\alpha_{0\lambda\beta\beta}$	−0.0311(13)	−0.0268(13)	−0.0226(31)	−0.0215(10)	−0.0190(9)
$k_{e\lambda\beta\beta}$	−1.850(140)	−2.943(165)	−3.332(93)	−3.994(264)	−4.655(318)
N	410	480	480	480	480
σ	$2.2 \cdot 10^{-3}$	$1.5 \cdot 10^{-3}$	$2.0 \cdot 10^{-3}$	$1.3 \cdot 10^{-3}$	$1.4 \cdot 10^{-3}$

Table 2.6. Parameters for the real-part model functions ${}^{20}f_p^{p'}(k, \lambda, \beta)$, ($\lambda \leq 3.5$).

	${}^{20}f_6^6$	${}^{20}f_8^8$	${}^{20}f_{10}^{10}$	${}^{20}f_{12}^{12}$	${}^{20}f_{14}^{14}$	${}^{20}f_{16}^{16}$
			$\beta \leq 1.0$			
a_{10}	−0.5539(32)	−0.5723(7)	−0.5986(83)	−0.586(2)	−0.6491(107)	−0.6716(118)
a_{11}	0.2527(64)	0.2521(100)	0.2724 (102)	0.292(6)	0.3053(103)	0.3187(110
$a_{1\beta}$	6.943(258)	6.806(515)	6.942(511)	6.921(246)	7.164(511)	7.256(513)
β_{1e}	0.822(16)	0.797(3)	0.796(3)	0.822(1)	0.794(3)	0.793(3)
α_0	0.4023 (85)	0.3925(250)	0.3254(204)	0.318(6)	0.2454(155)	0.2188(144)
$\alpha_{\lambda\beta\beta}$	0.1928(145)	0.3016(42)	0.2625(353)	0.170(11)	0.2207(27)	0.2082(25)
k_{e0}	2.846(27)	2.953(65)	2.976(76)	2.860(35)	2.984(97)	2.983(108)
$k_{e\lambda\beta\beta}$	0.4416 (201)	0.2546(368)	0.3008(423)	0.531(23)	0.3891(51)	0.4325(55)
N	2015	952	952	952	952	952
σ	$1.7 \cdot 10^{-3}$	$3.8 \cdot 10^{-3}$	$4.2 \cdot 10^{-3}$	$4.2 \cdot 10^{-3}$	$1.8 \cdot 10^{-3}$	$4.6 \cdot 10^{-3}$
			$\beta > 1.0$ ($k_{e0} = 0$)			
a_{10}	−0.6997(151)	−0.5979(129)	−0.5492(129)	−0.5214(129)	−0.4926(124)	−0.4745(122)
α_0	0.2400(74)	0.1938(54)	0.1652(50)	0.1317(39)	0.1321(45)	0.1220(44)
$\alpha_{\lambda\beta\beta}$	−0.0293(32)	−0.0255(12)	−0.0225(10)	−0.0189(80)	−0.0185(10)	−0.01711(8)
$k_{e\beta\beta}$	−1.7284(150)	−2.5918(130)	−3.2862(179)	-4.5735(213)	−4.429(285)	−4.876(342)
N	475	480	480	505	480	480
σ	$1.7 \cdot 10^{-3}$	$1.7 \cdot 10^{-3}$	$1.7 \cdot 10^{-3}$	$1.6 \cdot 10^{-3}$	$1.7 \cdot 10^{-3}$	$1.7 \cdot 10^{-3}$

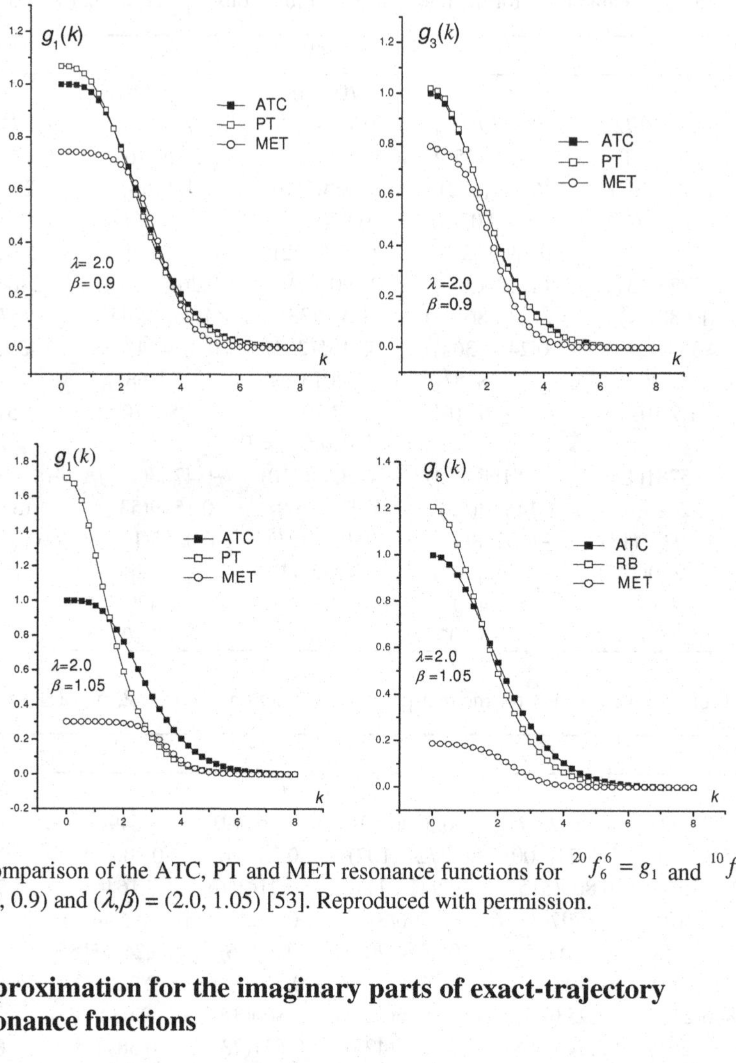

Fig. 2.4. Comparison of the ATC, PT and MET resonance functions for ${}^{20}f_6^6 = g_1$ and ${}^{10}f_7^7 = g_3$ at $(\lambda,\beta) = (2.0, 0.9)$ and $(\lambda,\beta) = (2.0, 1.05)$ [53]. Reproduced with permission.

2.12. Approximation for the imaginary parts of exact-trajectory resonance functions

The imaginary parts of resonance functions are defined by Eqs (2.7.13). For a straight-line trajectory the functions If_1, If_2, and If_3 can be found in Ref. [21] and the functions Ig can be calculated by Eq. (2.7.14). The imaginary parts of resonance functions in the RB formalism are determined in Ref. [36].

For the exact-trajectory model the imaginary parts of resonance functions were calculated in Refs [54, 55] using Eq. (2.7.13) for the same values of k_i, λ and β as for the real parts. The principal values of integrals were obtained as

$$If(k) = \frac{1}{\pi}\left[\int_{-A}^{0}\frac{f(-y)}{k-y}dy + \int_{0}^{k-\delta}\frac{f(y)}{k-y}dy + \int_{k+\delta}^{A}\frac{f(y)}{k-y}dy\right] \tag{2.12.1}$$

with $A = 400$, $\delta = 10^{-12}$. The model functions for their approximation were taken as

$$If^{model}(k,\lambda,\beta) = \mathrm{sh}[\alpha_0(\lambda,\beta)k]\left\{\frac{a_1(\lambda,\beta)}{\mathrm{ch}[\alpha_1(\lambda,\beta)(k-k_e(\lambda,\beta))]} + \frac{a_2(\lambda,\beta)}{\mathrm{ch}[\alpha_2(\lambda,\beta)(k-k_e(\lambda,\beta))]}\right\} \tag{2.12.2}$$

with the fitting parameters $a_n(\lambda,\beta)$, $\alpha_n(\lambda,\beta)$, $k_e(\lambda,\beta)$ defined like Eq. (2.11.6):

$$\begin{aligned} a_n(\lambda,\beta) &= a_{n0} + a_{n1}\{\mathrm{th}[a_{n\beta}(\beta-\beta_{ne})] + (\mathrm{th}[a_{n\beta}(\beta-\beta_{ne})])^2\},\ n = 1, 2; \\ \alpha_n(\lambda,\beta) &= \alpha_{n0} + \alpha_{n\lambda\beta\beta}\lambda\beta^2,\ n = 0, 1, 2; \\ k_e(\lambda,\beta) &= k_{e0} + k_{e\lambda\beta\beta}\lambda\beta^2. \end{aligned} \tag{2.12.3}$$

These parameters are obtained by the global fitting of Eq. (2.12.2) to the points $If(k_i, \lambda, \beta)$ calculated with Eq. (2.12.1) for various sets of (λ, β) (Tables 2.7–2.9).

The imaginary-part model resonance function If_1^{model} of Eq. (2.12.2) for the straight-line trajectory and exact trajectory are compared in Fig. 2.5 for $(\lambda, \beta) = (3.0, 1.0)$ and $(\lambda, \beta) = (3.0, 1.08)$.

Table 2.7. Parameters for the model functions $If_n^{model}(k,\lambda,\beta)$ [55]. Reproduced with permission.

	If_1	If_2	If_3
		$\beta \leq 1.0$ ($\alpha_2 = \alpha_0$)	
a_{10}	0.2623(63)	0.1437(41)	0.1198(31)
a_{11}	0.4223(105)	0.2851(72)	0.2569(71)
β_{1e}	8.941(842)	10.206(712)	10.704(572)
$a_{1\beta}$	1.056(9)	1.046(7)	1.044(5)
a_{20}	0.0205(7)	0.0120(26)	0.0108(5)
α_0	1.0363(272)	0.9196(192)	0.7532 (151)
$\alpha_{0\lambda\beta\beta}$	–0.0743(94)	–0.0369(74)	–0.0232(51)
k_{e0}	1.3468(82)	2.3014(121)	3.1530(142)
$k_{e\lambda\beta\beta}$	0.2885(121)	0.2222(195)	0.2059(219)
α_{10}	1.3134(265)	1.1319(184)	0.9238(146)
		$\beta > 1$ ($k_{e0} = 0$)	
a_{10}	0.7149(215)	0.7463(534)	0.7241(444)
α_0	0.8959(265)	0.4512(266)	0.3288(194)
$k_{e\lambda\beta\beta}$	–0.1376 (162)	–0.1018(164)	0.0506(210)
α_{10}	1.0902(163)	0.5786(222)	0.4083(150)
$\alpha_{1\lambda\beta\beta}$	0.0159(19)	0.0293(20)	0.0472(6)

It is worthy to note that for large k values an asymptotic representation of If^{model} is appropriate. It can be obtained in an analytical form if Eq. (2.11.4) is chosen for the corresponding real parts. Indeed, we have in this case [21]

$$If(k) \cong \frac{1}{k}\frac{2}{\pi}\int_0^{\infty} f(k')dk', \tag{2.12.4}$$

Table 2.8. Parameters for the model functions $I\,^{10}f_n^{model}(k,\lambda,\beta)$.

	$I\,^{10}f_7^7$	$I\,^{10}f_9^9$	$I\,^{10}f_{12}^{12}$	$I\,^{10}f_{13}^{13}$	$I\,^{10}f_{15}^{15}$
			$\beta \le 1.0$ ($\alpha_2 = \alpha_0$, $N = 1258$, $\sigma = 1.0\cdot 10^{-3}$)		
a_{10}	0.8686(361)	0.9589(762)	1.4151(819)	1.3273(663)	1.4305(685)
a_{11}	3.0165(1247)	3.2679(1605)	5.0427(287)	4.6717(.229)	5.0632(.237)
β_{1e}	1.051(2)	1.060(2)	1.069(2)	1.0633(26)	1.064(25)
$a_{1\beta}$	8.937(178)	7.664(169)	6.952(137)	7.257(149)	7.208(144)
a_{20}	0.0219(15)	0.0172(20)	0.03032(27)	0.0177(26)	0.01769(27)
α_0	0.6046(194)	0.6261(216)	0.3936(163)	0.5122(158)	0.4842(139)
$\alpha_{0\lambda\beta\beta}$	−0.0509(22)	−0.0653(26)	−0.0411(21)	−0.0723(22)	−0.0723(20)
k_{e0}	0.1392(403)	−0.1558(605)	−0.8004(1009)	−0.9093(906)	−1.1680(967)
$k_{e\lambda\beta\beta}$	0.5053(775)	0.6212(246)	0.9696(375)	0.9828(330)	1.1138(345)
α_{10}	0.7930(674)	0.7615(194)	0.5297(129)	0.6046(138)	0.5640(121)
			$\beta > 1$ ($k_{e0} = 0$, $N = 555$)		
a_{10}	0.3432(66)	0.2975(68)	0.2949(65)	0.2855(82)	0.2622(64)
α_0	0.6021(162)	0.4766(148)	0.4007(119)	0.3801(146)	0.3114(10)
$\alpha_{0\lambda\beta\beta}$	0.0243(15)	0.0228(17)	0.0190(15)	0.0229(19)	0.0179(16)
$k_{e\lambda\beta\beta}$	−0.6721(212)	−1.0418(369)	−1.2133(420)	−1.5143(645)	−1.8000(662)
α_{10}	0.7241(138)	0.5735(125)	0.4808(99)	0.4601(121)	0.3767(83)
σ	$3.9\cdot 10^{-4}$	$3.8\cdot 10^{-4}$	$3.8\cdot 10^{-4}$	$4.3\cdot 10^{-4}$	$3.3\cdot 10^{-4}$

Table 2.9. Parameters for the model functions $I\,^{20}f_n^{model}(k,\lambda,\beta)$.

	$I\,^{20}f_6^6$	$I\,^{20}f_8^8$	$I\,^{20}f_{10}^{10}$	$I\,^{20}f_{12}^{12}$	$I\,^{20}f_{14}^{14}$	$I\,^{20}f_{16}^{16}$
			$\beta \le 1.0$ ($\alpha_2 = \alpha_0$)			
a_{10}	0.680(31)	0.6961(295)	0.6835(292)	0.720(27)	0.7227(272)	0.7675(295)
a_{11}	1.9647(186)	2.2212(948)	2.2513(962)	2.4028(206)	2.4420(910)	2.6497(1004)
β_{1e}	1.074(1)	1.062(28)	1.063(30)	1.078(13)	1.061(2)	1.064(3)
$a_{1\beta}$	7.50(16)	7.353(166)	7.248(171)	7.200(144)	7.185(144)	7.041(139)
a_{20}	0.0326(11)	0.0220(16)	0.0179(18)	0.0249(17)	0.0129(19)	0.0115(21)
α_0	0.4378(22)	0.4497(171)	0.4776(177)	0.4317(32)	0.4748(144)	0.4624(137
$\alpha_{0\lambda\beta\beta}$	−0.0216(17)	−0.0262(22)	−0.0345(22)	−0.0314(21)	−0.0432(20)	−0.0450(19)
k_{e0}	1.328(16)	1.1936(261)	0.9001(354)	0.6680(273)	0.4813(478)	0.2578(578)
$k_{e\lambda\beta\beta}$	0.4401(146)	0.4324(180)	0.5512(232)	0.7751(235)	0.7704(275)	0.8752310
α_{10}	0.7016(28)	0.6541(132)	0.6336(145)	0.5721(23)	0.5767(125)	0.5482(121
N	1515	1258	1258	1515	1258	1258
σ	$7.8\cdot 10^{-4}$	$9.9\cdot 10^{-4}$	$1.2\cdot 10^{-3}$	$8.7\cdot 10^{-4}$	$1.2\cdot 10^{-3}$	$1.1\cdot 10^{-3}$
			$\beta > 1$ ($k_{e0} = 0$)			
a_{10}	0.3834(68)	0.3515(71)	0.3212(70)	0.2321(29)	0.2880(67)	0.2759(66)
α_0	0.5149(123)	0.4749(131)	0.3998(117)	0.3391(68)	0.3067(96)	0.2771(90)
$\alpha_{0\lambda\beta\beta}$	0.02481(11)	0.0207(14)	0.0193(15)	0.0252(9)	0.0176(15)	0.0170(15)
$k_{e\lambda\beta\beta}$	−0.5985(18)	−0.9191(29)	−1.1973(40)	−1.50(0.26)	−1.7504(62)	−2.007(73)
α_{10}	0.6307(102)	0.5707(110)	0.4817(97)	0.3947(65)	0.3714(78)	0.3355(73)
N	777	555	555	777	555	555
σ	$4.3\cdot 10^{-4}$	$4.2\cdot 10^{-4}$	$4.0\cdot 10^{-4}$	$5.1\cdot 10^{-4}$	$3.6\cdot 10^{-4}$	$3.4\cdot 10^{-4}$

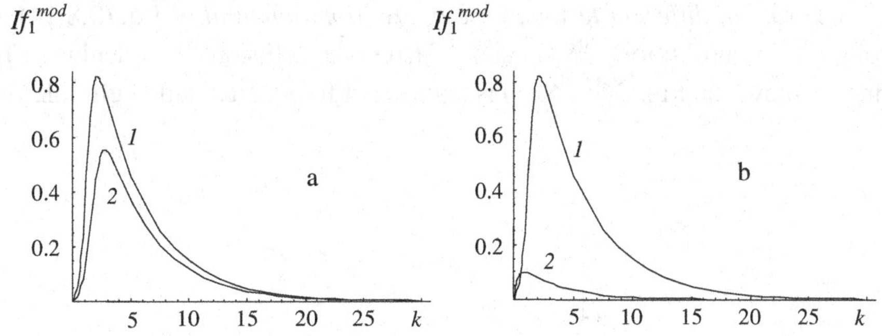

Fig. 2.5. Imaginary-part model resonance functions obtained with the straight-line trajectory (curve 1) and the exact trajectory (curve 2) models for $\lambda = 3.0$, $\beta = 1.0$ (a) and $\lambda = 3.0$, $\beta = 1.08$ (b).

and the substitution of $f(k')$ from Eq. (2.11.4) leads to the analytical expression

$$If^{\,model}(k) \cong \frac{2}{\pi k}\left[-\frac{a_1 + a_3}{\alpha}\ln(1 + e^{2\alpha k_e}) - \frac{a_3}{2\alpha \mathrm{ch}^2(\alpha k_e)}\right]. \qquad (2.12.5)$$

For $k > 9$ this formula gives the same values of $If^{\,model}(k)$ as Eq. (2.12.1).

2.13. Short-range forces and trajectory effects

As already mentioned in Sec. 2.9, one of the important advantages of the RB approach [34] is introducing realistic anisotropic short-range forces through the pair atom–atom interactions. The role of these forces for the collisional broadening as well as the trajectory effects can be conveniently demonstrated for HCl perturbed by Ar. Indeed, calculations of the cross-sections for pure rotational transitions [34] and the broadening coefficients for vibrotational transitions [57] were performed for this system by various theoretical approaches including the quantum mechanical CC method [44]. CC- and NG-calculated broadening coefficients for the 1–0 and 2–0 bands were compared to the experimental data in Ref. [57].

The influence of the trajectory model on the computed unaveraged HCl–Ar cross-sections for pure rotational transitions can be seen on Fig. 2.6. The using of parabolic trajectories [34] leads to larger cross-sections than the MET model.

The interval of integration on β has also a visible impact on the line-width value, as it can be seen on Fig. 2.7 for $0 \leq \beta \leq 1$ and $\beta > 1$. The interval $\beta > 1$ (close collisions) contributes significantly at any value of the rotational quantum number whereas including the interval $0 \leq \beta \leq 1$ is essential for low J only.

The role of different terms in the interaction potential of Eq. (2.8.14), i.e. the role of the contributions ${}^{10}S_2(b)$ and ${}^{20}S_2(b)$ in the collisional broadening of HCl–Ar lines is shown on Fig. 2.8: ${}^{10}S_2(b)$ is significant for small J and negligible for $J > 2$.

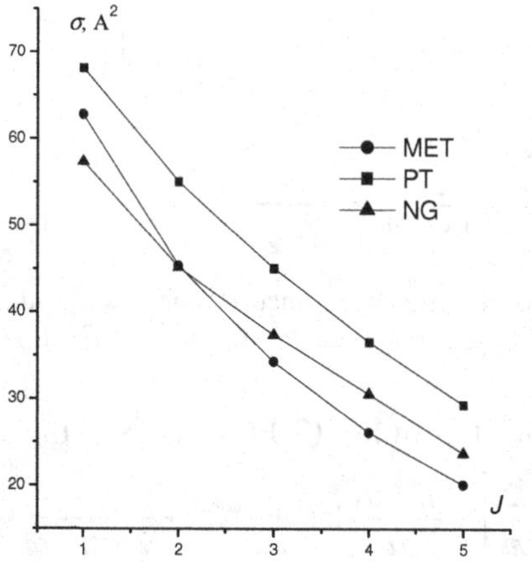

Fig. 2.6. Comparison of unaveraged HCl–Ar cross-sections for pure rotational transitions for reduced kinetic energies $E = m\mathring{v}^2/2 = 398$ K: MET, PT [34] and NG [45].

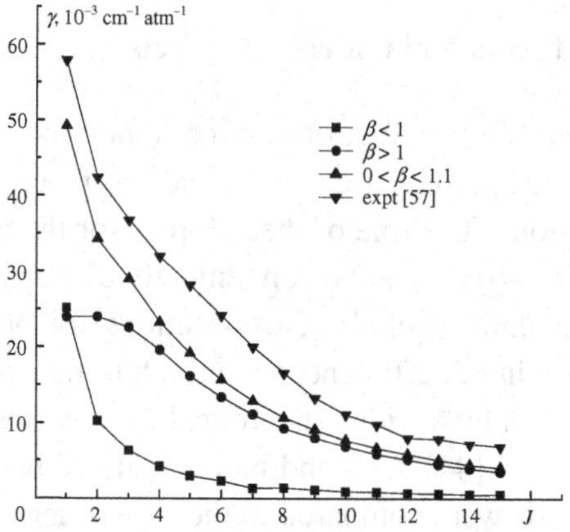

Fig. 2.7. Contributions of different integration intervals to HCl–Ar broadening coefficients (P-branch) obtained by the RB approach with MET resonance functions at 298 K.

The short- and long-range parts of the interaction potential influence the room-temperature HCl–Ar broadening coefficients in the manner shown in Fig. 2.9. For high J-values the long-range contributions are practically negligible. For low and middle J-values both kinds of contributions are indispensable.

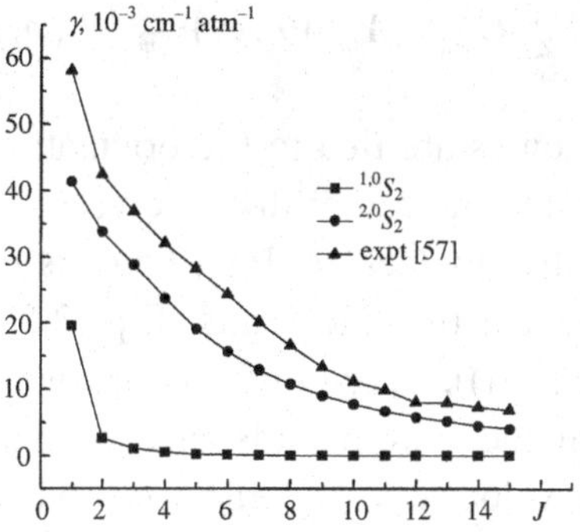

Fig. 2.8. Contributions of ${}^{10}S_2(b)$ and ${}^{20}S_2(b)$ to HCl–Ar P-branch broadening coefficients calculated by the RB formalism with MET resonance functions at 296 K.

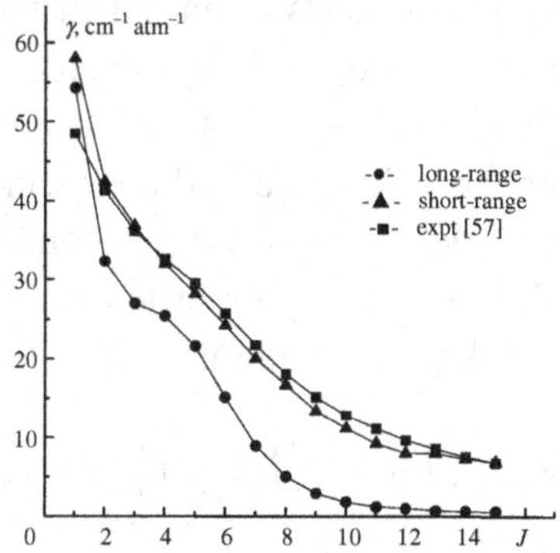

Fig. 2.9. Long-range ($R_1 = R_2 = 0$) and short-range ($A_1 = A_2 = 0$) contributions to the HCl–Ar broadening coefficients calculated within the RB formalism with MET resonance functions.

2.14. Robert and Bonamy formalism with exact trajectories

A systematic implementation of the exact classical trajectories into the framework of the original RB formalism [34] for linear, symmetric-top and asymmetric top molecules with both long-range and short-range interactions in the rotationally invariant form of Eq. (2.5.6) was realised in a series of works [41–43].

For any molecular system, the intermolecular potential is taken as a sum of the long-range electrostatic interactions given by Eqs (2.5.7) and (2.5.10) and pair atom–atom interactions described by Eq. (2.5.20) and is further developed in the laboratory-fixed frame in order to ensure the rotational invariance. In the case of linear or colliders, the development reads

$$V_{aniso}(\vec{r}) = \sum_{l_1 l_2 l} V_{l_1 l_2 l}(r) \sum_{m_1 m_2 m} C^{lm}_{l_1 m_1\, l_2 m_2} Y_{l_1 m_1}(\theta_1, \varphi_1) Y_{l_2 m_2}(\theta_2, \varphi_2) C^{*}_{lm}(\theta, \Psi), \tag{2.14.1}$$

where the spherical harmonics are tied to the orientations of the molecular axes ($Y_{l_1 m_1}$ and $Y_{l_2 m_2}$) and of the intermolecular distance vector $\vec{r}$ (C_{lm}). In the most general case of two asymmetric molecules, Eq. (2.5.6) is employed with the radial components including the electrostatic terms (Eqs (2.5.7) and (2.5.10)) and the atom–atom terms (Eq. (2.5.24)). These radial components are computed numerically for r varying from the minimal distance of the closest approach $r_{c\ min}$ (defined by the relative velocity v and the isotropic potential parameters) to (typically) 9–15 Å where the contributions to the line width become negligible. The numerical computation enables, in particular, immediate (that is, without line-width computation) testing and ensuring of the convergence on p and q summation indices in Eq. (2.5.22). (We remind that in the CRB approach this summation is automatically stopped at the eighth order.)

Since in Eq. (2.8.4) the factor $(1 - S_2^{middle}{}')$ can be inserted back in the exponent, the interruption function for the line-width calculation can be written as $S = 1 - \exp(-S_2)$ where the complete second-order contribution is $S_2 = S_{2,i} + S_{2,f} + S_2^{middle}$. In the case of a symmetric top colliding with a linear molecule (for which the rotational quantum number $K_2 = 0$), for example, it reads

$$S_2 = 2\hbar^{-2}\left(\frac{r_c}{\mathrm{v}}\right)^2 \sum_{l_1 l_2 l}\left[\sum_{j_i' j_2'}\left(C^{j_i' K}_{j_i K\, l_1 0}\right)^2 \left(C^{j_2' 0}_{j_2 0\, l_2 0}\right)^2 f_{l_1 l_2 l} + \sum_{j_f' j_2'}\left(C^{j_f' K}_{j_f K\, l_1 0}\right)^2 \left(C^{j_2' 0}_{j_2 0\, l_2 0}\right)^2 f_{l_1 l_2 l} \right.$$
$$\left. - \sum_{j_2'} (-1)^{j_2 + j_2'} \left(C^{j_2' 0}_{j_2 0\, l_2 0}\right)^2 D^{(l_1)}_{j_i j_f} f_{l_1 l_2 l}\right], \tag{2.14.2}$$

where the exact-trajectory resonance functions are defined by

$$f_{l_1 l_2 l}(r_c, k_c) = \sum_m \frac{(l-m)!(l+m)!}{2^{2l}((l-m)/2)!^2((l+m)/2)!^2}$$
$$\times \left\{ \frac{\int_1^{\infty} dz V_{l_1 l_2 l}(r_c z) z \cos\left[k_c A_0(z) + m\sqrt{1 - V^{\circ}_{iso}(r_c)} A_2(z)\right]}{\sqrt{z^2 - 1 + V^{\circ}_{iso}(r_c) - z^2 V^{\circ}_{iso}(r_c z)}} \right\}^2, \tag{2.14.3}$$

$$A_n(z) = \int_1^z \frac{dz'}{z'^{n-1}\sqrt{z'^2 - 1 + V^{\circ}_{iso}(r_c) - z'^2 V^{\circ}_{iso}(r_c z')}}.$$

In contrast to the electrostatic function of Eq. (2.10.12), these functions are not normalised to unity at $k_c = 0$ and are subscribed by the additional index l since for the atom–atom contributions l is not generally equal to $l_1 + l_2$. They are also computed by a numerical integration for r_c varying from $r_{c\,min}$ to 9–15 Å and k_c from 0 to 20–35 (tests of line-width convergence on k_c should be made). For a given temperature, the computations are typically made only with the mean thermal velocity $\overline{\mathrm{v}}$. In some cases, however, the velocity averaging for the differential cross-section is required and the resonance functions of Eq. (2.14.3) are to be additionally computed for other relative velocities (e.g. $0.5\,\overline{\mathrm{v}}$, $1.5\,\overline{\mathrm{v}}$, and so on). It needs to be reminded that in these cases the minimal r_c value should be properly modified since λ in Eq. (2.9.12) changes.

For asymmetric rotors the wave functions are expressed in terms of those of symmetric top [14]:

$$\left|J\tau M\right\rangle = \sum_{K} a_K^{J\tau}\left|JKM\right\rangle \tag{2.14.4}$$

with $-J \le K \le J$. The radial interaction potential components of Eq. (2.5.24) depend additionally on the projections k_1, k_2. As a result, the second order contributions have a more complex form [43]:

$$S_{2,i2} = \frac{2}{\hbar^2}\left(\frac{r_c}{\mathrm{v}}\right)^2 \sum_{J_i'\tau_i'J_2'\tau_2'} \sum_{l_1 l_2 l} \frac{1}{(2l_1+1)(2l_2+1)} \sum_m \frac{(l-m)!(l+m)!}{2^{2l}((l-m)/2)!^2\,((l+m)/2)!^2} \\ \times \left| \sum_{k_1 k_2} (-1)^{k_1+k_2} X_{-k_1}^{l_1}(J_i\tau_i J_i'\tau_i') X_{-k_2}^{l_2}(J_2\tau_2 J_2'\tau_2') I_{l_1 l_2 lm}^{k_1 k_2}(\omega_{i2,i'2'}) \right|^2, \tag{2.14.5}$$

$$S_{2,f2'i2'} = -\frac{4}{\hbar^2}\left(\frac{r_c}{\mathrm{v}}\right)^2 (-1)^{J_i+J_f+J_2+J_2'+\rho} \sqrt{\frac{(2J_2'+1)(2J_i+1)(2J_f+1)}{(2J_2+1)}} \sum_{l_1 l_2 l} \frac{(-1)^{l_2+l} W(J_i J_f J_i J_f;\rho l_1)}{(2l_1+1)(2l_2+1)} \\ \sum_m \frac{(l-m)!(l+m)!}{2^{2l}((l-m)/2)!^2\,((l+m)/2)!^2} \left[\sum_{k_1 k_2} (-1)^{k_1+k_2} X_{-k_1}^{l_1}(J_f\tau_f J_f\tau_f) X_{-k_2}^{l_2}(J_2'\tau_2' J_2\tau_2) I_{l_1 l_2 lm}^{k_1 k_2}(\omega_{2'2}) \right] \\ \times \left[\sum_{\eta_1 \eta_2} (-1)^{\eta_1+\eta_2} X_{-\eta_1}^{l_1}(J_i\tau_i J_i\tau_i) X_{-\eta_2}^{l_2}(J_2\tau_2 J_2'\tau_2') I_{l_1 l_2 lm}^{\eta_1 \eta_2}(\omega_{22'}) \right], \tag{2.14.6}$$

where

$$X_{k_1}^{l_1}(J_i\tau_i J_i'\tau_i') \equiv \sum_{K_i K_i'} a_{K_i}^{J_i\tau_i\,*} a_{K_i'}^{J_i'\tau_i'} C_{J_i K_i l_1 k_1}^{J_i' K_i'}, \tag{2.14.7}$$

$$X_{k_2}^{l_2}(J_2\tau_2 J_2'\tau_2') \equiv \sum_{K_2 K_2'} a_{K_2}^{J_2\tau_2\,*} a_{K_2'}^{J_2'\tau_2'} C_{J_2 K_2 l_2 k_2}^{J_2' K_2'}, \tag{2.14.8}$$

$$I_{l_1 l_2 l}^{k_1 k_2}(\omega) \equiv \int_1^\infty \frac{dz V_{l_1 l_2 l}^{k_1 k_2}(r_c z) z \cos[k_c A_0(z) + m\sqrt{1-V_{iso}^{\circ}(r_c)}A_2(z)]}{\sqrt{z^2 - 1 + V_{iso}^{\circ}(r_c) - z^2 V_{iso}^{\circ}(r_c z)}}. \tag{2.14.9}$$

The summation on m in Eqs (2.14.5) and (2.14.6) can not be made in order to introduce the resonance functions, and the integrals of Eq. (2.14.9) are preliminary computed for various r_c, k_c and $\bar{v}$ whereas the summation on m is made in the code computing S_2 contributions.

The exact-trajectory approach presented in this section is not limited to the Lennard–Jones form of the isotropic potential. In any formula, a numerical V_{iso} can be used (obtained, for example, from the atom–atom part of the interaction potential). The errors coming from the fitting are thus avoided. The numerical isotropic potential, however, needs the correction of Eq. (2.9.16) since now

$$b\,db = r_c\,dr_c\left[1 - V_{iso}^{\circ}(r_c) - \frac{r_c}{2} V_{iso}^{\circ\,\prime}(r_c)\right] \qquad (2.14.10)$$

and the first derivative $V_{iso}^{\circ}{}'(r)$ of $V_{iso}^{\circ}(r)$ should be known for the integration on the trajectory. Some examples of the RBE approach application to molecular systems of atmospheric interest are given in the last chapter.

Bibliography

1. Anderson P.W. (1949). Pressure broadening in the microwave and infrared region. *Phys Rev* 76: 647–661.
2. Tsao C.J. and Curnutte B. (1962). Line widths of pressure-broadened spectral lines. *J Quant Spectrosc Radiat Trans* 2: 41–91.
3. Fano U. (1963). Pressure broadening as prototype of relaxation. *Phys Rev* 131: 259–268.
4. Srivastava R.P. and Zaidi H.R. (1979). Intermolecular forces revealed by Raman scattering. In: Weber A. (ed.). *Raman spectroscopy of gases and liquids*, pp. 167–202, Springer-Verlag, Berlin, New York.
5. Ben-Reuven A. (1966). Impact broadening of microwave spectra. *Phys Rev* 145: 7–22.
6. Cherkasov M.R. (1976). To the theory of overlapping lines. *Opt Spectrosk* 40: 7–13 (in Russian).
7. Biedenharn L.C. and Louck J.D. (1981). *Angular momentum in quantum physics*. Addison-Wesley Publishing Company, Reading, Massachusetts.
8. Fiutak J. and van Kranendonk J. (1962). Impact theory of Raman line broadening. *Can J Phys* 40: 1085–1100.
9. Fiutak J. and van Kranendonk J. (1963). Theory of the impact broadening of Raman lines due to anisotropic intermolecular forces. *Can J Phys* 41: 21–32.
10. Herman R.M. and Tipping R.H. (1970). Impact theory for the noble gas pressure-induced HCl vibration-rotation and pure rotation line shifts – II. *J Quant Spectrosc Radiat Trans* 10: 897–908.
11. Hirschfelder J.O., Curtiss C.F. and Bird R.B. (1964). *Molecular theory of gases and liquids*. Wiley, New York.
12. Varshalovich D.A., Moskalev A.N. and Khersonskii V.K. (1988). *Quantum theory of angular momentum*. World Scientific, Singapore.
13. Landau L.D. and Lifshitz, E.M. (1977). *Quantum Mechanics (Non-Relativistic Theory)*. Pergamon Press, Oxford, England.
14. Leavitt R.P.(1980). Pressure broadening and shifting in microwave and infrared spectra of molecules of arbitrary symmetry: An irreducible tensor approach. *J Chem Phys* 73: 5432–5450.
15. Gray C.G. and Gubbins K.E. (1984). *Theory of molecular fluids, Volume 1: Fundamentals*. Clarendon press, Oxford.
16. Leavitt R.P. (1980). An irreducible tensor method of deriving the long-range anisotropic interactions between molecules of arbitrary symmetry. *J Chem Phys* 72: 3472–3482.

17. MacRury T.B., Steele W.A. and Berne B.J. (1976). Intermolecular potential models for anisotropic molecules, with applications to N_2, CO_2, and benzene. *J Chem Phys* 64: 1288–1299.
18. Yasuda H. and Yamamoto T. (1971). On the Two-center Expansion of an Arbitrary Function. *Prog Theor Phys* 45: 1458–1465.
19. Downs J., Gubbins K.E., Murad S. and Gubbins C.G. (1979). Spherical harmonic expansion of the intermolecular site-site potential. *Mol Phys* 37: 129–140.
20. Bykov A.D. and Lavrentieva N.N. (1991). Calculation of resonance functions in the impact line-broadening theory. *Atmos Oceanic Opt* 4: 718–729 (in Russian).
21. Boulet C., Robert D. and Galatry L. (1976). Shifts of the vibration-rotation absorption lines of diatomic molecules perturbed by diatomic polar molecules. *J Chem Phys* 65: 5302–5314.
22. Murphy J.S. and Boggs J.E. (1967). Collision Broadening of Rotational Lines. I. Theoretical formulation. *J Chem Phys* 47: 691–702.
23. Murphy J.S. and Boggs J.E. (1967). Collisional Broadening of Rotational Absorption Lines II. Self-broadening of Symmetric-top Molecules. *J Chem Phys* 47: 4152–4158.
24. Murphy J.S. and Boggs J.E. (1968). Collisional Broadening of Rotational Absorption Lines III. Broadening by Linear Molecules and Inert Gas and Determination of Molecular Quadrupole moments. *J Chem Phys* 49: 3334–3343.
25. Cattani M. (1972). On the calculation of the pressure line shape in the impact approximation. *Phys Lett A* 38: 147–148.
26. Cherkasov M.R. (1985). *Formalism of Liouville operator in line-broadening calculation.* Preprint, № 4, Institute of Atmospheric Optics, Tomsk.
27. Korff D. and Leavitt R.P. (1975). Calculation of impact broadened halfwidth parameters for HCl using a natural cutoff theory. *Phys Lett A* 53: 351–352.
28. Korff D. and Leavitt R.P. (1981). Cutoff-free theory of impact broadening and shifting in microwave and infrared gas spectra. *J Chem Phys* 74: 2180–2188.
29. Jarecki J. and Herman R.M. (1975). Widths and shifts of HF vibration-rotation absorption lines induced by pressure of rare gases. *J Quant Spectrosc Radiat Trans* 15: 707–726.
30. Jarecki J. (1976). Rare gas pressure broadening and shifting of HF vibration-rotation absorption lines — explanation of anomalous line shifts. *J Chem Phys* 65: 5318–5326.
31. Smith E.W., Giraud M. and Cooper J. (1976). A semiclassical theory for spectral line broadening in molecules. *J Chem Phys* 65: 1256–1267.
32. Davis R.W. and Oli B.A. (1978). Theoretical calculations of H_2O line-width and pressure shifts: comparison of the Anderson theory with quantum many-body theory for N_2 and air-broadened lines. *J Quant Spectrosc Radiat Trans* 20: 95–120.
33. Salesky E.T. and Korff D. (1979). Calculation of HCl line widths using a new impact parameter theory. *Phys Lett A* 72: 431–434.
34. Robert D. and Bonamy J. (1979). Short range force effects in semiclassical molecular line broadening calculations. *J Phys (Paris)* 40: 923–943.
35. Bonamy J., Bonamy L. and Robert D. (1977). Overlapping effects and motional narrowing in molecular band shapes: Application to the Q branch of HD. *J Chem Phys* 67: 4441–4453.
36. Lynch R. and Gamache R.R. (1996). Fully complex implementation of the Robert–Bonamy formalism: Half widths and line shifts of H_2O broadened by N_2. *J Chem Phys* 105: 5711–5721.
37. Ma Q., Tipping R.H. and Boulet C. (2007). Modification of the Robert–Bonamy formalism in calculating Lorentzian half-widths and shifts. *J Quant Spectrosc Radiat Trans* 103: 588–596.
38. Landau L.D. and Lifshitz E.M. (1976). *Mechanics*. Pergamon, Oxford.
39. Bykov A.D., Lavrentieva N.N. and Sinitsa L.N. (1992). Influence of trajectory model on the line shifts in the visible region. *Atmos Oceanic Opt* 5: 907–917.

40. Joubert P., Bonamy J. and Robert D. (1999). Exact trajectory in semiclassical line broadening and line shifting calculation test for H_2–He Q(1) line. *J Quant Spectrosc Radiat Trans* 61: 19–24.
41. Buldyreva J., Bonamy J. and Robert D. (1999). Semiclassical calculations with exact trajectory for N_2 vibrotational Raman linewidths at temperatures below 300 K. *J Quant Spectrosc Radiat Trans* 62: 321–343.
42. Buldyreva J., Benec'h S. and Chrysos M. (2001). Infrared nitrogen-perturbed NO line widths in a temperature range of atmospheric interest: An extension of the exact trajectory model. *Phys Rev A* 63: 012708.
43. Buldyreva J. and Nguyen L. (2008). Extension of the exact trajectory model to the case of asymmetric tops and its application to infrared nitrogen-perturbed linewidths of ethylene. *Phys Rev A* 77: 042720.
44. Green S. (1990). Theoretical line shapes for rotational spectra of HCl in Ar. *J Chem Phys* 92: 4679–4685.
45. Neilsen W.B. and Gordon R. (1973). On a semiclassical study of molecular collisions. II. Application to HCl–argon. *J Chem Phys* 58: 4149–4170.
46. Labani B., Bonamy J., Robert D., Hartmann J.-M. and Taine J. (1986). Collisional broadening of rotation-vibration lines for asymmetric top molecules. I. Theoretical model for both distant and close collisions. *J Chem Phys* 84: 4256–4267.
47. Neshyba S.P. and Gamache R.R. (1993). Improved line broadening coefficients for asymmetric rotor molecules with application to ozone lines broadened by nitrogen. *J Quant Spectrosc Radiat Trans* 50: 443–453.
48. Labani B., Bonamy J., Robert D. and Hartmann J.-M. (1987). Collisional broadening of rotation-vibration lines for asymmetric-top molecules. III. Self-broadening case; application to H_2O. *J Chem Phys* 87: 2781–2769.
49. Robert D. and Bonamy J. (1979). Internal report, Laboratoire de Physique Moléculaire, Faculté des Sciences. Besançon Cedex, France.
50. Bykov A.D., Lavrentieva N.N. and Sinitsa L.N. (1992). Calculation of resonance functions for real trajectory, *Atmos Oceanic Opt* 5: 1127–1132 (in Russian).
51. Bykov A.D., Sinitsa L.N. and Starikov V.I. (1999). *Experimental and theoretical methods in spectroscopy of water vapour*. Publishing House of the Institute of Atmospheric Optics, Novosibirsk (in Russian).
52. Bykov A.D., Sinitsa L.N. and Starikov V.I. (2004). *Introduction in ro-vibrational spectroscopy of polyatomic molecules*. Publishing House of the Institute of Atmospheric Optics, Tomsk (in Russian).
53. Lavrentieva N.N. and Starikov V.I. (2006). Approximation of resonance functions for exact trajectories in the pressure-broadening theory. Real parts. *Mol Phys* 104: 2759–2766.
54. Lavrentieva N.N. and Starikov V.I. (2005). Approximation of resonance functions for real trajectories in the pressure-broadening theory. *Atmos Oceanic Opt* 18: 729–734.
55. Starikov V.I. (2008). Calculation of the self-broadening coefficients of water vapor absorption lines using an exact trajectory model. *Opt Spectrosc* 104: 513–523.
56. Starikov V.I. and Lavrentieva N.N. (2007). Approximation of resonance functions for exact trajectories in the pressure-broadening theory. *Atmos Oceanic Opt* 20: 780–788.
57. Boulet C., Flaud J.-M. and Hartmann J-.M. (2004). Infrared line collisional parameters of HCl in argon, beyond the impact approximation: measurements and classical path calculations. *J Chem Phys* 120: 11053–11061.

Chapter 3

Collisional broadening of water vapour lines

Despite the fact that the main constituents of the Earth's atmosphere are nitrogen (~ 79%) and oxygen (~ 20%), other minor (~ 1%) pollutants and natural gases (CO_2, CO, H_2O, Ar, and so on) play a very important role. A particular place is occupied by water vapour which is the principle absorber of solar radiation concentrated mainly in the spectral region 0.3–15 μm. It is in this region that its strongest vibrotational bands are situated. Currently, about 50,000 spectral transitions of water vapour molecule are identified with the vibrational quantum numbers $v_1 + v_2 + v_3 \leq 10$ and the rotational quantum numbers $J \leq 42$, $K_a \leq 32$.

Let us specify the meaning of these quantum numbers. In the H_2O molecule, there are three normal vibrational modes, two of which refer to the variations in the OH bond length (stretching vibrations) and one describes the variations of ∠HOH angle (bending vibration). The symmetric and anti-symmetric stretching vibrations are characterised by the vibrational quantum numbers v_1 and v_3, whereas the bending vibration is associated with the vibrational quantum number v_2. Any vibrational state in the molecule is therefore defined by three quantum numbers: $V \equiv (v_1\ v_2\ v_3)$.

Since the H_2O molecule is an asymmetric top, its rotational energy levels $[J\ K_a\ K_c]$ are defined by three rotational quantum numbers J, K_a, K_c. The (pseudo-)quantum numbers K_a and K_c represent the projections of the angular momentum operator along the molecular axes a (axis of the smallest moment of inertia) and c (axis of the largest moment of inertia). They are not good quantum numbers but become them in the limit cases of prolate or oblate symmetric tops, respectively, reducing to the rotational quantum number K. It is for this reason that they are often referred to as pseudo-quantum numbers. The point group of symmetry is C_{2v}, so that all vibrotational states and wave functions belong to one of four symmetry types: A_1, A_2, B_1 or B_2. These types of symmetry (usually denoted by Γ) are presented in Table 3.1 [1]. The symmetry type Γ also defines the nuclear spin statistical weight w needed for the line-broadening and intensity calculations (Table 3.2).

The allowed transitions between the vibrotational states of H_2O molecule are: $A_1 \leftrightarrow A_2$, $B_1 \leftrightarrow B_2$, i.e. only the transitions between different A-states (A-bands) or B-states (B-bands) are possible. The strongest lines in an A-band ($\Delta v_3 = v_{3f} - v_{3i}$ is odd) correspond to the transitions with $\Delta K_a = |K_{af} - K_{ai}| = 0, 2$ and in a B-band (Δv_3 is even) they are due to $\Delta K_a = 1, 3$. The complete selection rules for vibrotational transitions in the H_2O molecule are collected in Table 3.3 [1].

Table 3.1. Symmetry types of vibrotational wave functions of X_2Y (C_{2v}) molecule in function of even (e) or odd (o) quantum number values [1]. Reproduced with permission.

v_3	K_a	K_c	Γ
e	e	e	A_1
o	o	e	A_1
e	o	o	A_2
o	e	o	A_2
e	o	e	B_1
o	e	e	B_1
e	e	o	B_2
o	o	o	B_2

Table 3.2. Nuclear spin statistical weight w in function of symmetry type Γ.

Γ	H_2O (C_{2v})	D_2O(C_{2v})	Γ	HDO (C_s)
A_1, A_2	1	6	A'	6
B_1, B_2	3	3	A''	6

Table 3.3. Selection rules for vibrotational absorption transitions in H_2O molecule [1]*. Reproduced with permission.

$\Delta J = 0, \pm 1$; $A_1 \leftrightarrow A_2$, $B_1 \leftrightarrow B_2$			
Band	Changing component of dipole moment $\vec{\mu}$	Selection rules for v_3	Selection rules for K_a and K_c
A	μ_a	$e \leftrightarrow o$	$ee \leftrightarrow eo$ or $oo \leftrightarrow oe$
B	μ_b	$e \leftrightarrow e$ or $o \leftrightarrow o$	$ee \leftrightarrow oo$ or $oe \leftrightarrow eo$

* Here μ_a and μ_b are the dipole moment components in the molecular frame with a, b, c axes corresponding to the moments of inertia $I_a < I_b < I_c$. The molecule lies in the XOZ plane and its symmetry axis OX is the permanent dipole moment axis (i.e. for the equilibrium configuration $\mu_x \neq 0$ and $\mu_z = 0$). Other notations for the rotational states can be found in Refs [1–3].

The knowledge of H_2O vibrotational line-broadening coefficients by main atmospheric gases N_2 and O_2 (γ_{N_2} and γ_{O_2}) allows the determination of these coefficients for the dry-air broadening (γ_{air}) according to the approximate relation

$$\gamma_{air} = 0.79\gamma_{N_2} + 0.21\gamma_{O_2}$$

(in the case of multi-component mixtures Eq. (1.3.4) must be used). For the Earth's atmosphere γ_{air} is to be known for the temperatures 200–300 K and for the rotational quantum numbers $J \leq 20$, $K_a \leq 15$ [4]. For the atmosphere of Venus

(95% of CO_2, 4% of N_2, and 1% of H_2O) the H_2O–CO_2 broadening coefficients are needed for 200–700 K and $J \le 35$, $K_a \le 20$. Finally, various industrial applications require N_2-, O_2-, CO_2-, H_2O- and Ar-broadening coefficients for 300–3000 K and $J \le 40$, $K_a \le 30$.

The precision of the broadening coefficients has a major influence on the accuracy of spectra calculations and, as a consequence, on the retrieval of temperatures and/or concentrations of atmospheric species from the recorded spectra. For most applications this precision should be better than 5% for both measurements and computations [4].

A large number of publications on experimental studies of H_2O spectral line broadening and shifting are currently available in the literature for the perturbation by linear molecules (CO, CO_2), air (N_2+O_2 mixture), rare gases (Ar, Kr, Ne, Xe) and water vapour itself. Such a wide choice of perturbing particles has allowed an investigation of the influence of various contributions in the intermolecular interaction potential and, in a few cases, the influence of the atom–atom potential contributions. An exhaustive analysis of these works up to 1989 is given in the monograph [2] and some results are considered in the monograph [3]. A survey of experimental data of 4,000 $H_2{}^{16}O$ line-broadening coefficients obtained before 1994 is presented in the review article [4]; the same authors reviewed the experimental results up to 2004 in Ref. [5].

Despite the fact that a lot of experimental results are available, the need for vibrotational H_2O line-broadening coefficients is much greater and their theoretical calculation is required. Before starting the presentation of various theoretical methods for H_2O line-width and line-shift calculations a precision on the molecular wave functions and operators is necessary.

3.1. Effective operators of physical quantities for X_2Y molecule and vibrotational wave functions for water vapour molecule

Line-width and shift calculations require the vibration-rotation matrix elements $A(i; i') = \langle \psi_i | A | \psi_{i'} \rangle$ of various physical quantities A (such as dipole moment $\vec{\mu}$ or polarizability α) in the basis of wave functions ψ_i of molecular vibration-rotation Hamiltonian H. In many cases, the vibrotational energies and wave functions are obtained not for the complete Hamiltonian H, but for an effective Hamiltonian $\tilde{H}$ describing the vibrotational molecular structure for a chosen range of vibrational quantum numbers. This effective Hamiltonian is obtained by perturbation theory methods: $\tilde{H} = T^{+}HT$ where T is the transformation operator. Usually, the contact transformations are used and T is chosen in the form $T = e^{-iS}$ where S is Hermitian operator, so that $T^{+} = T^{-1}$ is unitary. The operator T is introduced to remove the terms coupling the vibrational states of different polyads of interacting states from the initial Hamiltonian H. In the vibrational basis $|V_i\rangle$ (i = 1, 2,...) of harmonic oscillator, the effective Hamiltonian $\tilde{H}$ associated with a polyad of interacting vibrational states represents a $n \times n$ matrix:

$$\tilde{H} = \begin{bmatrix} H_{11} \; H_{12} \; \dots \; H_{1n} \\ \quad H_{22} \; \dots \; H_{2n} \\ h.c. \qquad \dots \\ \qquad\qquad H_{1n} \end{bmatrix}, \tag{3.1.1}$$

where n determines the number of interacting states within the polyads and *h.c.* denotes the Hermitian conjugation for the lower triangular part of the matrix. In a particular case, the operator matrix can be one-dimensional (n = 1), so that the Hamiltonian $\tilde{H}$ is associated with an isolated vibrational state $|V_i\rangle$ and the matrix element $H_{ii} = \langle V_i|\tilde{H}|V_i\rangle$ represents the effective rotational Hamiltonian for this given isolated vibrational state. In the matrix of Eq. (3.1.1), the diagonal matrix elements are the effective rotational Hamiltonians for each vibrational state and the off-diagonal elements account for the interactions between these states.

The rotational operators $H_{ik} = \langle V_i|\tilde{H}|V_k\rangle$ in Eq. (3.1.1) can take various forms. In general, these operators are defined by the components J_α ($\alpha = x, y, z$) of the total angular momentum $\vec{J}$. For the asymmetric top molecule they depend on the operators J^2, J_z and $J_+^2 + J_-^2 = 2(J_x^2 - J_y^2)$ where $J_\pm = J_x \mp iJ_y$; the presence of the last operator is due to the asymmetry of the molecule. In addition, the rotational operators H_{ik} include adjustable parameters obtained by the fitting of calculated energy levels (or transition frequencies) to their experimental values. The energy levels result from a numerical diagonalization of the Hamiltonian matrix of Eq. (3.1.1) in a suitable vibration-rotation basis.

Since the H_2O molecule allows a large amplitude motion (bending vibration), calculation of vibrotational energies with the effective Hamiltonian (3.1.1) encounters some difficulties. First, a power series expansion of H_{ik} does not converge rapidly (or even diverges), so that very high powers of J_α in the centrifugal distortion polynomial for H_{ii} should be introduced in order to insure a reliable fit to experimental data. Secondly, many accidental, non-regular resonance interactions are present in this molecule. For an accurate use of experimental data, model operators with improved J_α-convergence properties are developed for H_{ii}.

A review of these models can be found in the monographs [3, 6, 7]. In particular, to obtain some numerical results of the following sections of this chapter the effective operators H_{ii} of Eq. (3.1.1) were represented by the model of generating functions:

$$H_{ii} = F^{(i)}(J^2, J_z) + \frac{1}{2}\sum_{k=0}^{4}\{\chi_k^{(i)}(J^2, J_z), J_+^{2k} + J_-^{2k}\}, \tag{3.1.2}$$

where F and χ are generating functions dependent on the squared angular momentum operator J^2 and on its J_z component (being expanded into Taylor series on J^2 and J_z they give the usual polynomial representations for H_{ii}) and the braces denote the anti-commutator. For instance, this model yielded a standard deviation of 0.016 cm^{-1} when fitting 2039 experimental energy levels with rotational quantum numbers $J \leq 35$, $K_a \leq 32$ for five lowest interacting vibrational states (v_1 v_2 v_3) = {(000), (010), (100), (020), (001)} of H_2O [8]. As an example, Table 3.4 gives the wave function mixing coefficients for some rotational levels of (010) and (110) vibrational states of H_2O and D_2O molecules. Usually the state (010) is viewed as an isolated state and all mixing coefficients are put equal to zero.

Table 3.4. Mixing coefficients (in %) for some rotational energy levels E (in cm^{-1}) of (010) and (110) vibrational states of H_2O [8] and D_2O molecules.

			H_2O			
$J\ K_a\ K_c$	$E_{calc.}$	(000)	(010)	(020)	(100)	(001)
13 11 2	5654.774	0.0	76.9	17.5	5.6	0.0
19 8 12	7219.893	0.0	70.2	29.4	0.3	0.1
21 14 7	9897.558	0.0	61.2	35.6	3.1	0.1
			D_2O			
$J\ K_a\ K_c$	$E_{calc.}$	(030)	(110)	(011)		
8 0 8	4223.425	14.5	71.3	14.2		
8 1 7	4297.495	19.4	72.0	8.6		
8 2 6	4347.085	23.8	69.9	6.3		
8 3 5	4385.468	24.2	70.8	5.0		
8 4 5	4443.046	21.0	73.0	6.0		

In the majority of works devoted to γ and δ coefficients determination, a polynomial representation (see Eq. (4.1.1)) for the effective rotational Hamiltonian was used, which imposes some limits on the wave functions obtained for the rotational states [$J\ K_a\ K_c$] with high values of the quantum numbers J and K_a.

The wave functions φ obtained from the effective Hamiltonian $\tilde{H}$ are related with the wave functions ψ of the complete vibration-rotation Hamiltonian H by the equation $\varphi = T^{-1}\psi$ (which follows from $H\psi = E\psi = \tilde{H}\varphi = E\varphi$), so that in the matrix elements

$$A(i;i') = \langle \Psi_i | A | \Psi'_i \rangle = \langle \varphi_i | \tilde{A} | \varphi'_i \rangle \tag{3.1.3}$$

the effective operators $\tilde{A}$ of physical quantities should be used. These operators are to be constructed in the same way as the effective Hamiltonian $\tilde{H}$; so for $\tilde{A} = T^{-1}AT = e^{iS}Ae^{-iS}$ the same contact transformation operator S should be employed. The operators S are well known [7, 9, 10]. They depend on the elementary vibrational operators q_i, p_i (q_i are the normal coordinates and p_i are the corresponding conjugated moments for the i-th vibrational mode) and on the rotational J_α operators. To derive the transformed operator $\tilde{A}$ it is necessary to know the dependence of A on the internal coordinates, for example, on the normal coordinates q_i, i.e. the function $A(q_i)$. If this function $A(q_i)$ is expanded into Taylor series on q_i coordinates:

$$A(q_i) = A_0 + \sum_i A_i q_i + \frac{1}{2}\sum_{ij} A_{ij} q_i q_j + \cdots, \tag{3.1.4}$$

the derivatives A_i, A_{ij}, … are needed. In some cases, they are known from *ab initio* calculations or, as in the case of dipole moment or mean polarizability α, from an analysis of absorption or Raman scattering line intensities. Since the contact transformation operators S depend on the elementary vibrational (q_i, p_i) and rotational J_α operators, the operator $\tilde{A}$ is a vibration-rotational operator and can be written as

$$\tilde{A} = \tilde{A}_0(q_i, p_j) + \sum_\alpha \tilde{A}_\alpha(q_i, p_j)J_\alpha + \sum_{\alpha\beta} \tilde{A}_{\alpha\beta}(q_i, p_j)J_\alpha J_\beta + \cdots \tag{3.1.5}$$

where $\tilde{A}_0(q_i, p_j)$, $\tilde{A}_\alpha(q_i, p_j)$, … are functions of q_i and p_j. The diagonal matrix element for the transformed operator $\tilde{A}$ in the basis of vibrational wave functions $|V_i\rangle$ appears as

$$\tilde{A}(V_i) = \tilde{A}_0(V_i) + \sum_\alpha \tilde{A}_\alpha(V_i)J_\alpha + \sum_{\alpha\beta} \tilde{A}_{\alpha\beta}(V_i)J_\alpha J_\beta + \cdots, \tag{3.1.6}$$

where $\tilde{A}_0(V_i)$, $\tilde{A}_\alpha(V_i)$, … are the corresponding matrix elements. The operator $\tilde{A}(V_i)$ is a rotational operator called effective operator for a given vibrational state $|V_i\rangle$. The formulae for $\tilde{A}_0(q_i, p_j)$, $\tilde{A}_\alpha(q_i, p_j)$, … from Eq. (3.1.5) and for $\tilde{A}_0(V_i)$, $\tilde{A}_\alpha(V_i)$, … from Eq. (3.1.6) for the effective dipole moment of a molecule of arbitrary symmetry are given in Refs [9–12]. For the effective polarizability operator of X_2Y molecule these formulae can be found in Refs [13, 14].

For the X_2Y molecules the effective dipole moment operator $\tilde{\mu}$ (corresponding to the $\tilde{\mu}_x$ component along the molecular symmetry axis) can be written as [11]

$$\tilde{\mu}(V_i) = \sum_{nm} h_{nm0}(V_i) J^{2n} J_z^{2m} + \frac{1}{2} \sum_{nm} h_{nm2}(V_i) \{J_+^2 + J_-^2, J^{2n} J_z^{2m}\}. \qquad (3.1.7)$$

The components of the effective polarizability operator (in the molecular frame) are defined by the relation [13]

$$\tilde{\alpha}_{ln}^V = \alpha_{ln}^{(V)} + \Delta\alpha_{ln}^{(J)} J^2 + \Delta\alpha_{ln}^{(K)} J_z^2 + \Delta\alpha_{ln}^{(C)} (J_+^2 + J_-^2). \qquad (3.1.8)$$

In Eqs (3.1.7) and (3.1.8), only the terms giving contributions to the diagonal (in the basis of vibrotational wave functions) matrix elements are taken into account; the expansion coefficients determined in Refs [11–18] are collected in Table 3.5.

The values of coefficients h_{nmq} for the dipole moment operator show that the expansion (3.1.7) diverges, so that its use for the calculation of averaged matrix elements $\langle J\tau |\ \tilde{\mu}\ | J\tau\rangle$ for large values of rotational quantum numbers can lead to erroneous results. To resolve this problem of divergence, summation methods can be applied to Eq. (3.1.7). In Ref. [11], some non-polynomial forms for $\tilde{\mu}$ were tested, the simplest one being the Pade approximant

$$\tilde{\mu}_\lambda[1/1] = h_{000} + \frac{\left(h_{200} J^2 + h_{020} J_z^2\right)^2}{h_{200} J^2 + h_{020} J_z^2 - \left(h_{400} J^4 + h_{220} J^2 J_z^2 + h_{040} J_z^4\right)}. \qquad (3.1.9)$$

The dependence of this approximant on the parameters h_{nm2} from Eq. (3.1.7) is neglected since their contribution in the matrix element $\langle J\tau |\ \tilde{\mu}\ | J\tau\rangle$ is small.

The rotational dependence of effective operators $\tilde{A}$ can be important for calculation of line shifts for purely rotational transitions within the same vibrational state. Indeed, according to Eq. (2.6.3) the first-order contribution $S_1(b)$ is purely imaginary and is defined by

$$S_1(b) = -\frac{3\pi i}{8\hbar v b^5} \left\{ \left[\mu_{1i}^2 - \mu_{1f}^2 + \frac{3}{4} \overline{u} (\alpha_{1i} - \alpha_{1f}) \right] \alpha_2 + (\alpha_{1i} - \alpha_{1f}) \mu_2^2 \right\}, \qquad (3.1.10)$$

where the indices 1 and 2 refer to the absorbing and perturbing molecules, respectively. For the purely rotational transitions the differences $\mu_i - \mu_f$ = $\langle J_i\tau_i |\ \tilde{\mu}\ | J_i\tau_i\rangle - \langle J_f\tau_f |\ \tilde{\mu}\ | J_f\tau_f\rangle$ and $\alpha_i - \alpha_f$ are determined only by the rotational contributions of the effective operators. The effect of rotational dependence of $\tilde{\mu}$ and $\tilde{\alpha}$ is only significant if the intermolecular potential is due to the dispersion and induction interactions, i.e. for the systems of H_2O–A type where A is a rare gas atom.

Table 3.5. Parameters of effective operators for H_2O physical quantities*.

Parameter	(000)	(010)	(100)	(001)	(020)	(030)	(040)	(050)
$-h_{000}$	1.8543	1.8255	1.8601	1.8779	1.791	1.744	1.690	1.610
h_{200}	$0.14\cdot10^{-3}$	$0.16\cdot10^{-3}$	$0.14\cdot10^{-3}$	$0.14\cdot10^{-3}$	$0.19\cdot10^{-3}$	$0.23\cdot10^{-3}$		
$-h_{020}$	$0.11\cdot10^{-2}$	$0.16\cdot10^{-2}$	$0.11\cdot10^{-2}$	$0.11\cdot10^{-2}$	$0.26\cdot10^{-2}$	$0.46\cdot10^{-2}$		
h_{002}	$0.64\cdot10^{-4}$	$0.73\cdot10^{-4}$	$0.64\cdot10^{-4}$	$0.64\cdot10^{-4}$	$0.91\cdot10^{-4}$	$0.11\cdot10^{-3}$		
$-h_{400}$	$0.51\cdot10^{-7}$	$0.64\cdot10^{-7}$	$0.51\cdot10^{-7}$	$0.51\cdot10^{-7}$	$0.81\cdot10^{-7}$	$0.11\cdot10^{-6}$		
$-h_{220}$	$0.35\cdot10^{-6}$	$0.79\cdot10^{-6}$	$0.35\cdot10^{-6}$	$0.35\cdot10^{-6}$	$0.20\cdot10^{-5}$	$0.46\cdot10^{-5}$		
h_{040}	$0.62\cdot10^{-5}$	$0.17\cdot10^{-4}$	$0.62\cdot10^{-5}$	$0.62\cdot10^{-5}$	$0.55\cdot10^{-4}$	$0.23\cdot10^{-3}$		
$-h_{202}$	$0.53\cdot10^{-7}$	$0.74\cdot10^{-7}$	$0.53\cdot10^{-7}$	$0.53\cdot10^{-7}$	$0.11\cdot10^{-6}$	$0.16\cdot10^{-6}$		
$-h_{022}$	$0.11\cdot10^{-8}$	$0.43\cdot10^{-7}$	$0.11\cdot10^{-8}$	$0.11\cdot10^{-8}$	$0.12\cdot10^{-6}$	$0.42\cdot10^{-6}$		
$-h_{004}$	$0.28\cdot10^{-7}$	$0.42\cdot10^{-7}$	$0.28\cdot10^{-7}$	$0.28\cdot10^{-7}$	$0.66\cdot10^{-7}$	$0.11\cdot10^{-6}$		
α	1.4613[a]	1.4656[a]	1.5042[a]	1.5027[a]				
$\Delta\alpha^{(J)}$	$0.65\cdot10^{-4}$							
$\Delta\alpha^{(K)}$	$0.71\cdot10^{-4}$							
$\Delta\alpha^{(C)}$	$0.75\cdot10^{-5}$							
$\alpha_{2,0}$	0.071[b]							
$\Delta\alpha_{20}^{(J)}$	$0.56\cdot10^{-4}$							
$\Delta\alpha_{20}^{(K)}$	$-0.13\cdot10^{-3}$							
$\Delta\alpha_{20}^{(C)}$	$0.14\cdot10^{-4}$							
α_{22}	0.0267[b]							
$\Delta\alpha_{22}^{(J)}$	~ 0.0							
$\Delta\alpha_{22}^{(K)}$	$0.11\cdot10^{-3}$							
$\Delta\alpha_{22}^{(C)}$	~ 0.0							
Q_{xx}	−0.13							
Q_{vv}	−2.5							
Q_{zz}	2.63							
Ω	2.0							

* The parameters h_{nmq} for the effective dipole moment are given in D, the parameters for the polarizability operator are expressed in Å^3, the components of the quadruple moment Q are in ÅD, and the octupole moment Ω is in Å^{2}D. The spherical components $\alpha_{l,n}$ of the polarizability tensor are related with the Cartesian components $\alpha_{\alpha\beta}$ through Eq. (2.5.13). The mean polarizability α is defined as $\alpha = -\alpha_{00}/\sqrt{3}$, and its *ab initio* values in various vibrational states from Ref. [17] (denoted by the superscript «a») are usually modelled as $\alpha(v_1, v_2, v_3) = 1.4613 + 0.0043v_2 + 0.04v_1 + 0.042v_3$ (in Å^3). The spherical polarizability components (marked by the superscript «b») are determined from the components $\alpha_{xx} = 1.468$ Å^3, $\alpha_{yy} = 1.4146$ Å^3, $\alpha_{zz} = 1.5284$ Å^3 obtained from an analysis of Raman lines intensities [18]. The vibrational dependences of $Q_{\alpha\alpha}$ and Ω are unknown.

The local method of calculation of vibrotational energy levels and wave functions presented here is not unique: Ref. [19] describes a global method for H_2O energy level calculation which is based on the exact kinetic energy operator and *ab initio* intramolecular potential. In the framework of this method, the matrix elements for the physical quantity A can be calculated using the basis of wave functions ψ of the initial Hamiltonian H, but the dependence of A on the internal vibrational coordinates (or, in the first approximation, A-values for the equilibrium configuration) should be known. Like the local method, this global method includes the influence of vibrotational interactions for the calculation of matrix elements of Eq. (3.1.3). A very accurate and complete global calculation of H_2O energy levels was performed, for example, by Tennyson *et al.* [20] on the basis of variational methods.

3.2. Self-broadening of H_2O lines

For the remote sensing of terrestrial atmosphere the knowledge of H_2O line-broadening and shifting coefficients by air pressure is much more important than the knowledge of these coefficients in the case of self-broadening. Nevertheless, the self-broadening coefficients are indispensable for a reliable inversion of tropospheric water vapour spectra since the H_2O self-broadening line widths are in general five times larger than its air-broadening line widths [21] and for some transitions the self-induced line shift is greater than the air-induced line shift.

The absorption spectrum of water vapour represents an alternative sequence of absorption bands and transparency windows. The most intensive bands are the rotational band centred at 200 cm^{-1} (0.7–1650 cm^{-1}), the far- and mid-infrared bands situated near 6.2 μm (640–2820 cm^{-1}) and 2.7 μm (2040–4600 cm^{-1}) as well as the near-infrared bands at 1.8 μm (4600–6005 cm^{-1}), 1.4 μm (7250 cm^{-1}), 1.1 μm (8190–9555 cm^{-1}), followed by the bands in the visible region (13000–25000 cm^{-1}) [2].

3.2.1. Experimental studies

An analysis of experimental line-broadening and shifting coefficients of water vapour molecule by various gases (including self-broadening) obtained before 1994 is given in Ref. [4]. For the self-broadening case its list of citations contains more than 30 publications. Further intensive investigations conducted between 1994 and 2003 have completed this list by more than ten works given in the chronological order in Ref. [22]. A global review of experimental results obtained up to 2004 is made in Ref. [5]. The main works to be mentioned are Refs [21–24] where a great quantity of experimental data for γ and δ coefficients were obtained. Some new results were very recently published for vibrotational transitions [25–27] and for pure rotational transitions [28–32] (see also references cited therein).

In Ref. [21], the room-temperature (T = 296 K) self-broadened line widths and self-induced line shifts of $H_2{}^{16}O$, $H_2{}^{17}O$ and $H_2{}^{18}O$ were measured for 1900 transitions from (000)–(000), (010)–(010), (010)–(000), (020)–(010) and (100)–(010) vibrational bands ($0 \le J \le 18, 0 \le K_a \le 10$) in the region 590–2400 cm^{-1}.

In Refs [22, 23], about 5,000 line-width and shift coefficients at 296 K were obtained for more than ten vibrational bands of the same molecules ($0 \le J \le 15$, $0 \le K_a \le 8$) in the region 2800–8000 cm^{-1}.

In Ref. [24], the coefficients γ and δ and their temperature dependences were extracted for 270 H_2O absorption lines in the region of 0.7 μm (13000 cm^{-1}).

It was found that the observed line widths range from 0.52 to 0.1 cm^{-1} atm^{-1} and the line shifts vary between –0.05 and 0.05 cm^{-1} atm^{-1}. The precision of γ- and δ-values declared in Refs [21–23] is between 3 and 20%. It is, however, noteworthy that it is a controlled precision issued from γ and δ extraction by mean squares fitting of theoretical model profiles to experimental line shapes. The main uncertainty in their determination can be produced by the uncontrollable experimental conditions. As a result, γ- and δ-values obtained by different authors for the same lines can differ by up to 20% (in addition to the fact that different authors use different profile models for fitting, see Sec. 1.1).

As it was stated by Toth *et al.* [21] (Table 3.6), the difference between experimental self-broadening coefficients can be very significant and the ratio of HITRAN database values [35] to the measurements of Ref. [21] can reach a factor of two. A comparison of γ-values for the same lines of different isotopic species shows that the ratio of main H_2O isotope line widths to the line widths of other isotopes varies between 0.90 and 1.07 [21].

Table 3.6. Comparison of H_2O self-broadening coefficients obtained in [21] (γ) and by other authors (γ_{others}).

Reference	Region, μm	γ_{others}/γ
Mandin *et al.* [33]	5	0.67–1.9
Langlois *et al.* [34]	1.3	1.03–1.1
Grossman and Browell [24]	0.7	0.83–1.3
HITRAN [35]		0.5–1.9

A more detailed comparison can be realised with the help of Table 3.7 where the experimental γ-values for some absorption lines of the ν_2 band are given with the calculated values which will be discussed below.

The values of about 300 H_2O self-broadening and self-shifting coefficients for the region 1950–2750 cm^{-1} not studied by Toth *et al.* [21] are given in Ref. [36]; for some lines very important differences with δ-values of Ref. [21] are observed.

Table 3.7. Experimental and calculated H_2O self-broadening coefficients γ (10^{-3} cm^{-1} atm^{-1}) for some lines ($\Delta K_a = 3$) of the ν_2 band.

ν, cm^{-1}	$J_f K_{af} K_{cf}$	$J_i K_{ai} K_{ci}$	expt [33]	expt [21]	calc.
1805.147	5 3 2	5 0 5	460	433	430
1835.894	6 3 3	6 0 6	377	430	422
1852.405	5 4 1	5 1 4	416	415	409
1856.260	6 4 2	6 1 5	361	400	395
1876.633	7 3 4	7 0 7	420	419	415
1888.818	8 4 4	8 1 7	440	403	400
1904.355	5 3 3	4 0 4	511	465	450
1904.762	4 4 1	4 1 4	455	410	414
1919.688	9 4 5	9 1 8	323	370	368
1925.071	5 4 2	5 1 5	445	378	388
1926.727	8 3 5	8 0 8	411	406	417
1939.126	9 5 4	9 2 7	402	350	352
1946.365	6 3 4	5 0 5	452	400	388
1951.130	6 4 3	6 1 6	419	347	351
1956.235	4 4 1	3 1 2	463	428	417
1959.633	5 5 0	5 2 3	417	356	374
1976.199	5 4 2	4 1 3	426	432	424
1993.259	7 3 5	6 0 6	417	362	366
1998.924	6 4 3	5 1 4	488	413	413
2014.434	7 5 3	7 2 6	358	316	316
2020.536	8 4 5	8 1 8	346	323	329
2026.603	7 4 4	6 1 5	427	398	400
2027.025	5 4 1	4 1 4	415	438	420
2043.951	8 3 6	7 0 7	388	342	337
2060.484	8 4 5	7 1 6	416	371	370
2072.541	6 4 2	5 1 5	309	370	383
2090.362	5 5 1	4 2 2	366	365	375
2097.368	9 3 7	8 0 8	343	309	313
2100.433	9 4 6	8 1 7	381	336	339
2105.781	5 5 0	4 2 3	459	368	376
2106.347	6 5 2	5 2 3	441	377	382
2121.269	7 5 3	6 2 4	313	397	406
2124.888	7 4 3	6 1 6	315	385	379
2137.224	8 5 4	7 2 5	343	396	400
2145.468	10 4 7	9 1 8	380	310	310
2152.559	10 3 8	9 0 9	360	272	276
2156.565	9 5 5	8 2 6	306	368	375
2171.256	7 5 2	6 2 5	426	–	357
2181.344	10 5 6	9 2 7	308	345	350
2185.212	8 4 4	7 1 7	320	389	400
2208.737	11 3 9	10 0 10	314	220	239
2212.537	11 5 7	10 2 8	318	317	322
2246.031	12 4 9	11 1 10	276	236	242

New high-resolution room-temperature spectra of water vapour in the region 4200–6600 cm^{-1} were recorded for the $H_2{}^{16}O$, $H_2{}^{17}O$, $H_2{}^{18}O$, and HDO isotopologues [25] providing accurate measurements of positions, intensities, air- and self-broadening coefficients and air-induced line shifts for about 10,400 lines including 3,280 new weak lines. In particular, 7,832 self-broadening coefficients were determined (with uncertainty better than 3σ) which vary between 0.0415 and 0.847 cm^{-1} atm^{-1}. A comparison of these coefficients obtained for $H_2^{16}O$ with the smoothed values of Toth [22] ranging between 0.100–0.513 cm^{-1} atm^{-1} revealed that some measurements of Ref. [25] are outside this interval. However, the mean values of γ (0.341 cm^{-1} atm^{-1} for Ref. [25] and 0.342 cm^{-1} atm^{-1} for Ref. [22]) agreed very well. The self-broadening coefficients for other isotopologues were obtained for 382 lines of $H_2^{17}O$, 797 lines of $H_2^{18}O$ and 2291 lines of HDO which were very close to the results of Toth [22] and did not demonstrate significant differences between the isotopologues.

Self-broadening coefficients for two H_2O lines [6 6 0] (101) ← [6 6 1] (000) and [8 8 0] (101) ← [8 8 1] (000) located respectively at 7185.60 and 7154.35 cm^{-1} were obtained for room-temperature 296 K and pressures 6–830 Torr in the work of Li *et al.* [26]. The authors used Voigt and Galatry models to extract the collisional line shape parameters and found that the effect of collisional narrowing produces a relatively small error for γ (<1.8%) in particular. In comparison with previously published data, their results had a much better precision and improved the values of the HITRAN 2004 database.

In the region 12053–12281 cm^{-1}, 24 $H_2^{16}O$ lines of the $2\nu_1+\nu_2+\nu_3$ and $3\nu_1+\nu_2$ bands (at 296 K) were studied in Ref. [27] using the standard Voigt profile and modified Voigt profile with adjusted Doppler width. It was found that for the self-broadening case the difference between these profiles was about 15%. The obtained self-broadening coefficients were systematically lower (about 7%) than the data of HITRAN 2004 [35], with the mean ratio 0.934. The authors therefore evoked the importance of the collisional narrowing effect for the H_2O line shapes and recommended the using of corresponding profile models.

New measurements of H_2O self-broadening coefficients were also made for some purely rotational transitions mainly in the terahertz frequency domain [28–32]. In particular, in Ref. [28] the 183-GHz line was studied with the (actually best) relative accuracy of 0.3% for the line-broadening coefficient. Three lines of room-temperature $H_2^{16}O$ in the ground vibrational state were studied at low pressure in Ref. [29]: $10_{2,9}$–$9_{3,6}$ (321 GHz), $5_{1,5}$–$4_{2,2}$ (325 GHz) and $4_{1,4}$–$3_{2,1}$ (380 GHz). The self-broadening parameters were extracted using a normalised Lorentzian profile, in that number, the self-shifting coefficients were obtained for the first time. A comparative analysis with the previous experimental and theoretical results available in the literature for these lines was also made.

Cazzoli *et al.* [30] studied the $6_{1,6} \leftarrow 5_{2,3}$ (22.2 GHz) rotational transition in the temperature range 296–338 K by a frequency-modulation technique. The model of Voigt was used to extract the collisional line widths and shifts. The temperature exponents were also obtained. The results agreed with the HITRAN 2001 database but not with the HITRAN 2004 version [35]. The same authors also undertook a similar study of the transition $1_{1,1} \leftarrow 0_{0,0}$ (1.113 THz) at room temperature. The experimental value of the self-broadening coefficient 19.72(46) MHz Torr^{-1} agreed very well with its theoretical prediction of 19.8 MHz Torr^{-1}. They studied also the rotational transitions $7_{2,5} \leftarrow 8_{1,8}$ (1.147 THz), $8_{5,4} \leftarrow 7_{6,1}$ (1.168 THz), $7_{4,4} \leftarrow 6_{5,4}$ (1.173 THz), $8_{5,3} \leftarrow 7_{6,2}$ (1.191 THz), and $6_{3,3} \leftarrow 5_{4,2}$ (1.542 THz), and showed that the corresponding self-broadening parameters from HITRAN 2004 are characterised by large errors.

3.2.2. Calculations

Many calculations of H_2O self-broadening and self-shifting coefficients were performed by various semi-classical methods. It was shown that for the H_2O–H_2O system the most important contribution to γ- and δ-values is given by the electrostatic part of the interaction potential whereas the contributions from the dispersion and induction terms are negligible.

Mandin *et al.* [33] compared the γvalues of the ν_2 band computed by the ATC and QFT approaches (see Sec. 2.8). These calculations used an intermolecular potential including dipole–dipole, dipole–quadrupole and quadrupole-quadrupole interactions as well as induction and dispersion contributions. It was demonstrated that the ATC method gives a better agreement with experimental results.

In Ref. [37], the self-broadening coefficients for H_2O Raman lines were calculated by the semi-classical RB formalism with the polarization part of the interaction potential represented by atom–atom terms. A satisfactory agreement with experimental data was stated, in particular for high values of the rotational quantum number J. An estimation of the trajectory curvature influence on the calculated γ-values was also made. It was observed that for small J-values at $T = 296$ K the use of parabolic trajectory approximation for the long-range part of the potential instead of straight-line trajectories leads to the same results. However, it was supposed that for high values of the rotational quantum number the trajectory curvature can become important.

Calculations of H_2O self-broadening and self-shifting coefficients were also made by the ATC and RB′ approaches [8] (ν_2 band) as well as by the ET′ method [38] (ν_2, $3\nu_2$, $2\nu_1$, $2\nu_3$, $\nu_1+\nu_2$, $\nu_1+\nu_3$, and $\nu_2+\nu_3$ bands). The RB′ method differs from the standard RB approach presented in Sec. 2.8 by accounting for only long-range induction and dispersion terms proportional to r^{-6}, so that the straight-line trajectories and the corresponding interruption functions can be employed. The ET′ method is similar to the RB′ method except the resonance functions (see sections 2.11 and 2.12) which are obtained with exact trajectories. In Ref. [38], the real parts of these functions were first calculated numerically for the dipole–dipole, dipole–quadrupole, quadrupole-quadrupole interactions as well as for the dispersion and induction contributions in the intermolecular interaction potential. Their model representation was further developed and used for the calculation of the imaginary parts (approximated in their turn by model expressions). These model complex-valued resonance functions were then injected in the semi-classical RB formulae for the line widths and line shifts, for which analytical expression were also proposed. More details on the analytical representation of the resonance functions and line-broadening parameters are given in Sec. 3.3. The vibrotational wave functions were obtained according to the procedure described in Sec. 3.1 [8, 39, 40]. It was shown that the ATC calculations give overestimated line-broadening coefficients in comparison with the RB′, ET′ and experimental results. Figure 3.1 presents the RB′ [8] and experimental [21] γ-values for the ν_2 band (the lines are numbered in the increasing order of experimental values). A statistical analysis of theoretical results compared to experimental data is given in Table 3.8 (for the ET′ approach, two sets of isotropic Lennard–Jones parameters were tested).

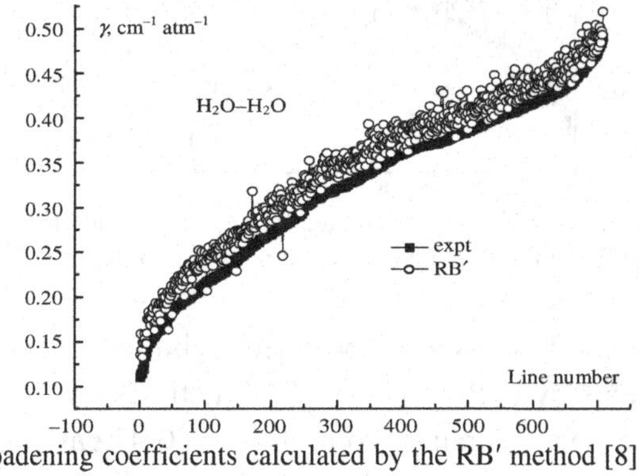

Fig. 3.1. H_2O self-broadening coefficients calculated by the RB′ method [8] and their experimental [21] values for the ν_2 band. The lines are numbered in increasing order of experimental values.

Table 3.8. Comparison of calculated [8, 38] and experimental [21] $H_2{}^{16}O$ self-broadening and self-shifting coefficients for the ν_2 band*.

Δ, %	δ				γ			
	ATC		RB′		ATC		RB′	
	N	N/N_{tot}, %	N	N/N_{tot}, %	N	N/N_{tot}, %	N	N/N_{tot}, %
0–10	128	18.1	139	19.7	13	1.8	505	71.1
10–20	112	15.8	104	14.7	502	70.7	189	26.6
20–30	108	15.3	92	13.0	191	26.9	14	2.0
30–50	106	15.0	120	17.0	3	0.4	1	0.1
> 50	253	35.8	252	35.6	1	0.1	1	0.1
N_{tot}	707				710			
	ET′				ET′			
	ε/k = 92.2 K, σ = 3.23 Å		ε/k = 356.0 K, σ = 2.72 Å		ε/k = 92.2 K, σ = 3.23 Å		ε/k = 356.0 K, σ = 2.72 Å	
0–10	135	19.0	150	21.1	541	76.2	693	97.6
10–20	114	16.1	110	15.5	154	21.7	15	2.1
20–30	93	13.1	85	12.0	13	1.8	1	0.15
30–50	104	14.6	110	15.5	1	0.15	0	0
> 50	264	37.2	255	35.9	1	0.15	1	0.15
N_{tot}	710		710		710		710	

* N is the number of coefficients γ(or δ) for which the discrepancies $\Delta = |(\gamma^{expt} - \gamma^{calc}) / \gamma^{expt}| \cdot 100\%$ (or $\Delta = |(\delta^{expt} - \delta^{calc}) / \delta^{expt}| \cdot 100\%$) are obtained; N_{tot} is the total number of these coefficients.

According to the Table 3.8 the RB′ calculations reproduce the experimental line widths with greater accuracy than the ATC calculations: for 98% of considered lines the discrepancies Δ are less than 20%. A good agreement is also observed with the experimental data of Mandin *et al.* [33] (Fig. 3.2), except for some lines with $\Delta K_a = |K_{a\,i} - K_{a\,f}| = 3$ (Table 3.7).

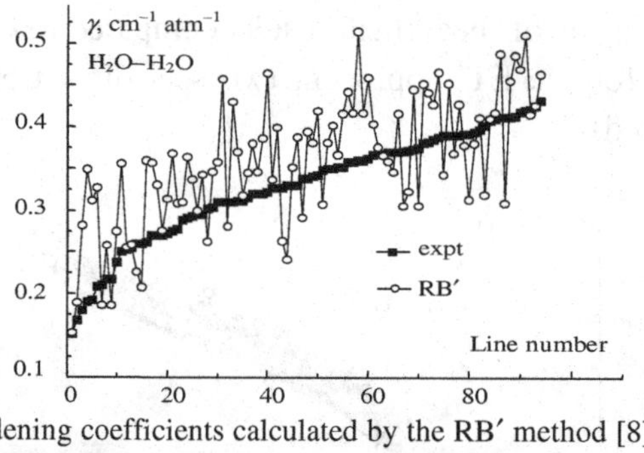

Fig. 3.2. H_2O self-broadening coefficients calculated by the RB′ method [8] and their experimental values [33] for the ν_2 band. The lines are numbered in increasing order of experimental values.

The line-shifting coefficients are reproduced with less accuracy than the line-broadening coefficients by both ATC and RB′ methods (with approximately the same quality). These δ-values range from −0.05 to 0.05 cm^{-1} atm^{-1}. Some calculated [38] and experimental [21] γ- and δ-values are given in Table 3.9.

Table 3.9. Calculated [38] and experimental [21] H_2O self-broadening and self-shifting coefficients for the ν_2 band (in cm^{-1} atm^{-1}).

(010) JK_aK_c	(000) JK_aK_c	δ expt	δ calc.	γ expt	γ calc.	(010) JK_aK_c	(000) JK_aK_c	δ expt	δ calc.	γ expt	γ calc.
1 1 1	0 0 0		–0.071		0.472	4 1 4	4 4 1	–0.040	–0.056	0.443	0.427
1 1 0	1 0 1		–0.072		0.460	4 3 2	4 4 1	0.012	–0.001	0.352	0.340
2 1 2	1 0 1		–0.035		0.455	5 1 4	4 4 1		–0.030		0.425
0 0 0	1 1 1		0.045		0.487	5 3 2	4 4 1	–0.022	–0.025	0.373	0.367
2 0 2	1 1 1	0.024	0.019	0.471	0.456	5 5 0	4 4 1	–0.022	–0.003	0.255	0.299
2 2 0	1 1 1		0.006		0.470	3 3 1	4 4 0		0.011		0.349
1 0 1	1 1 0		0.042		0.455	4 1 3	4 4 0		–0.037		0.434
2 2 1	1 1 0		0.047		0.467	4 3 1	4 4 0	0.001	–0.012	0.328	0.341
1 1 1	2 0 2	–0.026	–0.023	0.463	0.441	5 3 3	4 4 0	–0.007	–0.015	0.336	0.346
2 1 1	2 0 2		–0.036		0.432	5 5 1	4 4 0	0.026	–0.002	0.275	0.301
3 1 3	2 0 2		–0.000		0.427	4 1 4	5 0 5	–0.000	–0.010	0.468	0.460
3 3 1	2 0 2	0.030	0.024	0.413	0.418	4 3 2	5 0 5	0.034	0.039	0.412	0.413
2 2 1	2 1 2		0.016		0.484	5 1 4	5 0 5		0.003		0.467
3 0 3	2 1 2		0.000		0.501	5 3 2	5 0 5	0.021	0.022	0.433	0.430
3 2 1	2 1 2		0.018		0.475	5 5 0	5 0 5		0.039		0.390
2 0 2	2 1 1		0.037		0.429	6 1 6	5 0 5		–0.016		0.422
2 2 0	2 1 1	0.026	0.029	0.421	0.433	6 3 4	5 0 5	0.015	0.021	0.400	0.388
3 2 2	2 1 1	0.023	0.021	0.430	0.434	6 5 2	5 0 5		0.039		0.399
1 1 0	2 2 1	–0.043	–0.036	0.480	0.483	4 0 4	5 1 5	–0.046	–0.042	0.450	0.459
2 1 2	2 2 1		0.014		0.485	4 2 2	5 1 5	–0.040	–0.039	0.445	0.450
3 1 2	2 2 1	0.004	0.002	0.483	0.471	5 2 4	5 1 5	–0.004	–0.008	0.404	0.420
3 3 0	2 2 1		0.040		0.440	5 4 2	5 1 5	0.001	0.013	0.378	0.388
1 1 1	2 2 0	–0.010	–0.016	0.460	0.467	6 0 6	5 1 5	–0.035	–0.034	0.383	0.410
2 1 1	2 2 0		–0.017		0.451	6 2 4	5 1 5	–0.009	–0.009	0.460	0.452
3 1 3	2 2 0	0.015	0.012	0.483	0.481	6 4 2	5 1 5	–0.005	0.007	0.370	0.383
3 3 1	2 2 0		0.045		0.441	4 2 3	5 1 4	–0.000	0.003	0.464	0.467
2 1 2	3 0 3		–0.012		0.496	5 0 5	5 1 4		–0.006		0.462
3 1 2	3 0 3	–0.010	–0.017	0.470	0.487	5 2 3	5 1 4		–0.004		0.465
3 3 0	3 0 3	0.002	0.007	0.450	0.449	5 4 1	5 1 4	0.017	0.017	0.415	0.408
4 1 4	3 0 3		–0.015		0.499	6 2 5	5 1 4		0.002		0.427
4 3 2	3 0 3	0.002	0.010	0.483	0.466	6 4 3	5 1 4	0.020	0.020	0.413	0.412
2 0 2	3 1 3	0.006	0.019	0.410	0.427	4 1 3	5 2 4	–0.003	–0.006	0.470	0.471
2 2 0	3 1 3	0.004	0.004	0.456	0.445	4 3 1	5 2 4	0.018	0.014	0.409	0.401
3 2 2	3 1 3		0.011		0.436	5 1 5	5 2 4	0.005	0.000	0.406	0.429
4 0 4	3 1 3		–0.014		0.452	5 3 3	5 2 4	0.021	0.019	0.396	0.414

4 2 2	3 1 3	0.019	0.016	0.423	0.440	5 5 1	5 2 4	0.022	0.039	0.339	0.363
4 4 0	3 1 3		0.055		0.405	6 1 5	5 2 4	–0.011	–0.012	0.439	0.440
2 2 1	3 1 2	0.035	0.032	0.475	0.471	13 1 13	14 0 14	0.004	0.011	0.182	0.164
3 0 3	3 1 2		0.010		0.495	13 0 13	14 1 14	0.016	0.011	0.162	0.164
3 2 1	3 1 2		–0.002		0.461	15 0 15	14 1 14	0.011	–0.007	0.139	0.145
4 2 3	3 1 2	0.012	0.014	0.450	0.479	13 2 12	14 1 13	–0.007	0.001	0.146	0.166
4 4 1	3 1 2	0.065	0.069	0.428	0.417	14 0 14	14 1 13	0.020	0.015	0.161	0.161
2 1 1	3 2 2	–0.039	–0.041	0.420	0.427	14 2 12	14 1 13		–0.011		0.176
3 1 3	3 2 2		–0.015		0.426	15 2 14	14 1 13	–0.017	–0.007	0.142	0.159
3 3 1	3 2 2	0.025	0.028	0.392	0.405	13 1 12	14 2 13	0.002	0.001	0.153	0.165
4 1 3	3 2 2	0.000	–0.012	0.436	0.444	14 1 14	14 2 13	0.008	0.015	0.150	0.161
4 3 1	3 2 2	0.001	0.004	0.413	0.404	14 3 12	14 2 13		–0.012		0.173
2 1 2	3 2 1	–0.008	–0.003	0.467	0.475	15 1 14	14 2 13	–0.010	–0.007	0.147	0.159
3 1 2	3 2 1		–0.010		0.448	15 3 12	14 2 13		–0.024		0.194
3 3 0	3 2 1		0.041		0.423	13 3 11	14 2 12	0.000	–0.001	0.178	0.188
4 3 2	3 2 1	0.044	0.042	0.406	0.437	14 1 13	14 2 12	–0.010	0.005	0.168	0.176
2 0 2	3 3 1	–0.008	–0.012	0.435	0.440	15 3 13	14 2 12	–0.006	–0.003	0.163	0.175
2 2 0	3 3 1	0.003	–0.025	0.417	0.426	14 2 13	14 3 12	0.000	0.004	0.164	0.176
3 2 2	3 3 1	–0.008	–0.020	0.403	0.416	15 2 13	14 3 12	–0.008	–0.005	0.157	0.175
4 0 4	3 3 1	–0.040	–0.042	0.447	0.454	15 4 12	14 3 11		0.016		0.224
4 2 2	3 3 1	–0.018	–0.027	0.422	0.426	14 3 12	14 4 11	–0.013	0.004	0.199	0.202
4 4 0	3 3 1	0.022	0.020	0.363	0.374	15 1 14	14 4 11		0.014		0.190
2 2 1	3 3 0	–0.009	–0.008	0.427	0.434	15 3 12	14 4 11	–0.031	–0.014	0.194	0.202
3 0 3	3 3 0	–0.023	–0.023	0.462	0.471	15 5 10	14 4 11		–0.002		0.324
3 2 1	3 3 0		–0.027		0.415	14 3 11	14 4 10		0.008		0.318
4 2 3	3 3 0	–0.009	–0.018	0.411	0.428	14 4 11	14 5 10	–0.025	–0.011	0.236	0.237
4 4 1	3 3 0	0.019	0.022	0.337	0.374	14 6 9	14 5 10		–0.006		0.241
3 1 3	4 0 4	0.022	0.011	0.434	0.457	15 4 11	14 5 10		–0.027		0.267
3 3 1	4 0 4	0.034	0.039	0.433	0.436	15 6 9	14 5 10		0.036		0.296
4 1 3	4 0 4		0.021		0.478	15 6 10	14 5 9		–0.006		0.358
4 3 1	4 0 4	0.030	0.036	0.421	0.425	14 5 10	14 6 9		–0.021		0.234
5 1 5	4 0 4	0.034	0.030	0.440	0.452	15 5 10	14 6 9		0.000		0.336
5 3 3	4 0 4	0.044	0.045	0.465	0.450	13 5 9	14 6 8		0.003		0.306
5 5 1	4 0 4		0.061		0.399	14 6 9	14 7 8		–0.020		0.226
3 0 3	4 1 4	–0.018	0.002	0.490	0.501	14 8 7	14 7 8		–0.020		0.243
3 2 1	4 1 4	0.001	–0.002	0.456	0.466	14 6 8	14 7 7		0.020		0.271
4 2 3	4 1 4		0.019		0.445	14 7 8	14 8 7		0.005		0.216
4 4 1	4 1 4	0.044	0.064	0.410	0.413	14 1 14	15 0 15	–0.035	0.015	0.155	0.149
5 0 5	4 1 4		0.018		0.456	15 1 14	15 0 15		–0.018		0.147

5 2 3	4 1 4	0.012	–0.000	0.460	0.475	16 1 16	15 0 15	–0.003	–0.008	0.137	0.133
5 4 1	4 1 4	0.041	0.050	0.437	0.419	15 2 14	15 1 15		–0.018		0.147
3 2 2	4 1 3	0.012	0.007	0.441	0.464	15 2 13	15 1 14		–0.010		0.156
4 0 4	4 1 3	–0.015	–0.018	0.459	0.484	16 2 15	15 1 14	–0.019	–0.008	0.125	0.143
4 2 2	4 1 3	0.029	0.018	0.462	0.455	14 1 13	15 2 14		0.003		0.153
4 4 0	4 1 3	0.043	0.047	0.420	0.414	15 1 15	15 2 14	0.046	0.015	0.176	0.149
5 2 4	4 1 3	0.014	0.015	0.487	0.474	16 1 15	15 2 14	–0.007	–0.008	0.112	0.143
5 4 2	4 1 3	0.043	0.042	0.431	0.424	15 1 14	15 2 13		0.007		0.161
3 1 2	4 2 3	–0.011	–0.009	0.448	0.463	15 3 12	15 2 13		–0.018		0.187
3 3 0	4 2 3	0.025	0.027	0.406	0.409	16 3 14	15 2 13	–0.018	0.000	0.147	0.161
4 1 4	4 2 3		–0.007		0.456	16 2 14	15 3 13		–0.000		0.159
4 3 2	4 2 3		0.031		0.422	15 2 13	15 3 12	–0.018	0.009	0.189	0.190
5 1 4	4 2 3		–0.005		0.463	16 4 13	15 3 12		0.002		0.196
5 3 2	4 2 3	0.019	0.024	0.403	0.417	14 5 10	15 4 11		0.018		0.263
5 5 0	4 2 3	0.034	0.056	0.368	0.376	15 3 12	15 4 11		0.007		0.249
3 1 3	4 2 2	0.020	0.011	0.426	0.443	16 5 12	15 4 11		0.028		0.258
3 3 1	4 2 2	0.046	0.047	0.417	0.416	16 6 11	15 5 10		0.007		0.331
4 1 3	4 2 2	–0.023	–0.015	0.451	0.458	15 6 9	15 7 8		0.012		0.288
4 3 1	4 2 2	0.021	0.023	0.419	0.420	16 2 15	16 1 16		–0.018		0.135
5 1 5	4 2 2	0.042	0.036	0.457	0.455	17 0 17	16 1 16	–0.013	–0.007	0.114	0.119
5 3 3	4 2 2	0.048	0.044	0.427	0.438	17 2 16	16 1 15		–0.004		0.133
5 5 1	4 2 2	0.052	0.073	0.365	0.374	17 1 16	16 2 15	–0.003	–0.004	0.119	0.133
3 0 3	4 3 2	–0.008	–0.011	0.478	0.463	17 3 15	16 2 14		–0.002		0.149
3 2 1	4 3 2		–0.018		0.421	16 2 15	16 3 14		0.001		0.147
4 2 3	4 3 2	–0.024	–0.014	0.372	0.411	17 2 15	16 3 14		–0.003		0.149
4 4 1	4 3 2	0.026	0.026	0.358	0.358	17 3 14	16 4 13		–0.012		0.168
5 0 5	4 3 2	–0.004	–0.022	0.422	0.433	16 3 13	16 4 12		0.017		0.214
5 2 3	4 3 2		–0.016		0.425	16 4 13	16 5 12		–0.003		0.201
5 4 1	4 3 2	0.010	0.013	0.344	0.371	18 1 18	17 0 17		–0.002		0.109
3 2 2	4 3 1	–0.027	–0.028	0.391	0.406	18 4 15	17 3 14		–0.004		0.154
4 0 4	4 3 1	–0.047	–0.050	0.427	0.438	19 1 18	18 2 17		–0.006		0.118
4 4 0	4 3 1	0.025	0.023	0.345	0.358	19 2 17	18 3 16		–0.001		0.129
5 2 4	4 3 1	–0.008	–0.007	0.420	0.420	20 1 20	19 0 19		–0.001		0.094

Calculations in the framework of Modified CRB formalism were made by Anthony *et al.* [41] for 499 self-broadened H_2O lines from eight different vibrational bands. In comparison with the CRB results, the MCRB line widths were found to be generally larger, with the average absolute percent difference between the methods equal to 3.53 and the standard deviation equal to 3.86. For the line shifts characterised by small values the ratios were more informative. They ranged between −5 and +5 (maximal value of 148 and minimum value of 40.5), with the average ratio equal to 2.79 and the standard deviation equal to 69.2. These results showed that for strong interaction systems the effect of modification is very important. When compared with the measurements, the MCRB calculations demonstrated less agreement than the CRB values.

In the following work [42], Anthony *et al.* studied the self-broadening and self-shifting coefficients of 440 water vapour lines from 16 vibrational bands by the CRB formalism. The effects of the imaginary components on the line widths as well as vibrational, rotational and temperature dependences of γ were also studied. In particular, it was noted that for the H_2O–H_2O colliding pair the trajectory model (parabolic or exact) has little or no effect on the line widths and line shifts. Calculations of line widths made with the imaginary part of the second-order contribution S_2 and of the first-order contribution S_1 for the $3_{3,1}\leftarrow 3_{1,2}$ transition of the $2\nu_1+2\nu_2+\nu_3$ band showed that the difference with the case of the only real S_2 term included is about 9%. In comparison with the measurements for the H_2O self-broadening from Ref. [5], the CRB-calculated line-broadening coefficients demonstrated the (absolute) average percent difference between 0.04% ($2\nu_1+2\nu_2+\nu_3$ band) and 63.16% ($5\nu_1$ band).

Moreover, Antony and Gamache [43] calculated the H_2O–H_2O line-broadening and shifting coefficients for 5,442 rotational transitions in eleven vibrational bands ($J\leq 18$) in the range 3124.2–563.2 cm^{-1} (3.2–17.76 μm) at 296 and 225 K using the CRB approach using the mean thermal velocity approximation. Their line widths agreed with the experimental data reviewed in Ref. [5]. For the line shifts a good agreement with the measurements of Toth *et al.* [21] was also stated (the poorest agreement was observed for the rotational band). The authors established some general trends for 'families' of transitions concerning line-broadening and line-shifting coefficients as well as the temperature dependence of the line widths.

3.2.3. Temperature dependence of γ and δ coefficients

The line shape parameters γ and δ can be calculated for different temperatures T in the framework of any semi-classical method. (However, the temperature should not be too low so that the classical trajectory notion is valid.) For a small T interval the temperature dependence of the coefficient γ is usually modelled as

$$\gamma(T) = \gamma(T_0)(T_0/T)^n , \qquad (3.2.1)$$

where $\gamma(T_0)$ is the coefficient value for the reference temperature T_0 (usually $T_0 = 296$ K) and n is an adjustable parameter fitted for each line. The spectroscopic databases HITRAN and GEISA [35, 44] contain $\gamma(T_0)$- and n-values. Examples of calculated and experimental $\gamma(T)$-values for two lines are given in Table 3.10. Analogous results for four lines from the $\nu_1+\nu_3$ band can be found in Ref. [38].

Table 3.10. Calculated and experimental* temperature dependence of H_2O self-broadening and self-shifting coefficients for two lines of the ν_2 band (in 10^{-3} cm^{-1} atm^{-1}).

Line [7 7 0] → [8 8 1] (1320.868 cm^{-1})					Line [8 6 3] → [9 7 2] (1290.536 cm^{-1})				
T	δ, calc. [8]	γ, expt	γ, calc. [37]	γ, calc. [8]	T	δ, calc. [8]	γ, expt	γ, calc. [37]	γ, calc. [8]
296	–24.8	148±4[a]		168	296	–20.3	233±6[a]		257
408	–15.3	146±2	131	149	460	–8.5	227±7	191	204
522	–9.5	134±2	122	136	543	–5.4	197±6	176	186
616	–6.5	124±2	116	128	646	–2.9	180±5	160	169
708	–4.4	120±2	110	121	753	–1.3	160±4	147	155
839	–2.6	107±4	103	113	835	–0.5	150±7	139	145

*For the first line the experimental value of shifting coefficient is δ(296 K) = –23.6 cm^{-1} atm^{-1} and for the second line it is δ(296 K) = –16.9 cm^{-1} atm^{-1} [21]. The superscript «a» denotes the experimental data from Ref. [21], other experimental data are taken from Refs [37, 45]. Calculations [8] and [37] are performed with RB′ and RB methods.

Many experimental studies of the temperature dependence of H_2O self-broadening coefficients were realised (see, for example, the recent work [26] for vibrotational lines and Ref. [30] for the 22.2 GHz rotational line). However, in Ref. [30] the experimental value of the temperature exponent n = 1.23(54) was quite different from its theoretical estimation n = 0.76.

From a theoretical point of view, Eq. (3.2.1) is mainly used in the literature for historical reasons since it is valid for molecules interacting through a potential $r^{-\eta}$ [46, 47]. In the framework of this very rough treatment, the temperature exponent n is given by

$$n = \frac{1}{2} + \frac{1}{\eta - 1}.$$

As noted by Anthony *et al.* [42], Eq. (3.2.1) works well for the temperatures 50–100 K for which the principle contributions to the line width come from the real part of the second-order terms S_2. For higher temperatures, however, this formula is not valid, so that its use for combustion studies is recommended with caution. For 440 vibrotational transitions theoretically studied in Ref. [42] at 200, 250,

296, 350, and 500 K the power law did not work perfectly but provided a good representation of the temperature dependence. In contrast with the cases of N_2-, O_2- and air-broadening (see below), no negative values of n were obtained.

Equation (3.2.1) is also sometimes used in the literature to characterise also the temperature dependence of line shifts. A rough theoretical estimation of the temperature exponent for this case was obtained by Pickett [48]:

$$n = \frac{1}{2} + \frac{3}{2\eta - 2}.$$

In the general case, the power representation is not valid for the line shifts since for many H_2O lines the calculated δ-values change the sign with the temperature variation. An example of line-shift temperature dependence is given in Fig. 3.3.

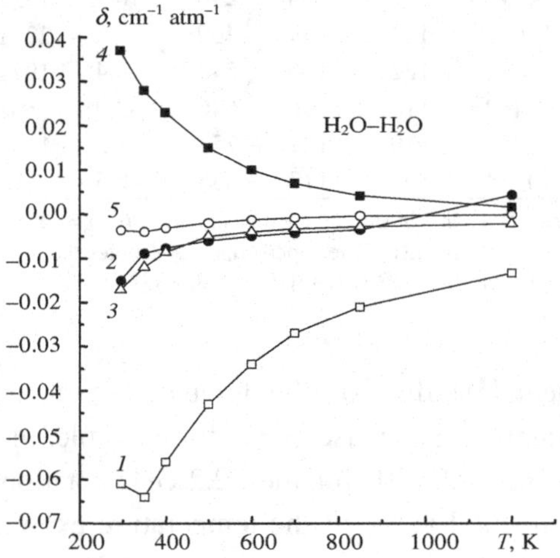

Fig. 3.3. Calculated temperature dependence of H_2O self-shifting coefficients for five lines of the ν_2 band. The lines *1–5* are: [0 0 0]→[1 1 1] (ν = 1634.967 cm^{-1}), [5 0 5]→[4 1 4] (ν = 1496.248 cm^{-1}), [5 5 0]→[6 6 1] (ν = 1991.885 cm^{-1}), [10 0 10]→[9 1 9] (ν = 1397.843 cm^{-1}), [10 1 01]→[9 9 0] (ν = 1292.372 cm^{-1}).

The changing of the line-shift sign with increasing temperature is not a particular feature of H_2O molecule: it is observed for other molecules as well. Various analytical expressions for $\delta(T)$ proposed in the literature are shortly reviewed in Ref. [49] (see also references cited therein):

$$\delta(T) = \delta(T_0)(T - T_0), \qquad (3.2.2)$$

$$\delta(T) = CT^{-m}[1 + A\ln(T)], \qquad (3.2.3)$$

$$\delta(T) = \delta(T_0)[S(T)/S(T_0)][T/T_0]^n, \tag{3.2.4}$$

$$\delta(T) = \delta(T_0)(T - T_0) + C(T - T_0) + D(T - T_0)^2, \tag{3.2.5}$$

where A, C, D, n are fitting parameters and the function $S(T)$ determines the sign of the shift.

3.2.4. Vibrational dependence of γ and δ coefficients

Vibrational excitation in the molecule provokes the changing of its molecular parameters and wave functions. The vibrational dependence of the H_2O dipole moment and polarizability is given in Table 3.5. A weak dependence of water vapour line widths on vibrational quantum numbers has been evoked in the literature many times. The γ- and δ-values computed for some vibrational states have a typical vibrational dependence presented in Figs 3.4 and 3.5 (the lines are numbered in the increasing order of calculated γ- or δ-values for the ν_2 band). For these figures the self-broadening and self-shifting coefficients were calculated by the RB′ method for the same rotational quantum numbers J, K_a, K_c of the lower and upper vibrational states but for different values of the vibrational quantum number v_2. Figure 3.4 shows that for 36 considered transitions the excitation of five bending quanta v_2 leads to the maximal γ variation of about 10%. The line-shift variations (Fig. 3.5) are even more significant.

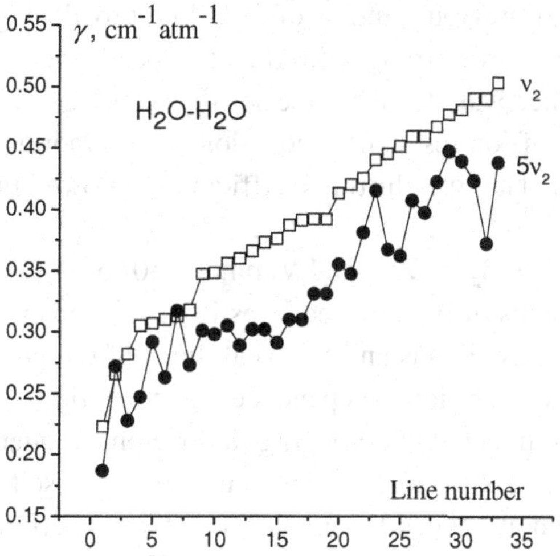

Fig. 3.4. Calculated dependence of H_2O self-broadening coefficient on the excitation of the bending mode ν_2.

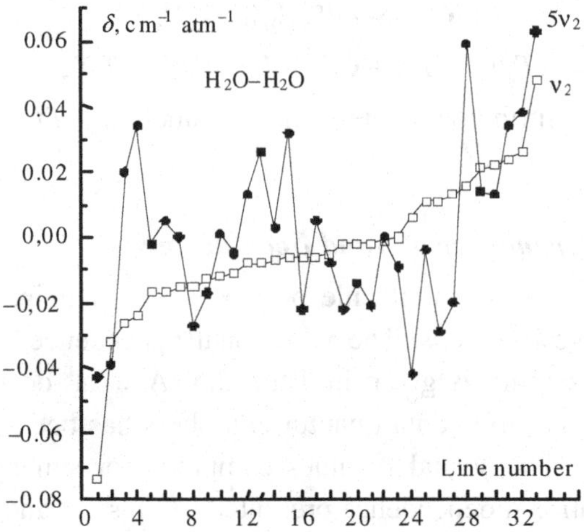

Fig. 3.5. Calculated dependence of the H_2O self-shifting coefficient on the excitation of the bending mode ν_2.

Additional theoretical analysis of the vibrational dependence of H_2O self-broadening and self-shifting coefficients can be found in Ref. [42]. The authors studied 71 vibrotational transitions for the purely rotational, $2\nu_3$, $4\nu_3$ and $6\nu_3$ vibrational bands: 36 transitions with $K_c = J$ and $K_a = 0$ and 35 transitions with $K_c = J$ and $K_a = 1$ (in previous studies of H_2O line broadening by nitrogen these lines were shown to have strong vibrational dependence). Only a very small vibrational dependence was stated for the self-broadening coefficients (due to the permanent presence of 'on resonance' collisions and important contribution from the rotational terms). The line-shifting coefficients followed the usual vibrational dependence.

The experimental study by Zou and Varanasi [50] of H_2O self-broadening and self-shifting coefficients of the infrared lines in the regions 610–2100 and 3000–4050 cm^{-1} (ν_1, ν_2, ν_3, $2\nu_2$ bands and the weak lines of the purely rotational spectrum) considered the vibrational dependence of transitions involving the same rotational quantum numbers but belonging to different frequency domains on the example of ν_1, ν_2, and $2\nu_2$ bands. They found that the self-broadening coefficients in the $2\nu_2$ band are generally smaller than these coefficients in the ν_2 band (except two values) with the average difference of about 10%. At the same time though large differences between the ν_1 and ν_2 bands and the ν_1 and $2\nu_2$ bands were stated, it was not possible to conclude which band had larger line widths.

3.2.5. *Influence of the rotational dependence of dipole moment and polarizability*

Since the effective dipole moment operator $\tilde{\mu}$ and the effective polarizability operator $\tilde{\alpha}$ are rotational operators for a given vibrational state, their matrix elements in the basis of vibrotational wave functions depend on the rotational quantum numbers.

Three forms of the matrix element $\mu(V,J,K) = \mu(V,J,K_a) \approx \mu(V,J,K_a,K_c) \approx \langle VJK|\tilde{\mu}|VJK\rangle$ dependence on the rotational quantum numbers J and K were tested in Ref. [8]. These forms correspond to different approximations used to obtain the effective operator $\tilde{\mu}$ of Eq. (3.1.7). The first approximation supposes that $\mu(V, J, K)$ does not depend on J nor on K; in this approximation, according to Eq. (3.1.7) $\mu(V, J, K) = h_{000}$. The second (most frequently used) approximation only keeps in Eq. (3.1.7) the quadratic dependence on the elementary rotational operators J_β (i.e. up to J_β^2, $\beta = x,y,z$): $\mu(v, J, K) = h_{000} + h_{200}J(J+1) + h_{020}K^2$. The third approximation defines the matrix element $\mu(V, J, K)$ by the diagonal in $|JK\rangle$ basis matrix element of operator $\tilde{\mu}$ given by Eq. (3.1.9). It was shown that the rotational dependence of the dipole moment does not significantly influence the line-broadening coefficients γ but can be very important for the line-shifting coefficients δ.

The dependence of calculated coefficients δ on the rotational quantum number K_a for the three abovementioned forms of $\mu(V, J, K)$ is presented in Fig. 3.6 [8] for a group of lines ($J = 20$, K_a, $K_c = 21 - K_a$) $\rightarrow$ ($J = 19$, $K_a - 1$, $K_c = 20 - K_a$). For $K_a \leq 10$ all forms of $\mu(V, J, K)$ give the same results, which means that for $K_a \leq 10$ the rotational dependence of the dipole moment operator can be neglected. Some differences between the calculated coefficients appear for $K_a > 10$. This value $K_a = 10$ corresponds approximately to the radius of convergence for the J_β-polynomial in the effective Hamiltonian of the water molecule for the ground (000) and vibrational (010) states.

According to Fig. 3.6 the usual polynomial form of $\mu(V, J, K)$ can produce significant errors in the line-shift calculation for large values of rotational quantum numbers. It means that it is rather preferable to not account for the rotational dependence of the dipole moment at all than to limit the expansion to the quadratic terms only.

However, it should be noted that the obtained dependence of the self-shifting coefficients does not allow the deduction of the real rotational dependence of the molecular dipole moment. This fact is due to the absence of experimental data for these rotational quantum numbers as well as the limits of the perturbation theory used for the construction of effective Hamiltonian for such a non-rigid molecule as H_2O. This theory allows the reconstitution of the polynomial coefficients in Eq. (3.1.7) but the expansion itself has a finite convergence radius.

Calculations show that the rotational dependence of the effective polarizability operator $\tilde{\alpha}$ does not influence the self-broadening and self-shifting coefficients, at least for $J \leq 25$.

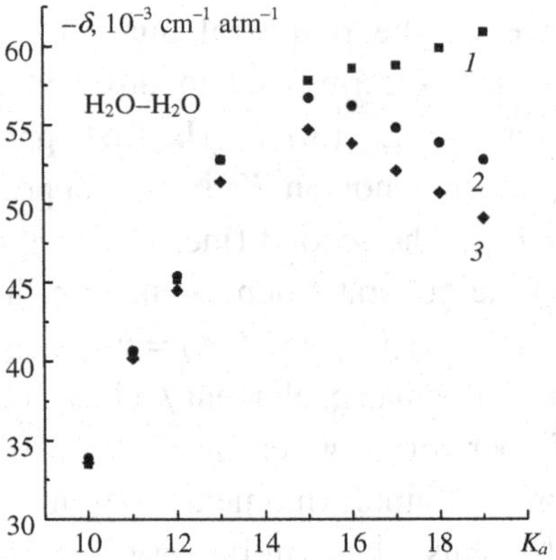

Fig. 3.6. Calculated dependence of H_2O self-shifting coefficients of the ν_2 band on the rotational quantum number K_a for different forms of the effective dipole moment operator: curve *1* corresponds to the polynomial form of Eq. (3.1.7) taken up to the quadratic terms, curve *2* is issued from the Pade expression (3.1.9) and curve *3* is obtained by using only the h_{000} contribution in Eq. (3.1.7). Reproduced with permission from Ref. [8].

3.2.6. Influence of accidental resonances

If the considered vibrotational state belongs to a group of interacting states, the wave functions contain non-zero mixing coefficients with other states of this group. Some examples are given in Table 3.4. In particular, the rotational energy level [13 11 2] (E = 5654.774 cm^{-1}) of the vibrational state (010) has a wave function mixing coefficient of about 17.5% with the rotational energy level [13 3 10] (E = 5654.761 cm^{-1}) of the vibrational state (020). The coefficients γ and δ were computed for all symmetry-allowed transitions from the rotational level [13 11 2] in the ν_2 band using the wave functions obtained for the isolated vibrational state (010) (mixing coefficient of 0%) and the wave functions found with the operator matrix of Eq. (3.1.1) (mixing coefficient of 17.5%). These computations showed that γ-values are practically identical for both cases whereas δ-values for some lines are quite different (up to 100% for δ-values close to zero). This difference can be ascribed to the precision of semi-classical computations.

Analogous calculations were performed [51] for the $\nu_2+\nu_3$ band of D_2O. For many upper vibrotational states involved in the considered transitions the mixing coefficient reached 50%, and the calculated γ and δ coefficients differed by 20% and 100% respectively from their values obtained without taking into account the resonance interactions (Fig. 3.7 shows an example for lines with $J_f = 8$).

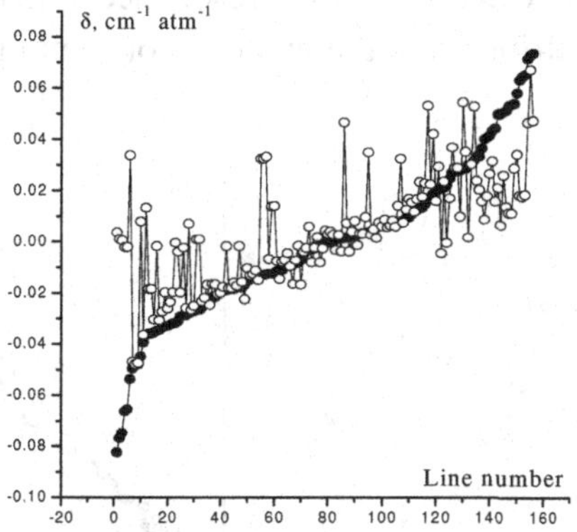

Fig 3.7. Calculated self-shifting coefficients for a group of lines with $J_f = 8$ in the $\nu_1+\nu_2$ band of D_2O [51] (reproduced with permission). Filled and open symbols correspond respectively to the wave functions obtained for the isolated and interacting (110) vibrational state (see Table 3.4).

3.2.7. *Influence of the trajectory model*

A comparison of H_2O self-broadening and self-shifting coefficients calculated within the CRB formalism with parabolic and exact trajectories [42] showed that there are no significant differences between these two models. Moreover, for the H_2O self-broadening case the intermolecular interaction is so strong that the real part of S_2 is large and the term $\exp(-S_2)$ in the expression for the line width becomes zero before the trajectories start to bend. It means that even a straight-line trajectory will give results similar to those obtained with more sophisticated parabolic or exact trajectories.

The self-broadening and self-shifting coefficients γ and δ calculated in the framework of the ET′ method with the exact-trajectory resonance functions of Eqs (2.11.4) and (2.11.5) [38] used two sets of isotropic H_2O–H_2O Lennard–Jones potential parameters: $\varepsilon/k_B = 92$ K, $\sigma = 3.23$ Å and $\varepsilon/k_B = 356$ K, $\sigma = 2.725$ Å. The last set was taken from a study of the second virial coefficient [52]; other parameter sets were found in Ref. [53]. These results are compared with ATC and RB′ calculations in Table 3.8 and in Fig. 3.8. The first set of isotropic potential parameters gives the same results for γ and δ coefficients as the straight-line trajectory model. The second set improves significantly the quality of calculations: the discrepancies Δ are less then 10% for about 98% of lines.

The self-shifting coefficients depend very weakly on the form of the isotropic potential. However, in some cases when calculated in the straight-line trajectory approximation δ-values are close to zero, the correction accounting for the trajectory curvature (using of exact-trajectory resonance functions) can reach 30%. This effect can also be attributed to the precision of semi-classical calculations.

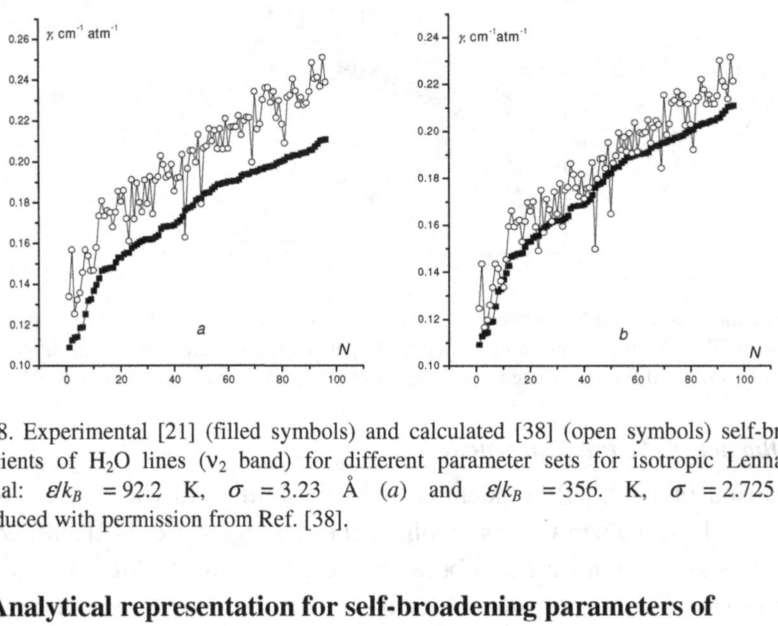

Fig. 3.8. Experimental [21] (filled symbols) and calculated [38] (open symbols) self-broadening coefficients of H_2O lines (ν_2 band) for different parameter sets for isotropic Lennard–Jones potential: ε/k_B = 92.2 K, σ = 3.23 Å (*a*) and ε/k_B = 356. K, σ = 2.725 Å (*b*). Reproduced with permission from Ref. [38].

3.3. Analytical representation for self-broadening parameters of water vapour

The calculation of a water vapour spectrum under real atmospheric conditions requires the knowledge of broadening and shifting coefficients for each line. For instance, the H_2O ν_2 band at T = 296 K and intensities superior to 10^{-27} cm^{-1}/(molecule·cm^{-2}) contains about 3,000 lines for which these coefficients should be estimated. The calculations can be made using semi-classical methods discussed in the second chapter and need the intermolecular interaction potential and vibrotational wave functions of the colliding molecules as input data.

Computational methods for self-broadening coefficients of water vapour dependent solely on adjustable parameters were proposed in Ref. [54]. Two model analytical formulae $\gamma(sur)$ ('*sur*' meaning 'surface') for the broadening coefficient were tested. After the determination of parameters by fitting the computed $\gamma(sur)$ values to a set of experimental data $\gamma(expt)$ these models enable the calculation of line-broadening coefficients for a wide range of rotational and vibrational quantum numbers and at various temperatures. The situation is analogous to that appearing in the analysis of experimental transition frequencies with effective Hamiltonians: first, the parameters of such a Hamiltonian are deduced from experimental transition frequencies; furthermore they are used for theoretical calculations of other transition frequencies.

The central point of $\gamma(sur)$ modelling is the representation of the dipole–dipole interaction resonance function $F_1(x)$ from Ref. [55]

$$F_1(x) = \frac{x^4}{4}\left[K_3(x)K_1(x) + 4K_2(x)K_0(x) - K_2^2(x) - K_1^2(x) - 3K_0^2(x)\right] \quad (3.3.1)$$

by a simple analytical expression

$$F_1^{mod}(x, n) = a_1/\mathrm{Ch}^n[\lambda(x - x_e)], \quad (3.3.2)$$

where n = 1, 2 and the parameters a_1, λ, x_e (Table 3.11) are obtained [56] by fitting $F_1^{mod}(x, n)$ to the exact function of Eq. (3.3.1) (see Fig. 3.9).

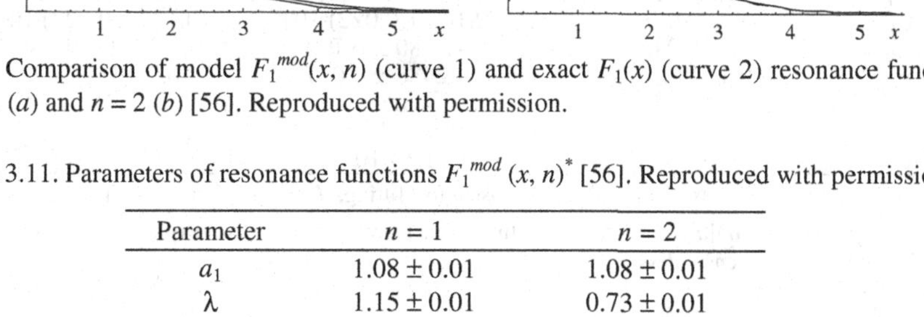

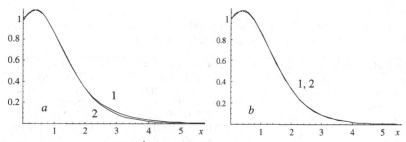

Fig. 3.9. Comparison of model $F_1^{mod}(x, n)$ (curve 1) and exact $F_1(x)$ (curve 2) resonance functions for n = 1 (*a*) and n = 2 (*b*) [56]. Reproduced with permission.

Table 3.11. Parameters of resonance functions $F_1^{mod}(x, n)^*$ [56]. Reproduced with permission.

Parameter	$n = 1$	$n = 2$
a_1	1.08 ± 0.01	1.08 ± 0.01
λ	1.15 ± 0.01	0.73 ± 0.01
x_e	0.40 ± 0.01	0.375 ± 0.005
σ	$1.7 \cdot 10^{-3}$	$8.0 \cdot 10^{-4}$

* Here $\sigma = \{\Sigma_i^N [F_1(x_i) - F_1^{mod}(x_i, n)]^2/(N-L)\}^{1/2}$, N = 28 is the number of points $x_i \in [0.0, 7.9]$ and L = 3 is the number of adjustable parameters.

Using the model function F_1^{mod} (x, n) with $n = 1$ leads to the following analytical expression for $\gamma(sur)$:

$$\gamma(sur) = x_1 + \left[\frac{x_2(J_i, K_i)}{\mathrm{Ch}\alpha(K_i - x_{ei})} + \frac{x_3(J_f, K_f)}{\mathrm{Ch}\alpha(K_f - x_{ef})} \right], \tag{3.3.3}$$

where the parameters x_1, x_2, x_3, α, x_e depend on the number density of the active molecules, relative velocity of colliders, interruption parameter b_0 and rotational quantum numbers J_i, J_f, $K_i \equiv K_{a\,i}$, $K_f \equiv K_{a\,f}$. To test Eq. (3.3.3) by fitting to experimental data of Ref. [21], the rotational dependences of parameters were chosen as

$$x_k = x_{k0} + x_{k1}(J_i + J_f),\ k = 1, 2, 3;$$

$$\alpha = \alpha_0 + \alpha_1(J_i + J_f),$$

$$x_{ei} = x_{ef} = x_{e0} + x_{e1}(J_i + J_f),$$

where x_{k0}, …, x_{e1} are adjustable parameters. An analysis of the fitting results showed that the maximal differences between experimental $\gamma(expt)$ and $\gamma(sur)$ broadening coefficients correspond to the lines with small values of J, K_i and K_f. As a consequence, 58 ν_2-band lines with $K_i = 0, 1$ and $K_f = 0, 1$ ($K_i + K_f = 1$) were fitted separately to their experimental values. Fitting of Eq. (3.3.3) to these data gave four parameters presented in the second column of Table 3.12 ($x_{20} = x_{30}$ was used since the parameters x_{20} and x_{30} were strongly correlated).

Table 3.12. Parameters of model function $\gamma(sur)$ obtained by fitting to experimental H_2O self-broadening coefficients [21] ($J < 17$) of the ν_2 band*.

Parameter	Data with $K_a = 0, 1$	Data with $K_a \neq 0, 1$	All data
x_{10}	0.1178 ± 0.0121	0.1014 ± 0.0116	0.1221 ± 0.0088
$x_{20} = x_{30}$	0.2121 ± 0.0049	0.2235 ± 0.0074	0.2093 ± 0.0059
$x_{21} = x_{31}$	0.0	$-(0.431 \pm 0.034)\cdot 10^{-2}$	$-(0.408 \pm 0.032)\cdot 10^{-2}$
α_0	1.331 ± 0.173	0.266 ± 0.029	0.2524 ± 0.029
α_1	0.0	$(0.107 \pm 0.022)\cdot 10^{-1}$	$(0.129 \pm 0.022)\cdot 10^{-1}$
x_{e0}	0.0	-1.280 ± 0.155	-1.291 ± 0.139
x_{e1}	0.1045 ± 0.0047	0.2488 ± 0.077	0.2487 ± 0.072
N	58	653	711
L	4	7	7
σ	$1.98\cdot 10^{-3}$	$1.19\cdot 10^{-3}$	$1.13\cdot 10^{-3}$

* N is the number of experimental values used for fitting, L is the number of parameters and $\sigma = \{\Sigma_i^N [\gamma_i(expt)\text{-}\gamma_i(sur)]^2/(N\text{-}L)\}^{1/2}$ (in cm^{-1} atm^{-1}). The parameters x_{10}, x_{20} and x_{21} are in cm^{-1} atm^{-1}; α_0, α_1, x_{e0} and x_{e1} are dimensionless.

The quality of this fitting can be seen in Fig. 3.10: for 93% of experimental data the differences $\gamma(expt) - \gamma(sur)$ are less then 10%.

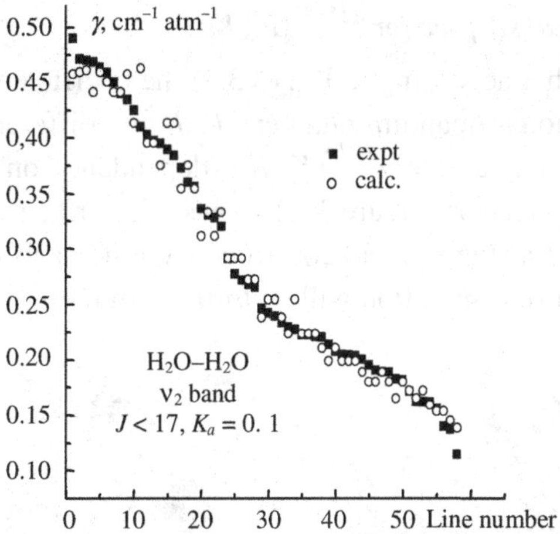

Fig. 3.10. Calculated by Eq. (3.3.3) and experimental [21] H_2O self-broadening coefficients for the lines $K_a = 0, 1$ $(K_{a\,i} + K_{a\,f}) = 1$ of the ν_2 band.

The next step was the fitting of Eq. (3.3.3) to two sets of experimental values of Ref. [21]: the first set excluded the previously tested data with $K_a = 0, 1$ $(K_i + K_f = 1)$ and the second set contained all the experimental data. The obtained parameters are given in the third and fourth columns of Table 3.12. Since the dependence of x_2 and x_3 with respect to the quantum numbers K_i, K_f was impossible to obtain separately, the relation $x_2(J_i, K_i) = x_3(J_f, K_f) = x_{20} + x_{21}(J_i + J_f)$ was used. Figure 3.11 shows the quality of this fitting.

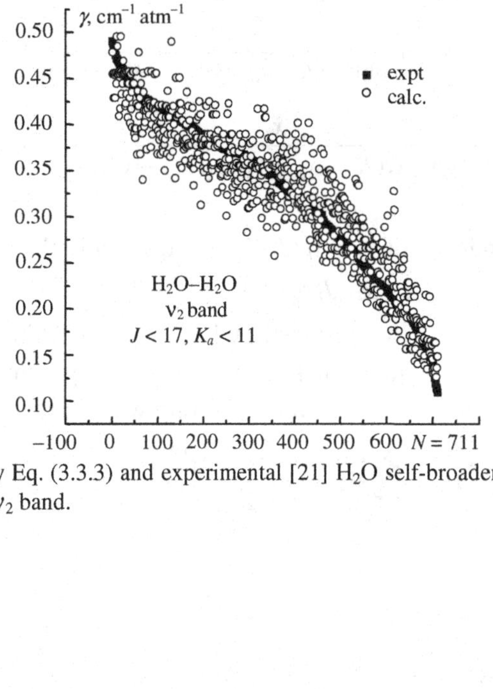

Fig. 3.11. Calculated by Eq. (3.3.3) and experimental [21] H_2O self-broadening coefficients for all the lines $K_a < 11$ of the ν_2 band.

3.3.1. Two-dimensional surface for $\gamma^{(J_i, J_f)}(K_i, K_f)$

It is worthy to note that according to Eq. (3.3.3) the coefficients $\gamma^{(J_i, J_f)}(K_i, K_f)$ for a given pair of rotational quantum numbers J_i, J_f and different $K_i = K_{a\,i}$, $K_f = K_{a\,f}$ form a two-dimensional surface $\gamma^{(J_i, J_f)}(K_i, K_f)$ (dependence on the quantum numbers $K_{c\,i}$ and $K_{c\,f}$ is neglected). Figure 3.12 shows this surface $\gamma^{(J_i, J_f)}(K_i, K_f)$ for the transitions $J_i = J_f = 15$ of the ν_2 band constructed on the basis of RB′ calculations (J_i, J_f indices in the surface notation will be further omitted for brevity).

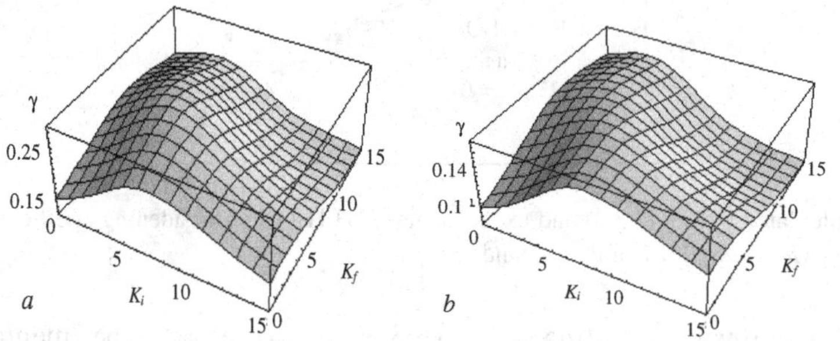

Fig. 3.12. H_2O self-broadening surfaces $\gamma(sur)$ for $J_i = J_f = 15$ lines of the ν_2 band at $T = 296$ K (*a*) and $T = 850$ K (*b*). Figure 3.12*a* is reproduced with permission from Ref. [54].

Two modified analytical expressions for $\gamma(K_i, K_f)$ surfaces were proposed in Ref. [54]:

$$\gamma(K_i, K_f) = x_1 + \frac{x_2 \mathrm{ch}[x_5(K_i - K_f)(K_i + K_f)]}{\mathrm{ch}[x_3(K_i - x_4)]\mathrm{ch}[x_3(K_f - x_4)]}, \tag{3.3.4}$$

$$\gamma(K_i, K_f) = x_1 + \frac{x_2}{\mathrm{ch}[x_3(K_i - x_4)]\mathrm{ch}[x_3(K_f - x_4)]} + \\ + \frac{x_5}{\mathrm{ch}[3x_3(K_i - x_4)]\mathrm{ch}[3x_3(K_f - x_4)]}, \tag{3.3.5}$$

The rotational dependence of $x_k = x_k(J_i, J_f)$ was chosen in the polynomial form

$$x_k = x_{k0} + x_{k1}(J_i + J_f) + x_{k2}(J_i + J_f)^2 \tag{3.3.6}$$

with the parameters x_{k0}, x_{k1}, x_{k2} deduced by least squares fitting of Eqs (3.3.4) and (3.3.5) to some known values $\gamma_i(u)$ (experimental or previously calculated) of self-broadening coefficients; the quality of fitting was characterised by

$$\sigma = \left\{ \sum_{i=1}^{N} \frac{[\gamma_i(u) - \gamma_i(sur)]^2}{N - L} \right\}^{1/2}, \tag{3.3.7}$$

Table 3.13. Quality of fitting of Eq. (3.3.4) to various sets of H_2O self-broadening coefficients $\gamma(u)$ of the ν_2 band [54] (σ in 10^{-4} cm^{-1} atm^{-1} and $L = 8$ for all sets). Reproduced with permission.

$\gamma(u)$	expt [21]	RB′							
T, K	296	296	296	350	400	450	600	700	850
$max\{J_i\}$	16	17	20	19	19	19	19	19	19
$max\{K_i\}$	10	16	19	10	10	10	10	10	10
N	705	3700	4020	980	980	980	980	980	980
σ	11.1	5.2	5.0	7.2	8.5	8.1	3.7	3.1	2.5

where N is the number of used data and L is the number of adjustable parameters. In Ref. [54], the model surfaces of Eqs (3.3.4) and (3.3.5) were fitted to both experimental data and theoretical RB′ values (purely rotational transitions at 296 K $\leq T \leq$ 1200 K and ν_2-band transitions at 296 K $\leq T \leq$ 850 K by step of 50 K) with approximately the same quality of fitting. Table 3.13 gives the results of fitting of Eq. (3.3.4) for the ν_2 band.

It can be seen from this table that the quality of fitting to experimental data [21] is worse than the quality of fitting to calculated RB′ coefficients and that with increasing temperature this quality is improved. Similar results were also obtained for the rotational band: $\sigma = 10.6 \cdot 10^{-4}$ cm^{-1} atm^{-1} for T = 296 K and $\sigma = 2.4 \cdot 10^{-4}$ cm^{-1} atm^{-1} for T = 800 K ($N \approx 1000$).

A statistical analysis of fitting to two sets of RB′ values at T = 296 K and T = 850 K [54] is given in Table 3.14 (the relative error Δ is defined as $\Delta = |\gamma(RB')-\gamma(sur)| / \gamma(RB')\cdot 100\%$). According to this table, for about 90% of lines at T = 296 K ($N \cong 4000$) the difference between $\gamma_i(sur)$ and $\gamma_i(RB')$ is less then 20%, and at T = 850 K there are practically 100% of such lines.

Table 3.14. Statistical analysis of fitting to $H_2O\nu_2$ band self-broadening coefficients $\gamma(RB')$ (N_L is the number of lines for which the given Δ-value is obtained) [54]. Reproduced with permission.

Δ, %	T = 296 K, $max\{J_i\}$ = 20, $max\{K_{a\,i}\}$ = 19				T = 850 K, $max\{J_i\}$ = 19, $max\{K_{a\,i}\}$ = 10			
	Eq. (3.3.4)		Eq. (3.3.5)		Eq. (3.3.4)		Eq. (3.3.5)	
	N_L	N_L/N, %	N_L	N_L/N, %	N_L	N_L/N, %	N_L	N_L/N, %
0–10	2483	62.35	2117	53.16	926	94.4	904	92.2
10–20	1081	27.15	1391	34.93	53	5.4	76	7.7
20–40	353	8.87	416	10.45	2	0.2	1	0.1
40–50	65	1.63	58	1.46	0	0	0	0
> 50	0	0	0	0	0	0	0	0
N	3984				981			

Table 3.15 presents $\gamma(RB')$ coefficients and differences $\Delta = \gamma(RB') - \gamma(sur)$ at $T = 850$ K [54]. Maximal values of Δ are obtained for J_i, J_f 2 and $K_i + K_f = 1$.

Table 3.15. Calculated H_2O self-broadening coefficients $\gamma(RB')$ for the ν_2 band and differences $\Delta = \gamma(RB') - \gamma(sur)$ (in $cm^{-1}\,atm^{-1}$) at $T = 850$ K [54]. Reproduced with permission.

$J_i K_{ai} K_{ci} \to J_f K_{af} K_{cf}$	$\gamma(RB')$	Δ	$J_i K_{ai} K_{ci} \to J_f K_{af} K_{cf}$	$\gamma(RB')$	Δ
10 0 10 9 1 9	0.167	–0.000	10 4 6 9 3 7	0.189	–0.006
10 0 10 10 1 9	0.161	–0.002	10 4 6 9 5 5	0.185	–0.007
10 0 10 10 3 7	0.188	–0.015	10 4 6 10 3 7	0.193	–0.012
10 0 10 11 1 11	0.153	–0.001	10 4 6 11 5 7	0.185	–0.009
10 0 10 11 3 9	0.161	0.006	10 5 6 9 2 7	0.188	–0.005
10 0 10 11 5 7	0.166	0.011	10 5 6 9 4 5	0.185	–0.008
10 1 10 9 0 9	0.167	–0.000	10 5 6 10 4 7	0.174	0.002
10 1 10 9 2 7	0.184	–0.007	10 5 6 10 6 5	0.167	0.000
10 1 10 10 2 9	0.158	0.013	10 5 6 11 4 7	0.190	–0.013
10 1 10 10 4 7	0.174	0.005	10 5 6 11 6 5	0.168	–0.000
10 1 10 11 0 11	0.153	–0.001	10 5 5 9 4 6	0.182	–0.004
10 1 10 11 2 9	0.171	–0.006	10 5 5 10 2 8	0.187	–0.007
10 1 9 9 2 8	0.167	0.010	10 5 5 10 4 6	0.183	–0.006
10 1 9 10 0 10	0.163	–0.004	10 5 5 10 6 4	0.172	–0.005
10 1 9 10 2 8	0.177	–0.006	10 5 5 11 6 6	0.173	–0.005
10 1 9 10 4 6	0.186	–0.007	10 6 5 9 5 4	0.161	0.005
10 1 9 11 2 10	0.155	0.011	10 6 5 9 7 2	0.146	0.003
10 1 9 11 4 8	0.169	0.006	10 6 5 10 5 6	0.157	0.010
10 2 9 9 1 8	0.169	0.008	10 6 5 10 7 4	0.148	0.002
10 2 9 10 1 10	0.163	0.009	10 6 5 11 7 4	0.151	0.002
10 2 9 10 3 8	0.165	0.014	10 6 4 9 5 5	0.158	0.008
10 2 9 10 5 6	0.168	0.012	10 6 4 9 7 3	0.148	0.001
10 2 9 11 1 10	0.154	0.012	10 6 4 10 5 5	0.164	0.003
10 2 9 11 3 8	0.184	–0.009	10 6 4 11 7 5	0.152	0.001
10 2 8 9 1 9	0.185	–0.008	10 7 4 10 6 5	0.138	0.013
10 2 8 9 3 7	0.181	0.001	10 7 4 10 8 3	0.128	0.001
10 2 8 10 1 9	0.175	–0.004	10 7 4 11 8 3	0.130	0.001
10 2 8 10 3 7	0.190	–0.011	10 7 3 10 6 4	0.139	0.012
10 2 8 10 5 5	0.180	0.000	10 7 3 10 8 2	0.128	0.001
10 2 8 11 1 11	0.171	–0.005	10 7 3 11 8 4	0.130	0.001
10 2 8 11 3 9	0.172	0.003	10 8 3 10 7 4	0.119	0.010
10 2 8 11 5 7	0.177	0.001	10 8 3 10 9 2	0.108	–0.005
10 3 8 9 0 9	0.179	–0.001	10 8 3 11 9 2	0.111	–0.006
10 3 8 9 2 7	0.186	–0.004	10 8 2 10 7 3	0.119	0.010
10 3 8 9 4 5	0.184	–0.001	10 9 2 10 8 3	0.101	0.001
10 3 8 10 2 9	0.168	0.011	10 9 2 10 10 1	0.090	–0.018
10 3 8 10 4 7	0.174	0.007	10 9 2 11 10 1	0.093	–0.019
10 3 8 10 6 5	0.164	0.012	10 10 1 9 9 0	0.083	–0.014
10 3 8 11 0 11	0.166	0.001	....		
10 3 8 11 2 9	0.169	0.005	16 1 16 16 2 15	0.104	0.002
10 3 8 11 4 7	0.186	–0.008	16 1 16 17 0 17	0.100	–0.015
10 3 7 9 2 8	0.190	–0.008	16 1 15 17 2 16	0.103	0.000
10 3 7 9 4 6	0.193	–0.010	16 2 15 17 1 16	0.103	0.000

10 3 7 10 2 8	0.192	−0.013	16 2 14 17 3 15	0.113	0.008
10 3 7 10 4 6	0.193	−0.012	16 3 14 16 2 15	0.111	0.013
10 3 7 10 6 4	0.178	−0.002	16 3 14 17 2 15	0.112	0.008
10 3 7 11 4 8	0.185	−0.007	16 4 13 17 3 14	0.125	0.011
10 3 7 11 6 6	0.179	−0.003	16 4 12 16 3 13	0.151	−0.011

3.3.2. Temperature dependence of $\gamma(sur)$

The temperature dependence of half-widths is usually modelled by Eq. (3.2.1). According to this formula it is necessary, for example, to know about 3,000 values of n for the lines of the ν_2 band. By analogy with Eq. (3.2.1), it was proposed in Ref. [54] to describe the temperature dependence of the broadening coefficients γ through the temperature dependence of the parameters x_{km} from Eqs (3.3.4)–(3.3.6):

$$x_{km}(T) = x_{km}(T_0)(T/T_0)^{n_{km}}, \qquad (3.3.8)$$

where $x_{km}(T_0)$ and n_{km} are adjustable parameters and $x_{km}(T_0)$ correspond to the temperature $T = T_0$. These parameters were obtained [54] on the basis of calculated $\gamma(RB')$-values for transitions with $max\{J_i\} = 17$, $max\{K_{a\,i}\} = 10$ for the rotational band and $max\{J_i\} = 20$, $max\{K_{a\,i}\} = 17$ for the ν_2 band (Table 3.16).

Table 3.16. Parameters $x_{km}(T_0)$ at $T_0 = 296$ K and n_{km} from Eqs (3.3.4) and (3.3.8) for H_2O self-broadening coefficients [54] (σ are given in 10^{-4} cm^{-1} atm^{-1}). Reproduced with permission.

Parameter	Values		Parameter	Values	
	Rotational band	ν_2 band		Rotational band	ν_2 band
x_{11}	$(0.473 \pm 0.042)\cdot 10^{-2}$	$(0.436 \pm 0.005)\cdot 10^{-2}$	n_{30}	-0.3847 ± 0.0143	-0.2792 ± 0.0032
n_{11}	-0.4060 ± 0.0007	-0.355 ± 0.015	x_{31}	$(0.447 \pm 0.099)\cdot 10^{-2}$	$(0.466 \pm 0.020)\cdot 10^{-2}$
x_{20}	0.549 ± 0.007	0.542 ± 0.002	x_{40}	-0.66 ± 0.15	-1.08 ± 0.05
n_{20}	-0.829 ± 0.003	-0.812 ± 0.006	n_{40}	0.493 ± 0.022	0.217 ± 0.034
n_{21}	-0.829 ± 0.003	-0.812 ± 0.006			
x_{21}	-0.0123 ± 0.0001	-0.0121 ± 0.0001	x_{41}	0.2143 ± 0.0073	0.2297 ± 0.0018
x_{30}	0.159 ± 0.004	0.149 ± 0.004	x_{50}	$(0.138 \pm 0.007)\cdot 10^{-1}$	$(0.137 \pm 0.002)\cdot 10^{-1}$
σ	1.82	2.2	N	9874	11430

The parameters of Table 3.16 can be used for the calculation of self-broadening coefficients for all rotational quantum numbers and temperatures from the studied intervals with the accuracy given in Tables 3.11–3.12 and visualized on Fig. 3.11. An example of calculation for the temperature dependence $\gamma(T)$ for two lines of the ν_2 band [54] is presented in Table 3.17 together with experimental data and theoretical results of other authors. The $\gamma(sur)$ coefficients calculated with Eqs (3.3.4) and (3.3.8) are close to the experimental values and theoretical results of Refs [37, 57].

Table 3.17. Calculated temperature dependence of H_2O self-broadening coefficients for two lines of the ν_2 band (in 10^{-3} cm^{-1} atm^{-1}) and two rotational lines (in MHz $Torr^{-1}$) [54]*. Reproduced with permission.

Line [7 7 0] → [8 8 1] (1320.868 cm^{-1})				Line [8 6 3] → [9 7 2] (1290.536 cm^{-1})			
T, K	expt [37]	calc. [54]	calc. [37]	*T*, K	expt [37]	calc. [54]	calc. [37]
408	146 ± 2	164	131	460	227 ± 7	196	191
522	134 ± 2	145	122	543	197 ± 6	178	176
616	124 ± 2	133	116	646	180 ± 5	161	160
708	120 ± 2	124	110	753	160 ± 4	147	147
839	107 ± 4	113	103	835	150 ± 7	138	139
Rotational line [3 1 3] → [2 2 0] (183 GHz)				Rotational line [4 1 4] → [3 2 1] (380 GHz)			
T, K	expt [57]	calc. [54]	calc. [57]	*T*, K	expt [58]	calc. [54]	calc. [58]
300	19.88 ± 0.1	19.09	18.80	300	20.54 ± 0.06	18.92	20.9
315	19.05 ± 0.08	18.36	18.09	324	19.23 ± 0.03	17.80	19.6
330	18.16 ± 0.1	17.55	17.43	348	17.95 ± 0.03	16.81	18.4
360	17.00 ± 0.06	16.50	16.27	373	17.00 ± 0.05	15.90	17.3
375	16.35 ± 0.06	15.97	15.76				
390	15.93 ± 0.06	15.47	15.28				

* Calculations [37] were performed by the RB method; calculations [54] correspond to Eq. (3.3.4) and parameters from Table 3.16; calculations [57] used Eq. (3.2.1) with γ(300 K) = 18.80 (MHz $Torr^{-1}$), n = 0.79 taken from Table 3 of Ref. [57]; calculations [58] were performed by the ATC theory; calculations with the parameters γ(300 K) = 18.80 (MHz $Torr^{-1}$), n = 0.82 (Ref.[57], Table 3) for this line give slightly underestimated (in comparison with experiment) results which are close to the values obtained with Eq. (3.3.4).

3.4. Semi-empirical approach to calculation of water vapour line widths and shifts

The semi-empirical approach to the calculation of spectral line widths and shifts includes various corrections to the approximate formulae of the ATC theory via a set of adjustable parameters [59]. Once determined by fitting calculated values to some experimental data, these parameters insure a sufficiently accurate prediction of line widths and shifts which have not been measured. Successful calculations by the semi-empirical method have been made, for example, for line widths, shifts and temperature exponents of H_2O–$N_2(O_2)$ and CO_2–$N_2(O_2)$ colliding systems [60–62]. Their results are included in a freely-available carbon dioxide spectroscopic data bank [63] and in the 'Atmos' Information System [64]. This method has recently been further developed by using anharmonic wave functions for the estimation of line shape parameters for H_2O–N_2, H_2O–O_2 and H_2O–H_2O

systems and resulted in a much better agreement of calculated line widths and shifts with their observed values [65–68]. In principle, these accurate wave functions resulting from extensive variational nuclear motion calculations extend the domain of applicability of the semi-empirical method up to the dissociation limit.

The ATC theory employs the simplest (straight path) representation of the relative translational motion of the colliding molecular pair and only the long-range interactions are retained in the anisotropic intermolecular potential. Distinguished by the relative simplicity of calculations, this method has proved its efficiency for HCl–DCl, H_2O–H_2O, H_2O–CO_2 etc. molecular systems characterised by strong interactions (when the distance of closest approach r_c is inferior to the interruption parameter b_0). For the case of weak interactions ($r_c > b_0$) this method generally fails. At the same time the alternative cut-off-free approaches, quite well describing the collisional broadening and shifting, do not enable, because of their complexity, a detailed analysis of various processes occurring in the colliding molecules (e.g. compensation of contributions from different scattering channels in the pressure-induced shift). The ATC method therefore seems worthy of some corrections to eliminate its main disadvantages.

Starting from Eq. (2.4.11) and taking into account Eqs (2.6.4) and (2.6.5) one can express the ATC half-width γ_{if} as

$$\gamma_{if} = B^{ATC}(i,f) + \sum_{l,i'} D^2(ii'|l)\, P_l^{ATC}(\omega_{ii'}) + \sum_{l,f'} D^2(ff'|l)\, P_l^{ATC}(\omega_{ff'}) + \cdots \tag{3.4.1}$$

where

$$B^{ATC}(i,f) = \frac{n}{c}\sum_k \rho(k)\int_0^{\infty} \mathrm{v}F(\mathrm{v})d\mathrm{v}b_0^2(k,i,f)$$

is the averaged 'interruption' part of the half-width and $D^2(ii'|l)$, $D^2(ff'|l)$ are the transition strengths (reduced matrix elements) for the scattering channels $i \to i'$, $f \to f'$ associated with the l-th order multipoles of the active molecule. The expansion coefficients ('interruption' or 'efficiency' functions)

$$P_l^{ATC}(\omega) = \frac{n}{c}\sum_k \rho(k) \sum_{l',k'} A_{ll'} D^2(kk'|l') F_{ll'}\left[\frac{2\pi c b_0(k,i,f)}{\mathrm{v}}(\omega + \omega_{kk'})\right] \tag{3.4.2}$$

depend on the properties of the perturbing molecule (energy levels, wave functions) and the interaction features (intermolecular potential, relative trajectory); $A_{ll'}$ are parameters for the ll'-type interaction: for example, for $l = 1$, $l' = 2$ (dipole–quadrupole) $A_{12} = 16Q^2/[45\hbar^2\mathrm{v}^2b_0^6(k,i,f)]$. In Eqs (3.4.1) and (3.4.2), n is the number density of perturbing molecules, k is the set of quantum numbers characterising their energy levels with thermal populations $\rho(k)$, and $F_{ll'}$ are the resonance functions. While the transition strengths are well known, the efficiency functions are determined with much less accuracy and are therefore worthy of refinement by means of a semi-empirical correction factor.

On the other hand, the general semi-classical expression for the pressure-broadened line half-width reads

$$\gamma_{if} = \mathrm{Re}\left\langle 1 - e^{-S(b)} \right\rangle, \tag{3.4.3}$$

where the average is to be made over the impact parameter, relative velocity and internal states of the perturber:

$$\langle ... \rangle = \frac{n}{2\pi c} \sum_k \rho(k) \int_0^\infty \mathrm{v}F(\mathrm{v})d\mathrm{v} \int_0^\infty 2\pi b db \; ..., \tag{3.4.4}$$

and the complex-valued interruption function $S(b)$ includes both short-range (atom–atom) and long-range (electrostatic) interactions.

The basic idea of the semi-empirical approach is to split the function $S(b)$ into terms $S_g(b)$ calculated with a high precision and terms $S_p(b)$ which are known only approximately (like short-range or high-order electrostatic contributions) and which will be further corrected using experimental data:

$$S(b) = S_g(b) + S_p(b). \tag{3.4.5}$$

The Taylor expansion of the exponent $e^{-S(b)}$ then allows the development of Eq. (3.4.3) as

$$\begin{aligned}\gamma_{if} &= \mathrm{Re}\left\langle 1 - e^{-S_g(b)} e^{-S_p(b)} \right\rangle = \mathrm{Re}\left\langle 1 - \left(1 - S_g(b) + S_g^2(b)/2! - ...\right) e^{-S_p(b)} \right\rangle = \\ &= \mathrm{Re}\left\langle 1 - e^{-S_p(b)} \right\rangle + \mathrm{Re}\left\langle S_g(b) e^{-S_p(b)} \right\rangle - \frac{1}{2!}\mathrm{Re}\left\langle S_g^2(b) e^{-S_p(b)} \right\rangle + ... \, .\end{aligned} \tag{3.4.6}$$

Using the notation

$$\begin{aligned} S_g(b) = &\sum_{l',i',k'} D^2(ii'|l) D^2(kk'|l') f_{ll'}(\omega_{ii'}, \omega_{kk'}) + \\ &+ \sum_{l',f',k'} D^2(ff'|l) D^2(kk'|l') f_{ll'}(\omega_{ff'}, \omega_{kk'}) \end{aligned} \tag{3.4.7}$$

with only dipole–dipole and dipole–quadrupole interactions included and taking into account that the transition strengths are independent from the averaging variables one can rewrite Eq. (3.4.6) (limited to two first terms) as

$$\gamma_{if} = \mathrm{Re}\, B(i,f) + \sum_{l,i'} D^2(ii'|l)\, \mathrm{Re}\, P_l(\omega_{ii'}) + \sum_{l,f'} D^2(ff'|l)\, \mathrm{Re}\, P_l(\omega_{ff'}), \tag{3.4.8}$$

where

$$B(i,f)=\left\langle 1-e^{-S_p(b)}\right\rangle, \qquad (3.4.9)$$

$$P_l(\omega)=\left\langle \sum_{l'}\sum_{k,k'} D^2(kk'|l')f_{ll'}(\omega,\omega_{kk'})e^{-S_p(b)}\right\rangle. \qquad (3.4.10)$$

Comparison of Eq. (3.4.8) with the ATC expression (3.4.1) yields

$$B(i,f)=B^{ATC}(i,f)C_0(i,f), \qquad (3.4.11)$$

$$P_l(\omega)=P_l^{ATC}(\omega)C_l(\omega), \qquad (3.4.12)$$

where $B^{ATC}(i,f)$ and $P_l^{ATC}(\omega)$ are the usual ATC terms including only dipole–dipole and dipole–quadrupole interactions. Equation (3.4.8) can therefore be seen as a sum of contributions related to different scattering channels with efficiency functions $P_l(\omega)$ given by Eq. (3.4.10). The form of correction factors $C_0(i,f)$, $C_l(\omega)$ in Eqs (3.4.11) and (3.4.12) and their adjustable parameters can be determined from fitting to experimental data.

The correction factor $C_l(\omega)$ is usually taken as

$$C_l(\omega_{ii'})=\frac{c_1}{c_2\sqrt{J_i}+1}, \qquad (3.4.13)$$

where c_1 and c_2 are adjustable parameters. However, their determination appears to be much more meaningful than simple curve fittings since their values obtained from fitting to only a few line-width data in a particular vibrational band correctly describe not only all the other line widths of this band but those of the whole variety of vibrational bands for the colliding pair. Results of line-width and line-shift calculations by this method are presented in sections 3.5 (H_2O–N_2, H_2O–O_2) and 3.7 (H_2O–H_2).

3.5. Broadening of water vapour lines by nitrogen, oxygen, air and carbon dioxide

The knowledge of spectral parameters of H_2O lines broadened by N_2, O_2 and CO_2 in a wide range of frequencies and temperatures is of great importance for atmospheric applications (characterisation of air pollution, study of ozone holes and greenhouse effect). Moreover, the H_2O–CO_2 case represents a particular interest for remote sensing of the atmospheres of Mars and Venus. Many papers devoted to line shape parameters of these molecular systems have been therefore published. A survey of experimental data obtained up to 2004 for the pressure broadening of H_2O absorption lines can be found in Ref. [5].

Very complete experimental investigations of H_2O lines broadened by air and nitrogen were performed in Refs [22, 69]. In Ref. [69], the line-broadening coefficients were obtained for more than 1300 vibrotational transitions of $H_2{}^{16}O$, $H_2{}^{17}O$ and $H_2{}^{18}O$ covering the region of 604–2271 cm^{-1} ($J \leq 18$, $K_a \leq 10$). In Ref. [22], these coefficients were determined for about 4000 H_2O–air and H_2O–H_2O absorption lines located between 2800 and 8000 cm^{-1}.

Valentin *et al.* [70, 71] recorded H_2O line shapes in the ν_2 band for perturbation by He, Ne, Ar, Kr and N_2. In particular, it was shown that for narrow lines neglecting the Dicke effect induces an error of several percent.

Five H_2O–$N_2(O_2)$ line-broadening coefficients for the region of 1.4 μm were obtained in Refs [61, 72]. The temperature dependence of line shape parameters for this band was studied in Ref. [73]. The same region was recently revisited [74] for the broadening coefficients of several lines of the ν_1+ν_3 and $2\nu_1$ bands broadened by N_2, O_2 and air pressure.

The room-temperature air-broadening and air-shifting coefficients were recently obtained in the spectral region 4200–6600 cm^{-1} in the abovementioned work of Jenouvrier *et al.* [25] for the isotopologues $H_2{}^{16}O$, $H_2{}^{17}O$, $H_2{}^{18}O$, a nd HDO. The authors reported very good agreement between their new measurements and the results listed in the database which improved significantly the description of the transmission in the windows situated on both sides of the band at 5400 cm^{-1}.

The air-broadening coefficients for 24 $H_2{}^{16}O$ lines (21 line form the $2\nu_1$+ν_2+ν_3 band and 3 lines from the $2\nu_1$+ν_3 band) in the region 12 053–12 281 cm^{-1} were recently measured by Ibrahim *et al.* [27] using the standard Voigt profile and its modified version with a variable Doppler contribution. For some lines studied the difference between the line widths extracted by the first and the second models was about 8% (modified profile giving larger values). The mean difference between the measurements (average of two experimental values for each line) and the air-broadening coefficients included in the HITRAN 2004 database was found to be about 1%, confirming thus that the HITRAN data are quite accurate for the considered spectral region.

For remote sensing of the terrestrial atmosphere an accurate experimental analysis of the rotational line $3_{1,3}$←$2_{2,0}$ (183 GHz) broadened by nitrogen, oxygen and air was realised by Tretyakov *et al.* [75] (see also the references cited therein for previous measurements and calculations for this line). In particular, the shifting coefficient values necessary for improved fitting of satellite data were measured in a laboratory for the first time. For the case of air-broadening this value (−0.07(2) MHz $Torr^{-1}$) agreed well with the existing theoretical prediction (−0.13 MHz $Torr^{-1}$). The air-broadening coefficient (3.84(4) MHz $Torr^{-1}$) was also found to be very close to the theoretical value (3.96 MHz $Torr^{-1}$). The N_2- and O_2-broadening and shifting coefficients for the 321, 325 and 380 GHz lines (transitions $10_{2,9}$–$9_{3,6}$, $5_{1,5}$–$4_{2,2}$, $4_{1,4}$–$3_{2,1}$) at room temperature were measured by Koshelev *et al.* [29]. For the 22.2 GHz line (transition $6_{1,6}\leftarrow 5_{2,3}$) they were studied both experimentally and theoretically by Cazzoli *et al.* [30] in the temperature range 296–338 K. Cazzoli and co-authors also considered the transition $1_{1,1}\leftarrow 0_{0,0}$ (1.113 THz) [31] and the transitions $7_{2,5}\leftarrow 8_{1,8}$ (1.147 THz), $8_{5,4}\leftarrow 7_{6,1}$ (1.168 THz), $7_{4,4}\leftarrow 6_{5,4}$ (1.173 THz), $8_{5,3}\leftarrow 7_{6,2}$ (1.191 THz), and $6_{3,3}\leftarrow 5_{4,2}$ (1.542 THz) [32] at room temperature.

Terahertz time-domain spectroscopy was used by Seta *et al.* [76] to determine the air-broadening coefficients for two lines $1_{1,0}\leftarrow 1_{0,1}$ (556.936 GHz) and $2_{1,1}\leftarrow 2_{0,2}$ (752.033 GHz). The values of 4.012(30) and 3.964(40) MHz $Torr^{-1}$ were obtained respectively. The first value was relatively small in comparison with the experimental result of Ref. [77] but close to the HITRAN value.

Golubiatnikov *et al.* [78] experimentally confirmed the N_2- and O_2-broadening coefficients obtained by Seta *et al.* [76] for the 556.936 GHz line of $H_2{}^{16}O$ and additionally reported the broadening and shifting coefficients for the isotopologues $H_2{}^{17}O$ and $H_2{}^{18}O$ (520.02 GHz and 547.676 GHz lines) as well as for the vibrationally excited state ν_2 of $H_2{}^{16}O$ (658.003 GHz line). Their values of the broadening parameters were very close to those of vibrotational lines 1_{10}–1_{01} in the region 1500–1700 cm^{-1} from Refs [21, 69].

The temperature dependence of the broadening parameters of the 556 GHz H_2O line by oxygen, nitrogen, carbon dioxide, hydrogen and helium pressure was recently examined by Dick *et al.* [79] in the temperature interval 17–200 K. In order to make a comparison with previous studies, the broadening by nitrogen and oxygen was also considered at room temperature. The room-temperature measurements agreed with recent works [76, 78] within one standard deviation for the nitrogen case and within two standard deviations for the oxygen case. The broadening coefficients extrapolated to 296 K with the temperature exponents n (see Eq. (3.2.1)) obtained for the experimentally studied range, however, showed a 13% difference for nitrogen and 8% difference for oxygen. For the N_2- and O_2-shifting coefficients an agreement within one standard deviation was stated with the values of Ref. [77] but the respective differences of 30% and 25% were observed with the results recently obtained by Golubiatnikov *et al.* [78]. The extrapolation of the line-shifting coefficients to the temperature of 296 K using the temperature exponents determined experimentally yielded a 40% disagreement for both perturbing gases. These results of extrapolation obtained for line widths and line shifts confirm that the temperature exponents should be used with caution when exceeding the temperature range for which they have been defined.

Extensive experimental and theoretical investigations of H_2O–CO_2 vibrotational lines were realised in Refs [80, 81]. Very recently Poddar *et al.* [82] measured the CO_2-induced broadening coefficients in the $2\nu_1+\nu_2+\nu_3$ overtone band using Voigt profile model and considered the dependence of the broadening coefficients on the rotational quantum numbers.

The pressure-broadening coefficients for thirty-two rotational line of H_2O–CO_2 system in the spectral range 550–3050 GHz (18–102 cm^{-1}) were studied for the first time by Sagawa *et al.* [83] using a terahertz time-domain spectrometer. The experimental values had a mean precision of 2.4% and their difference with the CRB calculations varied between −10.7% and +19.0%. The authors showed that their measurements had a significant impact on the retrieval of water vapour abundances in the atmosphere of Venus.

The semi-classical calculations for H_2O lines were mainly performed in the framework of the complex Robert–Bonamy formalism [80, 81, 84]. In order to obtain a complete set of air-broadening and air-shifting coefficients for $H_2{}^{16}O$ lines included in the HITRAN database from microwave to visible regions, Jacquemart *et al.* [85] realised semi-empirical calculations based on the fits of some recent measurements and calculations to the first-order terms in the expansion of the CRB equations. Semi-empirical vibrationally dependent coefficients were obtained for approximately 700 sets of rotational quantum numbers, with an accuracy of 5–10% for the air-broadening coefficients and 0.001–0.01 cm^{-1} atm^{-1} for the air-shifting coefficients.

The N_2- and O_2-broadening of H_2O lines was also considered by Anthony *et al.* [41] in the framework of the complex implementation of the MRB formalism (MCRB approach). For the $H_2{}^{16}O$–N_2 system noticeable differences were observed between the MCRB and CRB results, whereas for the $H_2{}^{16}O$–O_2 system these differences were small, in accordance with the strength of the intermolecular interaction. The CRB H_2O–N_2 line shifts especially showed a pronounced agreement with experimental values.

The vibrational dependence of selected transitions ($K_c = J$ or $K_c = J–1$) for the H_2O–N_2 colliding pair was recently studied by Gamache and Hartmann [86].

Their analysis confirmed that the bending and stretching vibrations have significantly different effects on the shifting coefficients due to the vibrational dependence of the interaction potential. At the same time it was found that the line widths in different vibrational bands are quite similar, with slightly smaller broadening for the bending band lines.

Wagner *et al.* [87] considered the temperature effect on the H_2O–air lines in the region of the ν_2 band. The air-broadening coefficients for the doublet lines with $K_c = J$ were measured at 296, 742 and 980 K and combined with previous experimental determinations to obtain the γ-values for the lines $0 < J < 16$ at these three temperatures. The temperature exponents corresponding to Eq. (3.2.1) were further extracted. The authors concluded that calculations are not able to reproduce experimental data at all temperatures simultaneously because of an inappropriate intermolecular potential and very strong dependence of the line-broadening coefficient on the relative velocity for high values of the rotational quantum number. For certain types of transitions the power law was found to be not valid (negative n-values) due to off-resonance energy jumps for collision-induced transitions in the colliding molecules and domination of the rotational contributions by the vibrational terms.

First calculations of H_2O–N_2, H_2O–O_2 and H_2O–CO_2 line widths and shifts by the semi-empirical method were attempted for the strongest ν_2 band [59, 60] where the lines corresponding to high values of the rotational quantum number are observable. Except some rotational energy levels (see Sec. 3.1), up to $J = 20$ the vibrational state (010) can be considered as isolated. The vibrotational wave functions and transition probabilities were therefore calculated by the well developed method of effective Hamiltonians. For the line-width calculations only dipole–quadruple and quadruple–quadruple interactions were taken into account.

A comparison of semi-empirical values with experimental data and CRB-computations of Ref. [84] for the $3\nu_1 + \nu_3$ band of H_2O–N_2 is given in Table 3.18). For the semi-empirical γ-values the mean square deviation from the experimental data is 0.00049 $cm^{-1}\,atm^{-1}$ whereas for the CRB values it is 0.00055 $cm^{-1}\,atm^{-1}$; for the line-shifting coefficients both theoretical methods result in the same value of 0.00037 $cm^{-1}\,atm^{-1}$. According to this comparison the semi-empirical and the semi-classical CRB methods have practically the same precision for the determination of line shape parameters.

Table 3.18. Comparison of semi-empirical (SE) line shape parameters with experimental and CRB-computed values of Ref. [84] (in cm^{-1} atm^{-1}).

$J_f K_{af} K_{cf}$	$J_i K_{ai} K_{ci}$	γ^{expt}	γ^{SE}	γ^{CRB}	δ^{expt}	δ^{SE}	δ^{CRB}
5 1 4	6 1 5	0.097	0.102	0.101	–0.014	–0.015	–0.015
5 2 4	6 2 5	0.091	0.091	0.089	–0.016	–0.017	–0.015
4 3 1	5 3 2	0.096	0.098	0.095	–0.012	–0.013	–0.011
4 2 2	5 2 3	0.100	0.106	0.105	–0.009	–0.010	–0.010
5 1 5	6 1 6	0.091	0.091	0.088	–0.019	–0.019	–0.018
5 0 5	6 0 6	0.090	0.093	0.089	–0.016	–0.019	–0.018
4 1 3	5 1 4	0.103	0.107	0.107	–0.016	–0.014	–0.014
4 2 3	5 2 4	0.098	0.099	0.097	–0.013	–0.014	–0.013
7 6 2	7 6 1	0.058	0.053	0.051	–0.026	–0.020	–0.028
6 6 0	6 6 1	0.055	0.049	0.046	–0.026	–0.035	–0.032
3 2 1	4 2 2	0.101	0.107	0.105	–0.011	–0.010	–0.010
3 1 2	4 1 3	0.107	0.109	0.109	–0.013	–0.011	–0.010
5 0 5	5 2 4	0.096	0.099	0.094	–0.021	–0.018	–0.016
3 2 2	4 2 3	0.102	0.104	0.101	–0.011	–0.012	–0.012
3 0 3	4 0 4	0.108	0.110	0.109	–0.013	–0.014	–0.013
3 1 3	4 1 4	0.104	0.108	0.106	–0.013	–0.013	–0.014
4 0 4	4 2 3	0.104	0.107	0.102	–0.016	–0.016	–0.017
2 2 0	3 2 1	0.107	0.108	0.104	–0.011	–0.012	–0.011
7 4 4	7 4 3	0.082	0.086	0.084	–0.021	–0.021	–0.018
8 4 4	8 4 5	0.088	0.091	0.090	–0.026	–0.015	–0.013
2 0 2	3 0 3	0.114	0.115	0.114	–0.013	–0.012	–0.013
2 1 2	3 1 3	0.110	0.111	0.110	–0.013	–0.012	–0.012
4 1 4	4 1 3	0.105	0.109	0.108	–0.009	–0.010	–0.008
1 1 0	2 1 1	0.114	0.119	0.116	–0.007	–0.009	–0.007
5 4 2	5 4 1	0.080	0.080	0.076	–0.016	–0.020	–0.017
5 4 1	5 4 2	0.084	0.080	0.076	–0.017	–0.019	–0.016
3 1 3	3 1 2	0.108	0.112	0.111	–0.010	–0.011	–0.009
4 3 2	4 3 1	0.096	0.095	0.091	–0.014	–0.015	–0.013
5 3 2	5 3 3	0.095	0.097	0.095	–0.012	–0.014	–0.013
6 3 3	6 3 4	0.099	0.098	0.097	–0.011	–0.014	–0.013
4 3 1	4 3 2	0.093	0.095	0.091	–0.012	–0.015	–0.013
3 2 2	3 2 1	0.105	0.108	0.105	–0.014	–0.013	–0.012
0 0 0	1 0 1	0.115	0.119	0.116	–0.009	–0.009	–0.009
4 2 2	4 2 3	0.106	0.105	0.104	–0.010	–0.010	–0.011
4 1 3	4 1 4	0.105	0.109	0.108	–0.010	–0.012	–0.013
2 1 2	1 1 1	0.110	0.113	0.112	–0.010	–0.009	–0.009
3 1 3	2 1 2	0.105	0.111	0.109	–0.010	–0.009	–0.009
3 0 3	2 0 2	0.111	0.114	0.113	–0.006	–0.007	–0.006
3 2 1	2 2 0	0.101	0.107	0.104	–0.009	–0.009	–0.011
4 1 4	3 1 3	0.100	0.107	0.106	–0.012	–0.010	–0.009
4 2 3	3 2 2	0.102	0.104	0.101	–0.013	–0.013	–0.013
4 0 4	3 0 3	0.102	0.109	0.109	–0.008	–0.007	–0.007
7 2 6	7 0 7	0.074	0.076	0.071	–0.021	–0.022	–0.024
6 1 6	5 1 5	0.088	0.088	0.086	–0.016	–0.016	–0.014
7 2 6	6 2 5	0.079	0.081	0.079	–0.020	–0.021	–0.018
8 1 8	7 1 7	0.063	0.065	0.059	–0.027	–0.026	–0.026
8 0 8	7 0 7	0.061	0.065	0.059	–0.027	–0.026	–0.026
7 3 5	6 3 4	0.085	0.087	0.084	–0.020	–0.020	–0.017
6 1 5	5 1 4	0.097	0.100	0.100	–0.011	–0.011	–0.009

Calculations of H_2O–N_2, H_2O–O_2, and H_2O–CO_2 line-broadening coefficients for the ν_2 band by the semi-empirical method were performed for J, $K_a \le 20$ at $T = 296$ K and for J, $K_a \le 10$ at $T = 200$, 300, 400 and 500 K in Ref. [59]. Figure 3.13 shows a comparison of these values with the experimental data [69] and the ATC computations for H_2O–N_2 mixture at 300 K. The semi-empirical results agree well with the measurements whereas the ATC theory overestimates the line broadening for small J- and K_a-values and underestimates it for high J- and K_a-values.

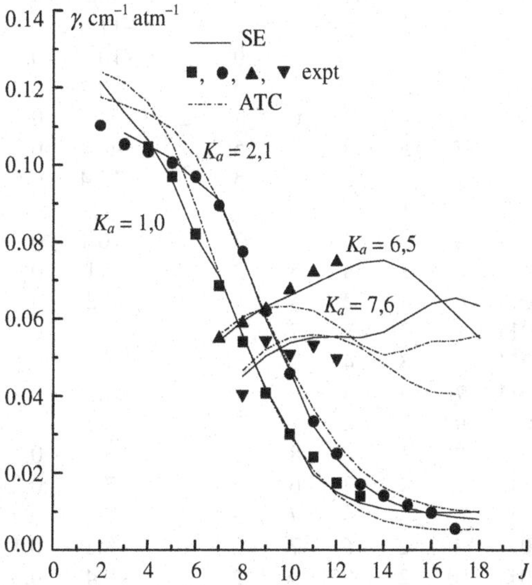

Fig. 3.13. Semi-empirical (SE), ATC and experimental [69] H_2O–N_2 line-broadening coefficients of the ν_2 band in function of rotational quantum numbers J, K_a at 300 K. Reproduced with permission from Ref. [59].

Semi-empirical calculations of line-shift coefficients for H_2O perturbed by N_2, O_2 and air were realised for various vibrational bands. Table 3.19 presents the results for $\nu_2 + \nu_3$, $2\nu_2 + \nu_3$, $\nu_1 + \nu_3$, $2\nu_1$, $\nu_1 + \nu_2 + \nu_3$ and $\nu_2 + 2\nu_3$ H_2O–N_2 lines compared to the experimental values of Ref. [88]. In these calculations, the mean dipole polarizability of the active molecule in the upper vibrational state (distorted by the application of the cut-off procedure and by the neglect of the repulsive part of the interaction potential) was considered as an additional fitting parameter.

Table 3.19. Semi-empirical (SE) and experimental [88] line-shifting coefficients for H_2O–N_2 lines (in 10^{-3} cm^{-1} atm^{-1}); σ stands for one standard deviation and B denotes the unresolved doublets.

Frequency, cm^{-1}	Band	J_i	K_{ai}	K_{ci}	J_f	K_{af}	K_{cf}	δ^{expt}	2σ	δ^{SE}
				2 μm						
5152.0993	011	7	1	6	8	1	7	−10.9	0.5	−10.6
5166.1931	011	7	5	3	8	5	4	−10.7	0.7	−11.0
5178.0992	011	6	4	2	7	4	3	−11.4	0.3	−8.6
5189.7198	011	5	3	2	6	3	3	−6.0	0.2	−5.3
5191.8791	011	6	5	1	7	5	2	−7.3	0.4	−11.8
5226.3515	011	4	1	4	5	1	5	−7.6	0.1	−8.0
5250.3091	011	4	0	4	4	2	3	−8.0	2.5	−11.3
5282.9623	011	6	2	5	6	2	4	−11.0	0.7	−4.8
5284.7803	011	1	0	1	2	0	2	−10.1	0.2	−7.9
5328.6310	011	6	3	4	6	3	3	−8.3	0.4	−5.5
5350.5344	011	6	4	3	6	4	2	−9.3	0.2	−10.7
5361.5245	011	7	5	3	7	5	2	−9.0	0.2	−12.6
5361.8928	011	7	5	2	7	5	3	−13.4	0.6	−11.1
5405.2906	011	5	1	4	5	1	5	−5.8	0.3	−7.9
5442.1522	011	4	3	1	3	3	0	−10.1	0.1	−10.5
5469.5269	011	6	2	5	5	2	4	−7.1	0.1	−6.7
5498.9953	011	6	4	2	5	4	1	−7.2	0.1	−9.4
5505.5546	011	8	2	7	7	2	6	−8.0	0.1	−8.0
5518.2874	011	7	3	4	1	3	3	−8.4	0.2	−6.0
5521.1385	011	7	4	4	6	4	3	−10.9	0.1	−10.2
5521.9039	011	9	2	8	8	2	7	−9.2	0.1	−8.7
5523.1329	011	9	1	8	8	1	7	−5.0	0.2	−7.2
5523.4538	011	7	4	3	6	4	2	−7.7	0.2	−6.9
5527.8448	011	8	3	6	7	3	5	−12.2	0.2	−9.6
5536.3074	011	8	2	6	7	2	5	−8.2	0.1	−5.5
5537.5414	011	10	2	9	9	2	8	−9.4	0.3	−9.7
5538.1290	011	10	1	9	9	1	8	−7.1	0.1	−8.4
5543.4219	011	8	4	5	7	4	4	−12.8	0.4	−10.0
5548.6557	011	8	4	4	7	4	3	−5.3	0.1	−5.7
5564.8228	011	9	4	6	8	4	5	−11.3	0.2	−10.5
5582.1670	011	11	3	9	10	3	8	−12.2	0.5	−11.9
5595.5578	011	10	3	7	9	3	6	−10.6	0.4	−7.8
				1.4 μm						
6705.0390	021	7	1	7	8	1	8	−10.7	0.7	−7.6
6726.2494	021	6	0	6	7	0	7	−12.2	0.1	−8.1
6727.2326	021	6	1	6	7	1	7	−12.2	0.5	−7.7
6748.8902	021	5	1	5	6	1	6	−8.3	0.3	−7.9
6753.5782	021	5	2	4	6	2	5	−3.7	0.5	−5.8
6799.3729	021	2	1	1	3	1	2	0.7	0.2	−1.7
6812.8111	021	2	1	2	3	1	3	−4.5	0.4	−5.8
6847.7267	021	0	0	0	1	0	1	−0.2	0.2	−0.6
6893.6485	021	2	2	0	2	2	1	−5.4	0.2	−1.0
6914.5496	021	4	3	2	4	3	1	−5.8	0.8	−5.7

6917.3693	021	2	0	2	1	0	1	1.5	0.2	1.7
6917.9525	021	2	1	2	1	1	1	–5.0	0.6	–3.5
6930.3670	021	2	1	1	1	1	0	–5.7	0.2	–6.5
6955.1576	021	4	0	4	3	0	3	0.2	0.1	–0.7
6956.3153	021	3	1	2	2	1	1	–5.3	0.4	–5.5
6963.1689	021	3	2	2	2	2	1	–3.8	0.3	–4.9
7004.2280	021	5	1	4	4	1	3	–0.6	0.6	–1.7
7006.1275	021	5	2	4	4	2	3	–6.9	0.2	–5.3
7025.3840	021	6	1	5	5	1	4	–2.5	0.3	–2.4
7034.4755	200	5	1	4	6	2	5	–10.1	0.4	–10.2
7063.1299	021	8	1	7	7	1	6	–5.8	0.9	–5.6
7070.7840	101	6	1	5	7	1	6	–11.6	0.1	–9.6
7079.1767	200	6	1	6	6	2	5	–5.8	0.9	–6.0
7080.5751	101	5	3	3	6	3	4	–4.4	0.1	–6.9
7104.6194	101	4	3	1	5	3	2	–3.8	0.2	–5.3
7108.7153	200	3	0	3	4	1	4	–10.9	0.2	–10.0
7120.3580	101	4	2	3	5	2	4	–8.0	0.2	–7.0
7127.0355	200	2	0	2	3	1	3	–8.0	0.4	–11.5
7131.9505	200	3	1	3	3	2	2	–8.7	0.3	–9.6
7133.9031	101	3	3	0	4	3	1	–5.1	0.2	–4.1
7134.9821	101	3	3	1	4	3	2	–5.7	0.9	–4.6
7136.0941	101	3	2	1	4	2	2	–5.2	0.2	–5.3
7165.8211	101	2	2	0	3	2	1	–5.1	0.2	–6.3
7202.9098	101	1	0	1	2	0	2	–10.5	0.1	–9.0
7216.1909	101	5	4	1	5	4	2	–8.0	0.3	–8.6
7227.9685	101	4	3	2	4	3	1	–8.8	0.1	–7.8
7236.4474	200	2	2	0	2	1	1	–4.0	0.3	–2.0
7240.4159	101	2	2	1	2	2	0	–7.8	0.0	–7.6
7249.9247	200	2	2	1	2	1	2	–6.1	0.3	–7.5
7266.6518	200	3	3	1	3	2	2	–5.8	0.5	–0.0
7281.0820	200	4	1	4	3	0	3	–1.3	0.1	0.9
7283.7319	101	6	2	4	6	2	5	–5.0	0.2	–7.7
7286.0516	200	6	2	5	6	1	6	–4.3	0.9	–7.1
7305.0814	200	6	0	6	5	1	5	–8.0	0.4	–4.9
7312.1963	101	3	0	3	2	0	2	–3.9	2.0	–1.9
7323.9579	101	4	1	4	3	1	3	–6.9	0.1	–4.3
7331.7156	200	8	1	8	7	0	7	–10.0	0.4	–6.8
7348.4037	101	5	2	4	4	2	3	–12.3	0.3	–7.8
7351.4852	101	5	3	2	4	3	1	–5.8	0.2	–8.3
7359.3343	101	6	4	2	5	4	1	–6.7	0.2	–8.9
7378.6791	101	7	4	4	6	4	3	–10.2	0.2	–9.9
7397.5754	101	8	1	7	7	1	6	–6.5	0.1	–6.4
7403.6163	101	8	4	4	7	4	3	–4.1	0.3	–6.6
7406.0282	101	9	2	8	8	2	7	–13.8	0.2	–10.4
7407.7830	101	9	1	8	8	1	7	–8.4	0.4	–8.1
7413.0192	101	9	4	6	8	4	5	–12.8	0.6	–11.6
7417.8213	101	10	1	9	9	1	8	–7.6	0.3	–10.0
7419.1750	101	8	3	5	7	3	4	–11.4	0.3	–7.3
				1 μm						
8636.7581	111	6	1	5	7	1	6	–12.3	0.8	–11.6

8665.1311	111	5	2	4	6	2	5	–7.0	0.4	–9.8
8675.7803	111	5	1	5	6	1	6	–14.3	0.2	–11.3
8680.2591	111	4	1	3	5	1	4	–6.8	0.2	–8.7
8696.9877	111	4	0	4	5	0	5	–10.8	0.2	–10.9
8713.6592	111	3	2	2	4	2	3	–5.2	0.3	–6.7
8717.9110	111	3	0	3	4	0	4	–9.9	0.3	–10.3
8730.1318	111	2	1	1	3	1	2	–4.1	0.4	–4.3
8733.8083	111	2	2	0	3	2	1	–7.1	0.1	–6.6
8742.9292	111	2	1	2	3	1	3	–9.6	0.4	–8.8
8754.9302	111	1	1	0	2	1	1	–3.8	0.4	–4.8
8760.1410	111	1	0	1	2	0	2	–10.3	0.4	–10.0
8765.0406	111	1	1	1	2	1	2	–4.5	0.1	–7.5
8811.0630	111	2	2	0	2	2	1	–4.9	0.0	–6.4
8812.0143	111	4	3	1	4	3	2	–6.7	0.2	–8.6
8821.9196	111	5	5	0	5	5	1*B*	–14.2	0.6	–17.8
	111	5	5	1	5	5	0*B*			
8830.2319	111	1	0	1	0	0	0	–8.0	1.2	–8.0
8848.0705	111	2	1	2	1	1	1	–6.6	0.4	–6.2
8866.1671	111	3	1	3	2	1	2	–6.5	0.2	–5.2
8869.8731	111	3	0	3	2	0	2	–3.0	0.2	–3.0
8879.1198	111	3	2	2	2	2	1	–7.0	0.1	–8.5
8882.8726	111	4	1	4	3	1	3	–5.2	0.2	–5.5
8885.5740	111	4	0	4	3	0	3	–3.9	0.1	–3.5
8898.1943	111	5	1	5	4	1	4	–5.8	0.3	–6.1
8899.1304	111	4	2	3	3	2	2	–8.7	0.4	–8.9
8912.2568	111	6	1	6	5	1	5	–9.8	0.3	–7.3
8912.9834	111	6	0	6	5	0	5	–7.9	0.1	–6.3
8917.6803	111	5	2	4	4	2	3	–11.3	0.2	–9.0
8925.2222	111	7	1	7	6	1	6	–8.2	0.2	–8.9
8928.4787	111	5	3	3	4	3	2	–9.9	1.2	–12.3
8933.4633	111	5	2	3	4	2	2	–7.2	0.5	–8.6
8934.7405	111	6	2	5	5	2	4	–12.0	0.6	–9.7
8948.4417	111	9	1	9	8	1	8	–12.4	0.5	–12.2
8950.3347	111	7	2	6	6	2	5	–13.3	0.3	–10.6
8956.2946	111	6	3	3	5	3	2	–7.5	0.4	–8.7
	012	7	7	1	6	6	0*B*	–27	3	–27

A statistical analysis of the data collected in Table 3.19 shows that the absolute error for the semi-empirical values ($\Delta = |\delta^{expt} - \delta^{SE}|$, in $cm^{-1}\,atm^{-1}$) does not exceed 0.0015 for 73% of lines. At the same time $0.0015 < \Delta \le 0.0030$ for 18%, $0.0030 < \Delta \le 0.0045$ for 6% and $0.0045 < \Delta \le 0.0060$ for 3% of lines.

3.5.1. Modelling of H_2O line widths broadened by N_2, O_2, air and CO_2

The SE-values of H_2O line-broadening coefficients obtained for these perturbing gases at various temperatures can be used together with the corresponding experimental data to determine the parameters of the analytical model $\gamma(sur)$

given by Eq. (3.3.4); in the case of CO_2 the CRB results [80] can be additionally taken into account. The temperature exponents n_k can be defined according to the relation

$$x_k(J, T) = x_k(J, T_0 = 296\text{ K})(T/T_0)^{n_k}, \quad (3.5.1)$$

with x_k $(J, T_0 = 296\text{ K})$ given by Eq. (3.3.6). The model parameters for these equations obtained by least squares fitting are listed in Tables 3.20–3.21. The quality of reproduction of experimental data by the model can be seen in Figs 3.14–3.16 for $T_0 = 296$ K. For $H_2O–N_2$ case $\gamma(T)$ is described by four parameters whereas for other systems there is only one parameter.

Table 3.20. Parameters $x_{km}(T_0 = 296\text{ K})$ and n_k for $\gamma(sur)$ model applied to $H_2O–N_2$ lines.

Parameter	$K_i, K_f = 0,1$ [69]]	All data [69]	All data [69] and SE
x_{20}	0.1195(0.0017)	0.1249(0.0016)	0.1363(0.0008)
x_{21}	–	$-0.237(0.011)\cdot10^{-2}$	$-0.3738(0.0074)\cdot10^{-2}$
x_{22}	–	0.0	$0.1034(0.0143)\cdot10^{-4}$
x_{30}	0.4182(0.0677)	0. 1836(0.0140)	0. 2064(00.0074)
x_{31}	–	$-0.400(0.077)\cdot10^{-2}$	$0.4511(0.0620)\cdot10^{-2}$
x_{32}	–	0.0	$-0.2011(0.0132)\cdot10^{-3}$
x_{40}	–	–0.7952(0.1116)	–0.8753(0.0423)
x_{41}	0.1737(0.0232)	0.2378(0.0063)	0.2525(0.0023)
x_{50}	–	0.0712(0.0048)	0.0230(0.0029)
x_{51}	–	$-0.4305(0.033)\cdot10^{-2}$	$-0.6252(0.0019)\cdot10^{-3}$
n_2	–0.7643(0.021)	–	–0.6476(0.0068)
n_3	–0.5253(0.047)	–	–0.2974(0.0194)
n_4	–	–	–0.4353(0.0182)
n_5	–	–	–1.3214(0.1927)
N	87	679	19100
σ, 10^{-4}	7.0	3.2	1.0

Table 3.21. Parameters $x_{km}(T_0 = 296\text{ K})$ and n_k of $\gamma(sur)$ model for $H_2O–O_2$, -air, and $-CO_2$ lines.

Parameter	$H_2O–O_2$ [a]	H_2O – air [b]	$H_2O–CO_2$ [c]
x_{20}	$0.7036(0.031)\cdot10^{-1}$	0.1136(0.0013)	0.2230(0.0034)
x_{21}	$-0.2348(0.0027)\cdot10^{-2}$	$-0.2322(0.0089)\cdot10^{-2}$	$-0.1016(0.004)\cdot10^{-1}$
x_{22}	$0.1773(0.0055)\cdot10^{-4}$	0.0	$0.1259(0.0099)\cdot10^{-3}$
x_{30}	0.3168(0.0036)	0. 2145(0.0129)	0. 3245(0.0014)
x_{31}	$-0.7440(0.0168)\cdot10^{-2}$	$0.2642(0.073)\cdot10^{-2}$	$-0.1016(0.0094)\cdot10^{-1}$
x_{40}	–0.5030(0.0363)	–0.6114(0.097)	–0.6834(0.1186)
x_{41}	0.2382(0.0024)	0.2244(0.0052)	0.2352(0.0102)
x_{50}	$0.1282(0.0066)\cdot10^{-1}$	$0.4556(0.0038)\cdot10^{-1}$	$0.4073(0.0262)\cdot10^{-1}$
x_{51}	$-0.3537(0.0294)\cdot10^{-3}$	$-0.1088(0.024)\cdot10^{-2}$	$-0.6419(0.1773)\cdot10^{-3}$
n_2	–0.6270(0.0079)	–	–0.5064(0.0212)
n_3	–0.2782(0.0256)	–	0.1592(0.07252)
n_4	–0.17365(0.0205)	–	
n_5	–0.544(0.072)	–	
N	12446	679	1583
σ, 10^{-4}	0.32	3.2	4.0

[a] Parameters deduced from SE calculations; [b] Experimental data [69] used without determination of n_k; [c] The SE-values used together with CRB results and some experimental data [80].

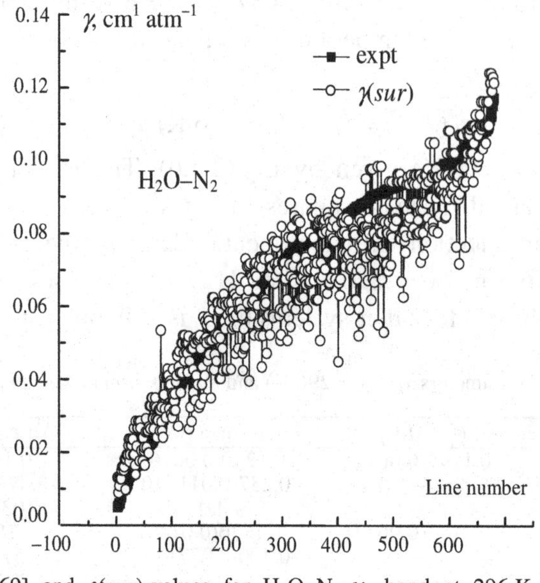

Fig. 3.14. Experimental [69] and $\gamma(sur)$-values for H_2O–N_2 ν_2 band at 296 K. The lines are numbered in increasing order of experimental values.

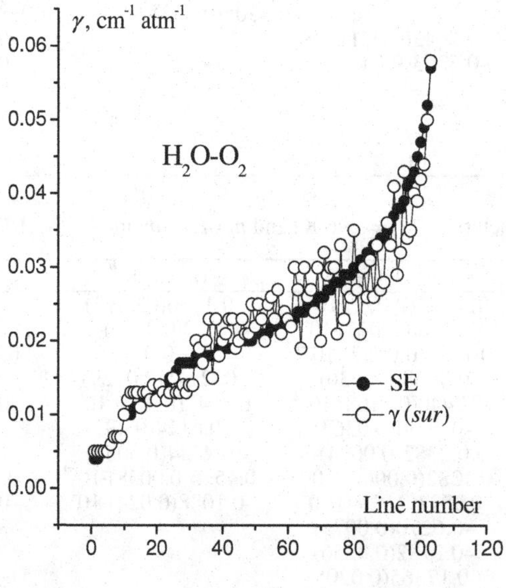

Fig. 3.15. Semi-empirical (SE) and calculated with $\gamma(sur)$ model H_2O–O_2 line-broadening coefficients in the ν_2 band at 296 K. The lines are numbered in increasing order of semi-empirical values.

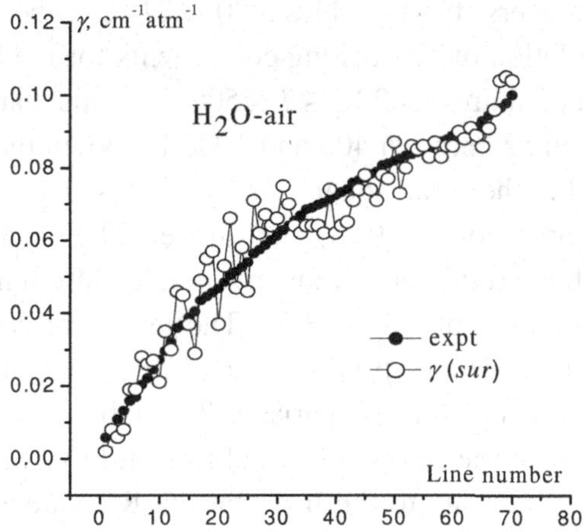

Fig. 3.16. Experimental [69] and calculated with $\gamma(sur)$ model H_2O–air line-broadening coefficients in the ν_2 band at 296 K. The lines are numbered in increasing order of experimental values.

The surfaces $\gamma(sur)$ of Eq. (3.3.4) for the H_2O–$N_2(O_2)$ and H_2O–CO_2 ν_2 bands are shown in Figs 3.17 and 3.18.

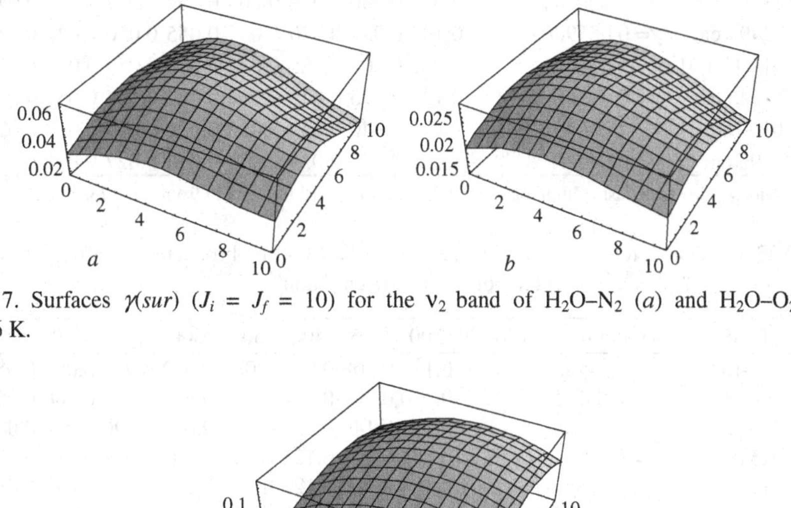

Fig. 3.17. Surfaces $\gamma(sur)$ (J_i = J_f = 10) for the ν_2 band of H_2O-N_2 (*a*) and H_2O-O_2 (*b*) at T = 296 K.

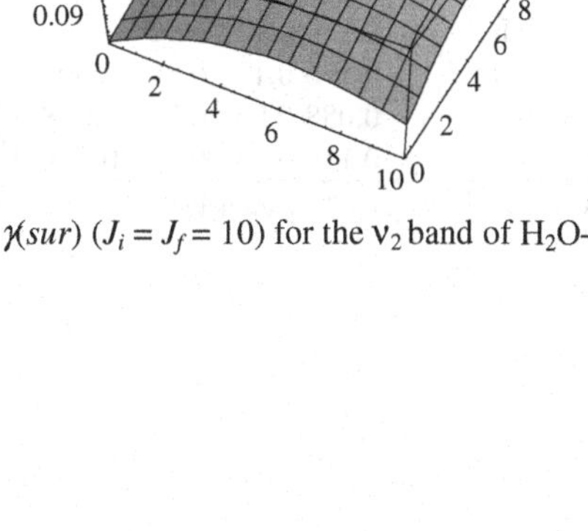

Fig. 3.18. Surface $\gamma(sur)$ (J_i = J_f = 10) for the ν_2 band of H_2O–CO_2 at T = 296 K.

The $\gamma(sur)$ parameters from Tables 3.20–3.21 can be further used, for example, in the calculation of broadening coefficients for the lines of the 1.4 μm band at various temperatures 200 K $\leq T \leq$ 500 K. This band was studied in Ref. [73] for temperatures between 400 and 1000 K. When the data of this reference are not included in the mean squares analysis, this calculation is a good test of the extrapolation properties of the $\gamma(sur)$ model. The calculated γ-values for some H_2O lines (with the minimal and the maximal quantum numbers K_a studied in Ref. [73]) are given in Tables 3.22, 3.23. The results of calculations with the usual temperature dependence $\gamma(T)$ of Eq. (3.2.1) with the parameters of Ref. [73] are also presented for comparison. The tables show a good general agreement of $\gamma(sur)$ with the results obtained in Ref. [73]. Slight deviations for the line 7185.596 cm^{-1} are due to the fact that the RB values [80] (T = 296 K) different from the data of Ref. [73] were used for fitting.

Table 3.22. Calculated temperature dependence of H_2O–N_2 line-broadening coefficients for the region of 1.4 μm (in cm^{-1} atm^{-1})*.

Line	Model for $\gamma(T)$, T(K)	200	296	400	500	600	700	800	900
[5 0 5]←[6 0 6]	$\gamma(sur)$	0.089	0.080	0.070	0.063	0.057	0.052	0.048	0.045
7117.2494 cm^{-1}	$\gamma = 0.0839(T_0/T)^{0.59}$	0.106	0.084	0.070	0.061	0.055	0.050	0.047	0.043
[4 1 3]←[5 1 4]	$\gamma(sur)$	0.129	0.102	0.085	0.074	0.065	0.059	0.054	0.050
7117.7546 cm^{-1}	$\gamma = 0.096(T_0/T)^{0.66}$	0.124	0.096	0.079	0.068	0.060	0.054	0.050	0.046
[6 6 0]←[6 6 1]	$\gamma(sur)$	0.058	0.046	0.039	0.035	0.032	0.029	0.028	0.026
7185.596 cm^{-1}	$\gamma = 0.047(T_0/T)^{0.64}$	0.060	0.047	0.039	0.034	0.030	0.027	0.025	0.023

* $\gamma(sur)$ parameters from the last column of Table 3.20, parameters of Eq. (3.2.1) from Ref. [73, 80]; T_0 = 296 K.

Table 3.23. Calculated temperature dependence of H_2O–CO_2 line-broadening coefficients for the region of 1.4 μm (in cm^{-1} atm^{-1})*.

Line	Model for $\gamma(T)$, T(K)	200	296	400	500	600	700	800	900
[5 0 5]←[6 0 6]	$\gamma(sur)$	0.133	0.108	0.091	0.080	0.072	0.066	0.062	0.058
7117.2494 cm^{-1}	$\gamma = 0.093(T_0/T)^{0.43}$	0.110	0.093	0.081	0.074	0.068	0.064	0.060	0.057
[4 1 3]←[5 1 4]	$\gamma(sur)$	0.171	0.140	0.120	0.107	0.098	0.090	0.084	0.080
7117.7546 cm^{-1}	$\gamma = 0.147(T_0/T)^{0.58}$	0.184	0.147	0.123	0.108	0.097	0.089	0.082	0.077
[3 2 2]←[3 2 1]	$\gamma(sur)$	0.184	0.150	0.127	0.112	0.102	0.093	0.087	0.081
7233.4233 cm^{-1}	$\gamma = 0.150(T_0/T)^{0.51}$	0.183	0.150	0.129	0.115	0.104	0.097	0.090	0.085
[4 3 2]←[4 3 1]	$\gamma(sur)$	0.154	0.124	0.104	0.092	0.083	0.076	0.070	0.065
7227.9829 cm^{-1}	$\gamma = 0.116(T_0/T)^{0.41}$	0.136	0.116	0.103	0.093	0.086	0.081	0.077	0.074
[6 6 0]←[6 6 1]	$\gamma(sur)$	0.088	0.068	0.056	0.048	0.042	0.038	0.034	0.031
7185.596 cm^{-1}	$\gamma = 0.086(T_0/T)^{0.74}$	0.115	0.086	0.069	0.058	0.051	0.045	0.041	0.037

* $\gamma(sur)$ parameters from the last column of Table 3.21, parameters for Eq. (3.2.1) from Ref. [73]; T_0 = 296 K.

3.6. Interference of water vapour spectral lines

The collision-induced interference of vibrotational spectral lines significantly influences the shape of absorption and Raman spectra of molecular gases at high densities. It is also observed in the atmospheric spectra of H_2O [89] composed of dense line manifolds. The analysis of transmittance in atmospheric windows and the description of spectral intensities in the line wings should therefore account for this interference effect. In the water vapour molecule, particular vibrational states are additionally mixed by various intermolecular resonances (Coriolis, Fermi, Darling–Dennison and more complex ones) [1–3], so that the relation between these resonances and the collisional interference should be carefully analysed.

In order to estimate the influence of the line interference on the atmospheric transmittance, calculations of broadening/shifting coefficients and cross-relaxation parameters were realised for two H_2O–N_2, H_2O–O_2 and H_2O–air lines in the 0.8 mm region ([6 3 4]←[5 4 1] line of the $8\nu_2$ band centred at 12414.2027 cm^{-1} and [6 5 2]←[5 4 1] line of the $3\nu_1+\nu_2$ band centred at 12413.9720 cm^{-1}) at atmospheric temperatures 200–300 K and pressures up to 10 atm. Both diagonal and off-diagonal elements of the relaxation matrix were calculated by the perturbative approach of Refs [90–92] within the impact approximation. In contrast with other studies of H_2O line interference, the resonance coupling of upper-state vibrotational wave functions was taken into account. Only the leading dipole–quadrupole interaction was retained in the interaction potential expression.

These calculations show that the line-mixing effect is in general small but for some lines involved in Coriolis or Fermi resonances becomes significant and results in a non-linear dependence of the line-shifting coefficient on pressure. The intramolecular resonances and the collisional line interference for these lines are therefore strongly related: the stationary vibrotational states always mixed by internal resonances are simultaneously perturbed by molecular collisions. As a result, the observed line shape deviates from Lorentzian profile, with a non-linear pressure-dependence of the line shift and an increased absorption in the atmospheric micro windows (up to several percent for the considered transitions). At high pressure (about 10 atm) the collisional broadening leads to significant overlapping and interference of almost all lines.

It should, however, be noted that the approximation of dominant dipole–quadrupole interaction is valid for transitions involving small values of the rotational quantum number. For the lines characterised by high J-values the corrections associated with the short-range forces can become significant. Since the expansion of the short-range potential over inverse powers of the intermolecular distance contains terms of various tensorial orders, the selection rules for the interfering lines should be completed by some new ones.

In the impact approximation, when the collision duration is short in comparison with the mean time interval between collisions, the absorption coefficient $K(\nu)$ defined up to a constant factor by Eqs (2.2.25) and (2.2.26) can be written as

$$K(\nu) = \frac{8\pi^2 \nu}{3\hbar c} n_a \operatorname{Im} I(\nu), \tag{3.6.1}$$

where n_a is the partial density of the molecular gas and

$$I(\nu) \approx Tr[\mu(\nu - L_0^a - \Lambda)^{-1} \mu\rho^a] \tag{3.6.2}$$

with $\Lambda \approx < [Tr(1 - S^{-1}S)\rho^b] >$. Here, as previously, μ, ρ^a and L_0^a are the dipole moment, density and Liouville operators of the active molecule, the relaxation operator Λ is averaged over the perturber states and molecular collisions, and the product $S^{-1}S$ involving the two-particle scattering matrix S is taken in the meaning of Sec. 2.3. To assess the absorption coefficient of Eq. (3.6.1) we should determine the scattering matrix, calculate the inverse of the relaxation matrix, and perform the trace operations. It should be noted that in the general case the line shape expression contains additional terms associated with the quadratic dependence of the line width on pressure. However, in Ref. [90] it was shown that these terms are small and can be omitted in Eq. (3.6.2).

In the case of the off-diagonal elements of the relaxation operator can be neglected, the matrix inversion is trivial and the absorption coefficient appears as a sum of contributions from individual lines of Lorentzian shape (cf. Eq. (2.2.29)). The line interference described by the off-diagonal elements can significantly modify the absorption at individual transitions (transfer of intensity between the lines). However, we shall consider the case of weak interference when the off-diagonal elements are much smaller than the diagonal ones, so that the matrix inversion in Eq. (3.6.2) can be performed using perturbation theory and series expansions. The spectral function $I(\nu)$ in this case can be written as a sum of contributions coming from individual lines numbered by the index k (transitions $i_k \to f_k$) [90, 91]:

$$\operatorname{Im} I_k(\nu) = \rho_k \mu_k^2 \frac{P_b \gamma_k + (\nu - \nu_k - P_b \delta_k - P_b^2 \operatorname{Re}\Delta_k) P_b Y_k}{(\nu - \nu_k - P_b \delta_k - P_b^2 \operatorname{Re}\Delta_k)^2 + (P_b Y_k)^2}, \tag{3.6.3}$$

where P_b is the partial pressure of the buffer gas (a binary mixture is considered), γ_k and δ_k are the broadening and shifting coefficients ($\Lambda_{kk} = iP_b\gamma_k - P_b\delta_k$),

$$Y_k = 2\sum_{l\neq k} \frac{\mu_l}{\mu_k} \frac{\operatorname{Re}\Lambda_{lk}}{\nu_k - \nu_l} \tag{3.6.4}$$

and

$$\operatorname{Re}\Delta_k = 2\sum_{l\neq k} \frac{\operatorname{Re}\Lambda_{kl}\operatorname{Re}\Lambda_{lk}}{\nu_l - \nu_k}.$$

In Eqs (3.6.3) and (3.6.4), ρ_k and $\mu_k = \langle i_k | \mu | f_k \rangle$ are the population of the lower transition level and the matrix element of the dipole moment operator, ν_k is the line position in vacuum, and $\Lambda_{kl} = \langle i_k i_l | \Lambda | f_k f_l \rangle$ is the off-diagonal matrix element of the relaxation operator connecting the lines k and l. From Eq. (3.6.3) it can be found that the observed shift of the absorption maximum of an individual line $\Delta\nu_k$ is defined by the relation

$$\Delta\nu_k = P_b\delta_k + P_b^2(\gamma_k Y_k / 2 + \operatorname{Re}\Delta_k) + O(P_k^3), \tag{3.6.5}$$

that is, in the presence of interference the line-shifting coefficient has a quadratic dependence on the buffer gas pressure which is defined by the off-diagonal elements of the relaxation operator.

In the calculation of matrix elements Λ_{kl} by perturbation theory for small values of the impact parameter b, the interruption approximation can be used. According to this approximation the relaxation matrix becomes diagonal for $b < b_0$ and, by analogy with the interruption functions $S_1(b)$, $S_2(b)$ of the ATC theory, we can introduce the interruption functions $\Theta_1(b)$, $\Theta_2(b)$ for the off-diagonal Λ_{kl} elements [92]. In particular, the function $\Theta_2(b)$ reads

$$\Theta_2(b) = \frac{1}{25\hbar^2 v^2 b^6} \sum_{2'} Q(22') \left\{ \sum_i D(ii'|1) D^*(ii''|1) \varphi(k_{ii'22'}, k_{ii''22'}) \delta_{ff'} \delta_{J_i J_{i'}} + \right.$$

$$\left. + \sum_f D(ff'|1) D^*(ff''|1) \varphi(k_{ff'22'}, k_{ff''22'}) \delta_{ii'} \delta_{J_f J_{f'}} \right\}, \tag{3.6.6}$$

where $D(ii'|1)$ are the reduced matrix elements of the H_2O dipole moment, $Q(22')$ are the strengths of quadrupole transitions in N_2 (or O_2) perturbing molecule, and $\varphi(k_{ii'22'}, k_{ii''22'})$ are the biparametric resonance functions of the resonance parameter

$$k_{ii'22'} = \frac{2\pi c b}{v}(\omega_{ii'} + \omega_{22'}) \tag{3.6.7}$$

dependent on the frequencies of virtual transitions in both molecules.

According to the numerical estimations of the interference effect for the water vapour lines [93], for the line broadening this effect is very small and can be neglected (at least, for the purely rotational spectrum). The weak influence of the interference is explained by the fact that the H_2O molecule is a light asymmetric top with low values of moments of inertia and, consequently, with large separations between the rotational energy levels. As a result, the transition frequencies $\omega_{ii'}$ and $\omega_{ii''}$ in Eq. (3.6.7) differ strongly, so that the resonance functions $\varphi(k_{ii'22'}, k_{ii''22'})$ and the Λ_{kl} elements have small values.

An analysis of Ref. [93] has, however, been performed for the purely rotational spectrum of H_2O, and its conclusions can not be directly applied to the transitions to excited vibrational states. Indeed, for the IR lines, and especially in the near infrared and visible regions, the rotational structure becomes denser. In addition, the vibrational dependence of molecular constants (rotational and centrifugal parameters, dipole moment) and the accidental Coriolis, Fermi, and Darling–Dennison resonances (arising between adjacent rotational levels of different vibrational states) can become very significant.

As a rule, the Λ_{kl} elements are defined by the anisotropic part of the interaction potential. The isotropic part and the term $\Theta_1(b)$ generally give a very small contribution in the interference of individual vibrotational H_2O lines but in some cases provoke an interference of vibrational bands [94]. Further analysis is consequently realised for the case when the term $\Theta_2(b)$ and the dipole–quadrupole interaction are dominant. For line-shift calculations, however, the isotropic part of the induction and dispersion interactions is also accounted for.

According to Eq. (3.6.6) the quantum numbers of interfering lines should satisfy some selection rules, namely: the upper states of the corresponding transitions should have the same quantum numbers J whereas the other vibrational (v_1, v_2, v_3) and rotational (K_a, K_c) quantum numbers can be different. Interference can be observed for transitions from the same lower state. The interference effects related with the fundamental vibrational state can be neglected for H_2O, so that only the second term should be kept in the braces of Eq. (3.6.6). Moreover, the following conditions should be satisfied.

i) Since the upper states should be coupled by dipole transitions to a common virtual state they must have the same symmetry and $J_{f'} = J_{f''}$.
ii) Since the resonance functions $\varphi(k, k')$ decrease rapidly with $|k - k'|$ increasing, the frequencies $\omega_{ff'}$ and $\omega_{ff''}$ should be close; that is, the difference $\Delta E = E_{f'} - E_{f''}$ should be small.

iii) The reduced matrix elements $D(ff'|1)$, $D^*(ff''|1)$ corresponding to the transitions from the common level f to the interfering levels f' and f'' should have rather large values to provoke interference. The wave functions Ψ_f, $\Psi_{f'}$ and the dipole moment should therefore have a large overlapping area.

In light polyatomic molecules like H_2O (with high vibrational frequencies and large values of rotational constants), such an overlapping of close levels of the same symmetry is possible only if they belong to different vibrational states and are coupled by some accidental resonance strongly mixing the wave functions. For two states coupled by a resonance their wave functions can be written as

$$\Psi_1 = c\varphi_1 + s\varphi_2,$$
$$\Psi_2 = -s\varphi_1 + c\varphi_2, \tag{3.6.8}$$

where $c = \cos\beta$, $s = \sin\beta$, $\beta = \mathrm{arctg}\,[2H_{12}/(H_{11}-H_{22})]$ and H_{11}, H_{22}, H_{12} are the matrix elements of the molecular Hamiltonian calculated with the wave functions φ_1, φ_2 of the isolated states; in the case of an exact resonance $c = s = 1/\sqrt{2}$. The calculation of the reduced matrix elements with the wave functions of Eq. (3.6.8) and of their product in the braces of Eq. (3.6.6) gives

$$D(ff'|1)D^*(ff''|1) = cs\mu^2\langle\varphi_1|K_{xx}|\varphi_1\rangle, \tag{3.6.9}$$

where K_{xx} is the element of the cosine matrix of Eq. (3.6.8). The conditions i–iii are fulfilled, for example, for the rotational levels schematised on Fig. 3.19 [95].

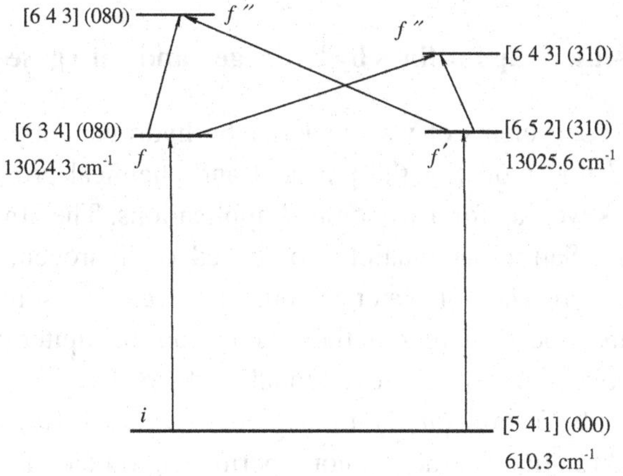

Fig. 3.19. Upper rotational levels of two H_2O lines [6 5 2] (310) ← [5 4 1] (000) and [6 3 4] (080) ←[5 4 1] (000) are coupled by accidental resonances with the levels [6 4 3] (080) and [6 4 3] (310), respectively [95]. Reproduced with permission.

To calculate the diagonal and off-diagonal elements of the relaxation matrix for two considered transitions, the rotational energy levels obtained by Partridge and Schwenke [19] were used for the vibrational state (080); for the vibrational state (310) they were determined through the diagonalization of the effective rotational Hamiltonian matrix. The wave functions of the isolated states were calculated as the eigenfunctions of the effective rotational Hamiltonian without molecular resonances. The mixing coefficient was determined from the observed line intensities. The polarizabilities α_{080} and α_{310} in the excited vibrational states were estimated through the equation

$$\alpha(v_1, v_2, v_3) = \alpha(0,0,0) + \alpha_1 v_1 + \alpha_2 v_2 + \alpha_3 v_3 . \tag{3.6.10}$$

The relaxation parameters for perturbation by air were calculated as

$$X(air) = 0.79\, X(N_2) + 0.21 X(O_2) \tag{3.6.11}$$

with $X = \gamma_k$, δ_k, Y_k, or W_{kl}. Other molecular parameters used in the calculations can be found in Table 3.4 and Appendix F. The obtained values of the line broadening and line-shifting coefficients as well as of the non-linear pressure dependence coefficients Y_k and W_{kl} elements [95] are presented in Table 3.24. As can be seen from this table, two studied lines are characterised by a strong interference effect: the coefficients Y_k have values about 0.02 cm^{-1} atm^{-1} at room temperature. These large values are due to a strong mixing of the wave functions for the upper states (mixing coefficient $s^2 = 0.4$) and a small separation of the line positions (1 cm^{-1}).

3.7. Broadening of water vapour lines by hydrogen and rare gases

The line-broadening coefficients of water vapour by hydrogen, argon or helium pressure are required for studies of the physical and chemical processes in the Earth's atmosphere as well as for astronomical applications. The atmospheres of giant planets (Jupiter, Saturn) are mainly composed of hydrogen, helium and methane, whereas the presence of water vapour is detected in some frequency regions of their infrared spectrum (e.g. in the 5μm region for Jupiter [96]).

Besides the practical interest, the collisional partners He, Ne, Ar, Kr, and Xe are worthy of study from a purely fundamental point of view. They have neither an internal vibrotational structure nor a permanent dipole moment, so that the long-range part of the interaction potential is completely defined by the induction and dispersion terms which can be examined in detail. The analysis of the short-range part, in its turn, enables an estimation of the trajectory curvature influence on the line shape parameters.

Table 3.24. Relaxation parameters (in 10^{-3} cm^{-1} atm^{-1}) and temperature exponents for the H_2O lines [6 3 4] (080)←[5 4 1] (000) and [6 5 2] (310)←[5 4 1] (000) broadened by N_2, O_2 and air [95]*. Reproduced with permission.

Perturber	Parameter	Temperature, K 200	230	260	297	330	n
		Line [6 3 4] (080) ← [5 4 1] (000)					
N_2	W_{kl}	–4.6	–4.0	–3.5	–3.1	–2.8	0.97
	Y_k	41	35	31	27	24	1.01
	γ_k	79	75	71	67	63	0.47
	δ_k	–49	–40	–34	–28	–24	1.43
O_2	W_{kl}	–3.6	–3.0	–2.6	–2.3	–2.0	1.11
	Y_k	31	26	23	20	17	1.17
	γ_k	34	32	30	28	27	0.47
	δ_k	–73	–63	–55	–48	–43	1.05
Air	Y_k	38	33	29	25	22	1.04
	γ_k	69	66	62	59	55	0.47
	δ_k	–54	–45	–38	–32	–28	1.30
		Line [6 5 2] (310) ← [5 4 1] (000)					
N_2	γ_k	87	81	78	73	70	0.44
	δ_k	–26	–21	–17	–14	–12	1.52
O_2	γ_k	37	34	33	31	30	0.39
	δ_k	–47	–40	–35	–30	–27	1.10
Air	γ_k	76	71	68	64	62	0.40
	δ_k	–30	–25	–21	–17	–15	1.43

* For the line [6 5 2] (310)←[5 4 1] (000) the parameters W_{kl} are the same as for the line [6 3 4] (080)←[5 4 1] (000) while the parameters Y_k have opposite signs.

3.7.1. Broadening of H_2O vibrotational lines by hydrogen

Experimental investigations of H_2O–H_2 line-broadening coefficients in the regions of 5, 2.7and 1.4 μm were realised in Refs [61, 72, 96, 97]. In Ref. [97], their room-temperature values are given for 630 lines ($J \leq 11$, $K_a \leq 6$) of the rotational spectrum (55–328 cm^{-1}) and the vibrational bands ν_2 (1260–2041 cm^{-1}), ν_1 (3320–3975 cm^{-1}) and ν_3 (3480–4045 cm^{-1}). These measured broadening coefficients range from 0.101 to 0.033 cm^{-1} atm^{-1}. When passing from the purely rotational transitions to those of the ν_2 band, the variation of γ is about 3%, but the excitation of the symmetric stretching vibration ν_1 provokes a changing of γ by up to 18%.

The broadening coefficients of rotational transitions from the region of 380–600 cm^{-1} were measured by Steyert *et al.* [98]. For the hydrogen-broadened H_2O

lines these coefficients vary between 0.033 and 0.076 cm^{-1} atm^{-1} with the mean value of 0.056 cm^{-1} atm^{-1}.

The most complete theoretical calculations of γ coefficients were performed in Ref. [96] in the framework of the semi-classical RB formalism. In total, 386 lines of the rotational H_2O band with $J \leq 14$ and $K_a \leq 8$ were considered. Moreover, for 33 lines the calculations were made for $T = 200$, 296, 400 and 750 K, in order to determine the temperature exponents of Eq. (3.2.1). A very good agreement (2–4%) was obtained with the experimental data of purely rotational and ν_2 bands. However, the calculated values show large discrepancies in comparison with the experimental values of Ref. [99]. The mean theoretical value was estimated at about 0.7 (in function of the interaction type) whereas the calculated values varied strongly from 0.19 to 0.71. Since a changing of the temperature exponent from 0.5 to 0.7 induces a 5% error in the gas concentration retrieval for some altitudes [96], the use of the mean n-value or the extrapolation of n obtained for some lines to other lines can hardly be justified.

For many applications the γ-values should be known for the lines with $J > 14$ and $\Delta K_a > 3$. A polynomial approximation for them was proposed in Ref. [96]:

$$\gamma = b_0 + b_1[(J_i + J_f)/2] + b_2[(J_i + J_f)/2]^2 + b_3[(K_{a\,i} + K_{a\,f})/2], \qquad (3.7.1)$$

where the adjustable parameters $b_0 = 0.093121(0.002)$, $b_1 = -0.71920(0.06)\cdot 10^{-2}$, $b_2 = 0.24511(0.04)\cdot 10^{-3}$, $b_3 = 0.43104(2)\cdot 10^{-4}$ (in cm^{-1} atm^{-1}, within one standard deviation given in parentheses) were obtained by least-squares fitting to calculated values. It should be mentioned that like any polynomial representation Eq. (3.7.1) has a limited application domain and leads to erroneous results for asymptotics of high J- and K_a-values.

For five H_2O–H_2 lines from the vibrational bands $\nu_1 + \nu_3$ and $2\nu_1$ (region of 1.4 μm) calculations of line-broadening and line-shifting coefficients were made [59] by the semi-empirical method described in Section 3.4. The results are presented in Table 3.25 together with the corresponding experimental data [61].

Table 3.25. Semi-empirical (SE) and experimental [61] values of H_2O–H_2 line-broadening and line-shifting coefficients (in 10^{-3} cm^{-1} atm^{-1}).

$J_i K_{a\,i} K_{c\,i}$	$J_f K_{a\,f} K_{c\,f}$	$(v_1 v_2 v_3)_f$	γ^{expt}	δ^{expt}	γ^{SE}	δ^{SE}
1 0 1	1 1 0	(200)	84 ± 8	−14 ± 1	84.7	−14.5
2 1 2	3 1 3	(101)	73 ± 8	−15 ± 1	74.0	−12.3
3 0 3	3 2 2	(101)	74 ± 8	−15 ± 1	76.3	−15.8
5 1 5	5 1 4	(101)	69 ± 8	−12 ± 1	69.1	−11.4
6 6 0	6 6 1	(101)	43 ± 8	−10 ± 1	43.0	−10.5

3.7.2. Broadening of H_2O vibrotational lines by rare gases

Line shape parameters of water vapour perturbed by He, Ne, Ar, and Kr were experimentally studied in Refs [26, 61, 71, 72, 79, 96, 98, 100–104].

The most complete results for the line-broadening coefficients were obtained for the ν_2 band [71] (H_2O–Ar, H_2O–He, H_2O–Ne, H_2O–Kr) and the regions of 0.72 μm [101] (H_2O–Ar) and 1.4 μm [61, 72] (H_2O–He). It was found that the broadening of H_2O lines by argon is much weaker that their broadening by nitrogen and oxygen, whereas the line shifts induced by these perturbers in the ν_2 band have the same order of magnitude. Recently, Steyert *et al.* [98] obtained the helium-broadened coefficients for pure rotational $H_2{}^{16}O$ lines in the region 380–600 cm^{-1} which range from 0.006 to 0.017 cm^{-1} atm^{-1}, with the mean value of 0.013 cm^{-1} atm^{-1}. The temperature dependence of the 556 GHz line in the temperature range 17–200 K was recently studied by Dick *et al.* [79].

A remarkable feature of these studies consisted in the systematic observation of line narrowing (Dicke effect). In Ref. [100], this effect was clearly observed for the doublet [14 1 14]→[15 0 15]/[14 0 14]→[15 1 15] with argon pressure increasing from 4 to 85 mbar (the line interference was excluded as forbidden by the symmetry of the doublet lines). The observed dependence of the line-width γ on the perturber pressure P followed the relation [100]

$$\gamma = AT^{\alpha}/P + BT^{-\beta}P, \tag{3.7.2}$$

where the first and second terms respectively describe the Dicke narrowing and the collisional broadening; the adjustable parameters A, B, α and β are determined by fitting to experimental data. The minimal line width is given by

$$\gamma_{min} = 2(AB)^{1/2}T^{-(\alpha-\beta)/2}. \tag{3.7.3}$$

The authors analysed the experimental line shapes using the Voigt model and separately determined the dependences of Doppler (γ_D) and collisional (γ_c) line widths on pressure. It was shown that the Doppler width decreases rapidly with increasing pressure and the line width has its minimum at P = 85 mbar. No numerical estimation of the narrowing coefficient was made.

Quantitative results for these coefficients were obtained in Ref. [71] for R-branch lines of ν_2 band (1850–2140 cm^{-1}) at He, Ne, Ar, Kr, and N_2 pressures corresponding to collisional narrowing regime (between 50 and 300 Torr, in function of the perturber); the H_2O pressure was about 1 Torr. Under these conditions a simple Voigt profile is not able to reproduce the experimental line shapes. The line-narrowing coefficients β^0 were determined for the model profiles of

Galatry (soft collisions) and Rautian–Sobel'man (hard collisions), but with the additional inclusion of the dependence on the velocity v_a of the absorbing molecule. Namely, the line shape was supposed to be defined by a weighted manifold of Lorentzian profiles corresponding to different classes of velocity v_a, so that the absorption coefficient $K(\nu)$ was written as

$$K(\nu) = \frac{S_0 P_{H_2O}}{\pi} \int_0^\infty \frac{\gamma(v_a)}{\gamma(v_a)^2 - (\nu - \nu_0)^2} f(v_a)\, dv_a , \tag{3.7.4}$$

where S^0 is the integral line intensity per unit pressure, $f(v_a)$ is the Maxwell–Boltzmann distribution of velocities v_a and the collisional line-width

$$\gamma(v_a) = \int_0^\infty f(v|v_a) C v^n dv$$

depends on the relative velocity v as Cv^n (C is a constant) and on the distribution $f(v|v_a)$ of relative velocities of perturbing particles with respect to the absorbing molecule moving at v_a. The function $\gamma(v_a)$ was chosen as

$$\gamma(v_a) = \gamma(0)\, M(3/2 + n/2,\, 3\eta^2/2)\exp(-\eta^2), \tag{3.7.5}$$

where $M(\ldots)$ is a hypergeometric function with $n = (q-3)/(q-1)$ for a collision interaction dependent on the intermolecular distance r as r^{-q} and $\eta = v_a[m_b/(k_B T)]^{1/2}$ (m_b is the perturber mass and k_B is the Boltzmann constant). For $v_a = 0$ this function is related with the usual thermal average γ_c by

$$\gamma(0) = \gamma_c[m_a/(m_a + m_b)]^{n/2}. \tag{3.7.6}$$

The profile of Eq. (3.7.4) was further convoluted with the soft-collision (Galatry) and hard-collision (Rautian–Sobel'man) profiles (with $\gamma_c = 0$), and γ_c, S_0, β_0 were deduced by fitting to experimental line shapes. The collisional narrowing coefficients β_0 were compared to the dynamic friction coefficient β_{diff} related to the diffusion coefficient D. It was found that the coefficients $\beta_0{}^G$ are systematically greater than $\beta_0{}^R$ by about 10%, whereas the differences between γ-values are approximately 1%. The mean (over all lines) β_0-values from Ref. [71] are given in Table 3.26. A good agreement can be stated between $\beta_0{}^G$, $\beta_0{}^R$, and β_{diff}. An analysis of γ-values showed that they strongly depend on the model profile used for fitting: when passing from He to Kr the difference between γ_V (Voigt profile) and γ_G or γ_R can reach 50%.

Line-width and line-shift calculations were performed for frequency ranges where experimental data were available. The most complete results concern the case of Ar-broadening. In Ref. [103], γ and δ CRB calculations were made for ν_1,

ν_2, $3\nu_3$, $3\nu_1 + \nu_3$, and $2\nu_1 + 2\nu_2 + \nu_3$ vibrational bands, leading to a good general agreement with experimental results. The influence of the truncation procedure for the atom–atom potential and the influence of the imaginary part of $S(b)$ on the line widths and line shifts were also discussed. The vibrationally dependent matrix elements $\langle i|V_{iso}|i\rangle$ and $\langle f|V_{iso}|f\rangle$ needed for the line-shift calculation were evaluated with the long-range induction and dispersion potentials given by Eq. (2.8.13), with the vibrational dependence coming from the dipole moment and polarizability of the active molecule. The trajectories were governed by the isotropic part of the pair-wise atom–atom potential similar to Eq. (2.8.12) fitted to a Lennard–Jones form. It was mentioned that the isotropic induction-dispersion contributions formally correspond to the same forces as the attractive part of the isotropic atom–atom potential (so that they are tacitly included in the trajectory calculations) and their magnitude is about half the value of this isotropic atom–atom term. For the H_2O–Ar system the atom–atom potential was expanded to the 12th order to insure its convergence and a good agreement of calculations with experimental results. The imaginary components of the Liouville scattering matrix strongly affected both line-shift and line-width calculations: the contribution from the imaginary part varied from line to line reaching 37%. For some lines with high J- and small K_a-values the CRB widths were two or three times greater than the experimental ones, and the authors ascribed this fact to the line-narrowing effect. A need for corrections for velocity-changing collisions and more accurate determination of the intermolecular potential parameters were evoked.

Table 3.26. Mean values of the narrowing coefficients for H_2O ν_2 lines in comparison with the dynamic friction coefficient (in 10^{-3} cm^{-1} atm^{-1}) [71]. Reproduced with permission.

Coefficient	He $P = 101$ Torr	Ne $P = 100$ Torr	Ar $P = 50$ Torr	Kr $P = 68$ Torr
β_{diff}	8.0	18	32	42.0
β_0^R	10.8	21.8	32.4	45.6
β_0^G	12.0	24.3	36.2	50.5

Calculations of H_2O–Ar line-shift coefficients for the $\nu_1 + \nu_2$ and $\nu_2 + \nu_3$ bands were realised in Ref. [102]. The ATC approach used by the authors led to a satisfactory agreement with experimental data.

Line-broadening coefficients for H_2O–He were calculated in Refs [96, 104]. The case of ν_2 band lines broadened by Ar, He, Ne, and Kr was studied in Ref. [105] by ATC, RB and RB′ approaches. For the H_2O dipole moment the approximation $\mu(V, J, K_a, K_c) = \mu(V, J, K_a) = \mu(V, J, K)$ was used, and the function $\mu(V, J, K)$ was calculated by Eq. (3.1.9). The mean polarizability $\alpha(V, J, K_a, K_c)$ in a given vibrotational state was approximated by $\alpha(V, J, K_a, K_c) = \alpha(V, J, K_a) = \alpha(V, J, K) = \alpha(V) + \Delta\alpha^{(J)}J(J+1) + \Delta\alpha^{(K)}K^2$ with $\alpha(V)$ of Ref. [17] (*ab initio* values) and the parameters $\Delta\alpha^{(J)}$ and $\Delta\alpha^{(K)}$ of Table. 3.4. These calculations showed that the line-broadening coefficients obtained by the ATC and RB′ methods depend weakly on the rotational quantum numbers. Figure 3.20 gives an example of their comparison for the ν_2 band of H_2O–Ar system (the lines are numbered in decreasing order of experimental γ-values).

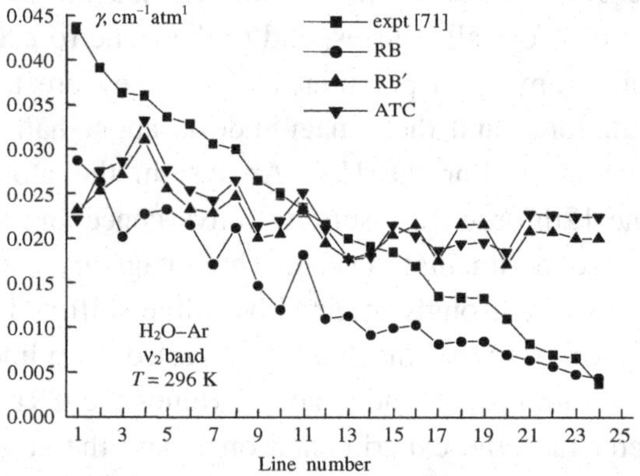

Fig. 3.20. Comparison of H_2O–Ar line-broadening coefficients calculated by the semi-classical ATC, RB and RB′ methods for the ν_2 band [105]. Reproduced with permission.

According to this figure the RB method provides much better results than the other approaches but significantly underestimates the line broadening for small values of the rotational quantum numbers (corresponding to high γ-values). This underestimation can be explained by the fact that in the RB method the γ coefficients are strongly dependent on the input Lennard–Jones parameters σ and ε which govern the relative molecular trajectory. When the literature value $\sigma = 3.45$ Å [52] (used together with $\varepsilon/k_B = 108$ K for Fig. 3.20) is reduced to 3.4 Å, the RB curve suffers a parallel shift up, so that for small rotational quantum numbers a good agreement with the experimental data is achieved and for high transition numbers an overestimation of γ-values is stated. These calculations also showed that the rotational dependence of the mean H_2O polarizability does not influence the line-broadening coefficients. Analogous conclusions applied to the broadening by He, Ne, and Kr. The RB-calculated γ-values are compared with the corresponding experimental data in Table 3.27.

Table 3.27. RB-computed line-broadening coefficients for the ν_2 band of H_2O in comparison with their experimental values [71] (in 10^{-3} cm^{-1} atm^{-1})*. Reproduced with permission from Ref. [105].

JK_aK_c		H_2O–Ar			H_2O–He			H_2O–Ne			H_2O–Kr		
f	*i*	expt	RB	*sur*	expt	RB	*sur*	expt	RB	*sur*	expt	RB	*sur*
3 1 3	2 2 0		42.6	33.7		22.7	19.5		27.4	25.5		47.7	41.7
4 1 4	3 2 1		40.4	32.5		21.4	18.7		25.8	23.7		45.5	41.3
4 4 1	3 1 2	36.0	32.0	34.5		17.5	18.4		20.7	22.0	40.5	36.2	42.9
5 4 1	4 1 4	44.1	38.7	32.4		20.1	17.5		24.7	20.1		43.9	39.8
5 4 1	4 3 2		27.1	25.3		15.8	15.9		17.9	17.6	23.7	30.0	28.6
5 4 2	4 1 3	34.6	29.0	32.4		18.1	17.5		19.2	20.1	45.3	32.7	39.8
6 1 6	5 2 3		37.8	30.1		19.7	17.2		23.2	19.9		42.9	37.0
6 4 2	6 1 5		28.0	29.6		16.2	16.2		18.1	18.2	37.0	32.3	34.3
7 3 5	6 0 6	31.9	31.5	29.4		16.6	16.2		20.0	18.8		36.2	34.7
7 3 4	7 0 7	39.1	36.9	28.8		18.4	15.8		23.2	18.3		42.2	33.6
7 4 4	6 1 5	30.5	26.0	29.4		15.5	15.8		16.8	17.7	33.2	29.9	32.8
8 2 6	7 1 7	34.8	33.9	28.8		16.9	15.8		21.2	17.6		39.1	32.6
8 5 4	7 2 5		21.6	27.3		13.1	14.5		13.8	16.1		25.3	28.9
9 2 7	8 1 8	30.0	31.7	26.6		15.7	15.0		19.8	16.5		36.7	31.5
9 3 7	8 0 8	23.2	28.2	25.2		14.5	14.5		17.7	16.6	24.3	32.7	31.5
9 6 4	8 5 3	16.9	16.5	16.4	11.1	11.2	10.6	11.2	11.4	12.2		18.2	22.4
9 7 3	8 6 2	13.4	13.8	12.6	9.2	10.3	9.5	9.8	10.1	11.4			20.4
9 2 7	10 3 8	18.5[a]	20.2	24.6		12.1	13.8		12.7	14.0		23.6	28.7
9 3 6	10 4 7	24.3[a]	21.9	23.5		11.6	12.8		13.2	13.0		26.0	26.8
9 5 5	10 6 4	14.4[a]	15.5	16.7		10.7	10.2		10.6	11.3		17.1	23.2
10 3 7	9 2 8		28.9	24.6		14.1	13.8		10.7	14.0	33.1	34.0	28.7
10 4 6	9 3 7	26.5	23.8	23.5		12.1	12.8		14.5	13.0		28.0	26.8
10 5 5	9 4 6	20.0	18.3	20.6	13.2	11.3	11.5	12.5	11.9	12.1		20.8	25.0
10 6 5	9 5 4	18.5	16.3	16.7	11.7	10.7	10.2	11.8	10.9	11.3		18.4	23.2
10 7 4	9 6 3		13.3	12.8		10.0	9.1			10.5	15.2		21.5
11 2 10	12 1 11	7.7[a]	13.4	14.8		9.8	12.4			11.0			24.2
11 4 8	10 3 7	25.2	20.5	22.3		10.9	12.3		12.3	11.4	28.1	24.5	25.4
11 5 6	10 4 7		18.6	20.4	13.4	10.8	11.0	12.8	11.7	10.6		21.4	23.9
12 2 10	11 3 9	13.5	14.2	18.2	11.5	10.1	12.4	9.5	9.6	10.2	14.2	15.8	22.8
12 4 9	11 3 8	21.6	18.8	20.1		10.2	11.7		11.4	9.5	24.0	22.3	21.4
13 2 12	12 1 11	8.2	12.7	10.3	9.0	9.1	11.4	6.0	9.1	8.7	9.6	13.6	16.5
13 2 11	12 3 10	11.0	12.6	13.9	10.0	9.3	11.6	7.8	8.7	8.1	11.0	13.9	15.4
13 1 12	14 2 13	5.7[a]	11.9	6.6		8.5	10.3			6.9			8.4
13 2 12	14 1 13	5.9[a]	11.9	6.6		8.4	10.3			6.9			8.4
14 1 13	13 2 12		11.9	6.6		8.5	9.4			6.9	4.7		8.4
14 2 13	13 1 12	7.0	12.0		8.4	8.4	8.5	5.4		6.9	4.7		8.4
14 0 14	15 1 15	3.4[a]	11.9	2.2			9.4			5.9			4.6
14 1 14	15 0 15												
15 0 15	14 1 14	6.6	11.8	2.2	7.8	8.1	8.5	4.7	9.2	5.9			4.6

* For the RB calculations the parameters $\sigma = 3.4$Å (H_2O–Ar), $\sigma = 3.05$Å (H_2O–He), $\sigma = 3.1$Å (H_2O–Ne) and $\sigma = 3.48$Å (H_2O–Kr) were used, other parameters can be found in Refs [52, 53]; [a] the experimental data are from Ref. [100].

3.7.3. Modelling of calculated and experimental H_2O line widths

The RB line-broadening coefficients and their available experimental values for H_2O-Ar, -He, -Ne and -Kr systems were used for the determination of the $\gamma(sur)$ parameters (Eq. (3.3.4)). For the ν_2 band of H_2O-Ar the parameters n_{kl} describing the temperature dependence of $\gamma(sur)$ (Eq. (3.3.8)) were also obtained.

3.7.3.1. H_2O–Ar system

The model parameters for Eqs (3.3.4), (3.3.6) and (3.3.8) were determined for the bands ν_2, $3\nu_1+\nu_3$, and $2\nu_1+2\nu_2+\nu_3$. For the combination bands the experimental data [101] (transitions with $\Delta K_a = 0$) were used as $\gamma_i(u)$ in Eq. (3.3.7) ($N = 79$ and $N = 31$, respectively). The obtained $\gamma(sur)$ parameters are listed in Table 3.28, and the calculated γ-values for the $3\nu_1+\nu_3$ band are shown in Fig. 3.21 (for the $2\nu_1+2\nu_2+\nu_3$ band the results are similar).

Table 3.28. Parameters of $\gamma(sur)$ model for H_2O–Ar at 296 K* [105]. Reproduced with permission.

	ν_2	$3\nu_1+\nu_3$	$2\nu_1+2\nu_2+\nu_3$		ν_2
x_{20}	$0.454(7)\cdot10^{-1}$	$0.646(12)\cdot10^{-1}$	$0.625(16)\cdot10^{-1}$	n_{20}	$-0.6395(46)$
x_{21}	$-0.1089(36)\cdot10^{-2}$	$-0.1329(130)\cdot10^{-2}$	$-0.1827(180)\cdot10^{-2}$	n_{21}	$-1.250(21)$
x_{30}	$0.767(80)\cdot10^{-1}$	$0.1989(120)$	$0.1891(207)$	n_{30}	$-1.217(35)$
x_{31}	$0.750(39)\cdot10^{-2}$			$n_{31}=n_{30}$	
x_{40}	$-3.925(290)$	$-0.793(270)$	$-0.793(270)^{\#}$	n_{40}	$0.699(42)$
x_{41}	$0.335(13)$	$0.2254(220)$	$0.2014(220)$	n_{41}	$0.0^{\#}$
x_{50}	$0.552(19)\cdot10^{-1}$			n_{50}	$-0.894(27)$
x_{51}	$-0.160(11)\cdot10^{-2}$			$n_{51}=n_{50}$	
N	1141	79	31	N	8450
L	8	5	4	L	13
σ	$7.4\cdot10^{-5}$	$3.4\cdot10^{-4}$	$4.8\cdot10^{-4}$	σ	$2.4\cdot10^{-5}$
	$\Delta K_a \le 5$	$\Delta K_a = 0$	$\Delta K_a = 0$		$\Delta K_a \le 5$

* Parameters x_{20} and x_{21} as well as σ are in cm^{-1} atm^{-1}, parameters denoted by # were kept fixed.

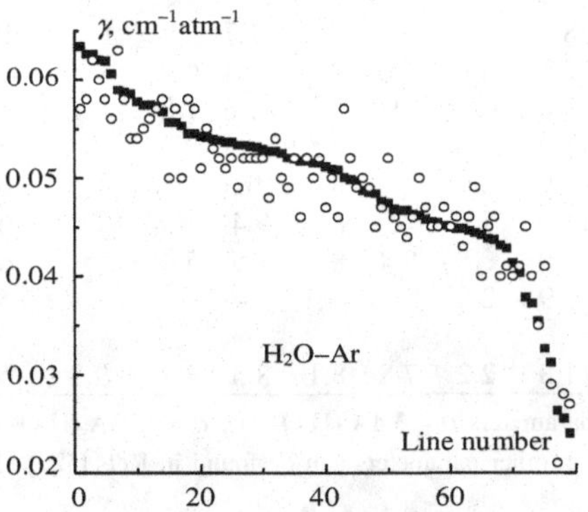

Fig. 3.21. Comparison of H_2O–Ar $\gamma(sur)$ coefficients (empty circles) with experimental data [101] (dark squares) for the $3\nu_1 + \nu_3$ band (parameters of Table 3.28) [105]. Reproduced with permission.

For the ν_2 band the available experimental data of Refs [100, 103, 106] (N~30) were not sufficient to extract all the parameters of Eqs (3.3.4) and (3.3.6), so that the theoretical line-broadening coefficients obtained by the RB method were added to 25 experimental values to form the set $\gamma(u)$ used for fitting. The corresponding model parameters are given in Table 3.28 and the quality of fitting to the available experimental data can be seen from Table 3.27. The set of parameters from Table 3.28 enables the calculation of line-broadening coefficients for about 1200 transitions ($J \le 15$) at $T = 296$ K.

For the same ν_2 band, RB line widths were also calculated for the temperature range 300–2500 K. These values were further used to determine the parameters of Eq. (3.3.8) which are responsible for the temperature dependence of $\gamma(sur)$ (Table 3.28). In Ref. [107], the γ-values for the ν_2 [12 3 10]←[11 2 9] line were experimentally studied at 1300–2300 K, and the dependence

$$2\gamma(T) = 0.027(T/1300)^{-0.90} \tag{3.7.7}$$

was obtained. Figure 3.22 shows a comparison of the line-broadening coefficients calculated by Eq. (3.7.7) and by the $\gamma(sur)$ model with parameters of Table 3.28. As can be seen from this figure, the $\gamma(sur)$-values are much greater than the experimental ones, like the conclusions of Ref. [107].

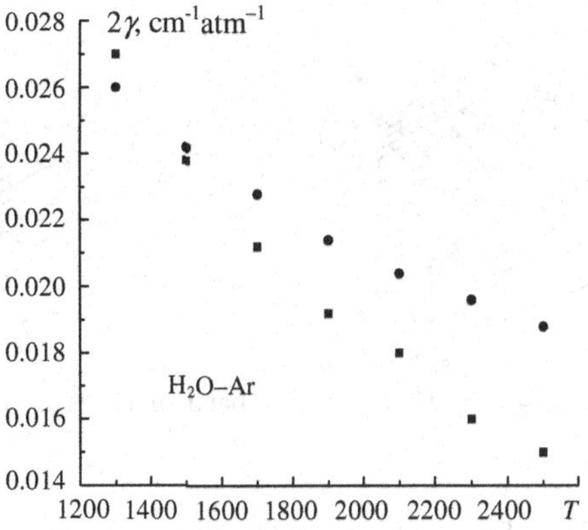

Fig. 3.22. Comparison of H_2O–Ar line-broadening coefficients calculated by $\gamma(sur)$ model (dark circles) and experimental data [107] (dark squares) for the line [12 3 10]←[11 2 9] of the ν_2 band. Reproduced with permission from Ref. [105].

3.7.3.2. H_2O–He, H_2O–Ne, and H_2O–Kr systems, the ν_2 band

The number of experimental data [71] was insufficient for the determination of $\gamma(sur)$ parameters, so they were completed by RB calculations. The obtained parameters are given in Table 3.29, and the quality of fitting can be easily seen from Table 3.27. This quality is quite good for H_2O–He but is less so for H_2O–Kr. The general agreement between $\gamma(sur)$, RB-calculated and experimental line-broadening coefficients is very satisfactory. Figure 3.23 shows the surface $\gamma(sur)$ for the ν_2 band of H_2O–Ar and H_2O–Kr.

Table 3.29. Parameters of $\gamma(sur)$ model for H_2O lines of ν_2 band broadened by He, Ne [105] and Kr at 296 K*.

	H_2O–He	H_2O–Ne	H_2O–Kr
x_{10}	$0.648(14)\cdot10^{-2}$	$0.493(31)\cdot10^{-2}$	$0.163(8)\cdot10^{-1}$
x_{20}	$0.160(3)\cdot10^{-1}$	$0.2094(42)\cdot10^{-1}$	$0.974(39)\cdot10^{-1}$
x_{21}	$-0.4323(96)\cdot10^{-3}$		$-0.394(15)\cdot10^{-2}$
x_{30}	$0.687(140)\cdot10^{-1}$	0.261(15)	0.181(12)
x_{31}	$0.849(72)\cdot10^{-2}$	$-0.186(14)\cdot10^{-1}$	
x_{32}		$0.477(39)\cdot10^{-3}$	
x_{40}	−1.50(39)	−4.71(23)	−4.956(35)
x_{41}	0.152(15)	0.765(38)	−0.130(34)
x_{42}			$0.150(18)\cdot10^{-1}$
x_{50}	$0.278(92)\cdot10^{-1}$	$0.239(88)\cdot10^{-1}$	$0.887(42)\cdot10^{-1}$
x_{51}			$-0.329(23)\cdot10^{-2}$
N	2427	1148	1140
L	8	8	9
σ	$3.2\cdot10^{-5}$	$5.3\cdot10^{-5}$	$1.1\cdot10^{-4}$

* Parameters x_{10}, x_{21}, x_{21} and σ are given in $cm^{-1}\,atm^{-1}$, other parameters are dimensionless.

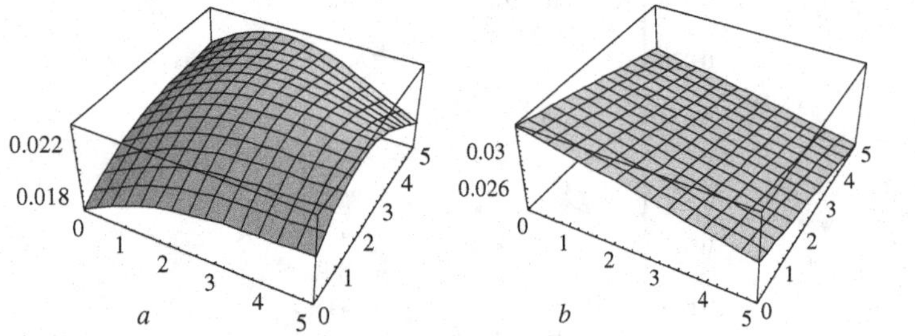

Fig. 3.23. Surfaces $\gamma(sur)$ ($J_i = J_f = 10$) for the ν_2 band of H_2O–Ar (*a*) and H_2O–Kr (*b*) [105]. Reproduced with permission.

Line-shift calculations for the water vapour molecule perturbed by rare gases are of particular interest in two cases. The first one concerns the shifts of purely rotational lines. As mentioned above, the first-order line shifts are defined by the difference of the matrix elements of the interaction potential in the final and initial vibrotational states. This difference determines the form of the interruption function

$$S_1(b) = -\frac{3\pi i}{8\hbar v b^5}\left[\mu_{1i}^2 - \mu_{1f}^2 + \frac{3}{2}\bar{u}(\alpha_{1i} - \alpha_{2f})\right]\alpha_2 \quad (3.7.8)$$

issued from the long-range part of the interaction potential. If the transitions occur between the rotational levels of the same vibrational state and if the rotational dependence of μ and α is neglected, one has $S_1(b) = 0$. Accounting for this rotational dependence by Eqs (3.1.7) and (3.1.8) leads to $S_1(b) \neq 0$. The influence of the rotational dependence of the H_2O polarizability components α_{ln} on the purely rotational line shifts induced by Ar was studied in Ref. [13]. It is shown in Fig. 3.24 for three cases: without rotational dependence (*a*), with rotational dependence (*b*) and with rotational dependence of the mean polarizability only (*c*). Analogous results were obtained for the H_2O-Kr system [13] (Fig. 3.25) and for the H_2O–He and H_2O–Ne systems. These figures show that the rotational dependence of $\alpha_{l,n}$ has a significant influence on the line-shift calculations for the purely rotational transitions. However, the existing experimental data (see Refs [28, 79, 98]) do not allow drawing definite conclusions on this influence.

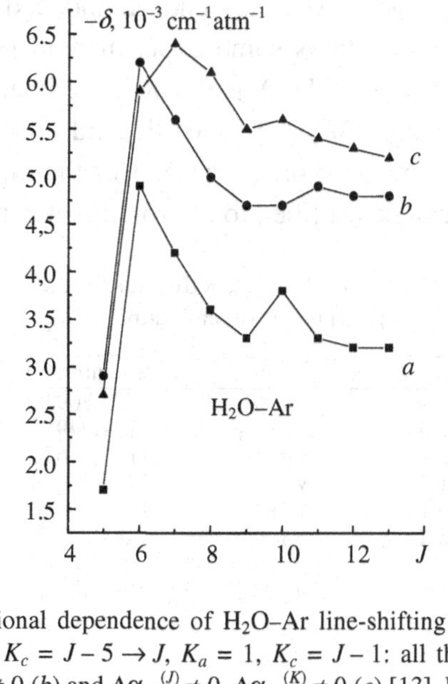

Fig. 3.24. Calculated rotational dependence of H_2O–Ar line-shifting coefficients for purely rotational lines $J+1$, $K_a = 6$, $K_c = J-5 \rightarrow J$, $K_a = 1$, $K_c = J-1$: all the parameters $\Delta\alpha_{ln}^{(\ldots)} = 0$ in Eq. (3.1.8) (*a*), all $\Delta\alpha_{ln}^{(\ldots)} \neq 0$ (*b*) and $\Delta\alpha_{00}^{(J)} \neq 0$, $\Delta\alpha_{00}^{(K)} \neq 0$ (*c*) [13]. Reproduced with permission.

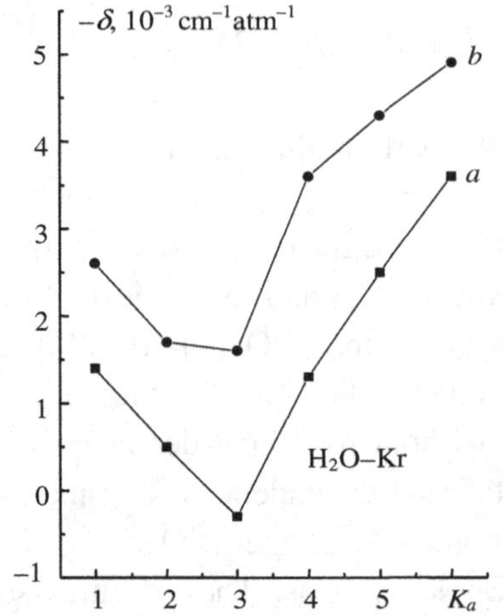

Fig. 3.25. Calculated rotational dependence of H_2O–Kr line-shifting coefficients for purely rotational lines $J = 10$, $K_a + 3$, $K_c = J - K_a - 3 \rightarrow J = 9$, K_a, $K_c = 10 - K_a$: all $\Delta\alpha_{ln}^{(\ldots)} = 0$ (*a*) and all $\Delta\alpha_{ln}^{(\ldots)} \neq 0$ (*b*) in Eq. (3.1.8) [13]. Reproduced with permission.

The second case of particular interest concerns the line shifts induced by helium atoms. In contrast with Ar-, Ne- and Kr-induced shifts, they can take positive values. Table 3.30 contains some experimental δ coefficients [61, 104] for different vibrational bands of H_2O–He. The experimental line shifts for the lines $[J\ K_a\ K_c]$ (011)←$[J\ K_a + \Delta K_a\ K_c]$ (000) depend on the sign of ΔK_a [104]; δ-values depend strongly on the short-range part of the interaction potential and their unusual behaviour can be ascribed to the rotational dependence of V_{iso}.

Table 3.30. Experimental H_2O–He line-shifting coefficients for the $2\nu_1$, $\nu_1 + \nu_3$ [61] and $\nu_2 + \nu_3$ [104] bands (in $cm^{-1}\ atm^{-1}$).

$J_f\ K_{af}\ K_{cf} \leftarrow J_i\ K_{ai}\ K_{ci}$	Band	Line centre, cm^{-1}	δ
1 0 1 ← 1 1 0	$2\nu_1$	7178.20910	0.0020
2 1 2 ← 3 1 3	$\nu_1 + \nu_3$	7182.94955	0.0002
3 0 3 ← 3 2 2	$\nu_1 + \nu_3$	7175.98675	−0.0008
5 1 5 ← 5 1 4	$\nu_1 + \nu_3$	7165.21504	0.0023
6 6 0 ← 6 6 1	$\nu_1 + \nu_3$	7185.59600	0.0056
3 1 3 ← 4 3 2	$\nu_2 + \nu_3$	5090.6257	-0.0035
8 4 4 ← 9 4 5	$\nu_2 + \nu_3$	5119.6539	0.0013
3 1 2 ← 4 3 1	$\nu_2 + \nu_3$	5122.4661	-0.0031
6 3 4 ← 7 3 5	$\nu_2 + \nu_3$	5173.4818	0.0012
4 4 0 ← 5 4 1	$\nu_2 + \nu_3$	5231.6828	0.0021
7 3 5 ← 7 3 4	$\nu_2 + \nu_3$	5314.1420	0.0021
7 4 4 ← 6 4 3	$\nu_2 + \nu_3$	5521.1340	0.0017
8 3 6 ← 7 3 5	$\nu_2 + \nu_3$	5527.8407	0.0018
9 4 6 ← 8 4 5	$\nu_2 + \nu_3$	5564.8187	0.0019

3.8. Tabulation of H_2O line-broadening coefficients for high temperatures

Systematic experimental and theoretical studies of H_2O line broadening by various foreign gases were mainly realised at room temperature. The available experimental data are limited by the accessible temperature range, and the computations are typically restricted by the J- and K_a-values which allow the use of effective Hamiltonians.

The most complete information on the semi-classically calculated line-broadening coefficients of H_2O molecule is presented in the HITRAN [35] and GEISA [44] databases. Their temperature dependence is usually taken in the form of Eq. (3.2.1) with $T_0 = 296$ K. The line-broadening coefficients calculated for different temperatures by one of semi-classical methods are fitted to Eq. (3.2.1) in order to obtain the temperature exponent n for each line. Values of $\gamma(T_0)$ and n are listed in the databases. Since the H_2O line widths depend only slightly on the vibrational states, these $\gamma(T_0)$ and n can be used for different vibrational bands. For high values of the rotational quantum number J Eq. (3.2.1) gives a 10% precision in the reproduction of experimental data. In the HITRAN database, the line-width calculations are generally made by the semi-classical RB formalism, which gives a good agreement with measurements except in the case of broadening by rare gases for high $J \sim 15$ (for these J-values the RB-calculated line widths are greater than the experimental results by a factor of two or three; cf. Sec 3.7). The temperature dependence of γ coefficients can also be described by the semi-empirical model of Sec. 3.4.

In order to extrapolate the H_2O line-broadening coefficients to strongly excited rotational states and high temperatures, their modelling can be realised according to the following schemes.

3.8.1. Surface $\gamma(sur)$

This model is described in Sec. 3.3 of the present chapter. The parameters of the analytical expression for $\gamma(sur)$ are obtained by fitting to experimental and/or calculated values of $\gamma(T)$. These parameters presented in Secs. 3.3, 3.5, 3.7 are determined for J, $K_a \leq 20$ and 296 K $\leq T \leq$ 1200 K for the self-broadening case, for J, $K_a \leq 20$ and 296 K $\leq T \leq$ 500 K for the case of broadening by N_2, O_2, CO_2, and for J, $K_a \leq 17$ and 296 K $\leq T \leq$ 1200 K for the case of broadening by Ar. In some cases (for example, for H_2O–Ar system), the model $\gamma(sur)$ gives better results than the semi-classical methods since its parameters can be deduced directly from the experimental data.

3.8.2. Interpolation procedure of Delaye–Hartmann–Taine

A very simple scheme for the calculation of the temperature dependence of H_2O line widths broadened by H_2O, N_2, O_2 and CO_2 for temperatures up to 2000 K was proposed in Ref. [108]. The line-broadening coefficients $\gamma_{0J\tau-0J\tau}(300)$ (for $T_0 = 300$ K) and the temperature exponents n for the hypothetical rotational transitions $0J\tau \rightarrow 0J\tau$ in the ground vibrational state (000) (i.e. transitions $[J\,K_a\,K_c]$ (000) $\rightarrow$ $[J\,K_a\,K_c]$ (000)) are needed as input parameters. For each buffer gas about 200 values were thus needed. The necessary input line-broadening coefficients were obtained by the RB method for the lines with rotational energies below 2900 cm^{-1}. These coefficients (for $J \le 16$, $K_a \le 10$) as well as the n-values can be found in Ref. [108]. The line-broadening coefficients $\gamma_{VJ\tau-V'J'\tau'}$ for an arbitrary transition $[J\,K_a\,K_c]$ $(v_1\,v_2\,v_3) \rightarrow [J'\,K_a'\,K_c']$ $(v_1'\,v_2'\,v_3')$ are then calculated by the relation

$$\gamma_{VJ\tau-V'J'\tau'} \approx \gamma_{0J\tau-0J'\tau'} \approx [(\gamma_{0J\tau-0J\tau}^{(\varepsilon-1)} + \gamma_{0J'\tau'-0J'\tau'}^{(\varepsilon-1)})/2]^{\frac{1}{\varepsilon-1}}, \tag{3.8.1}$$

where the parameter ε depends on the perturbing gas and has values $\varepsilon = 3$ for H_2O, $\varepsilon = 4$ for N_2, $\varepsilon = 13$ for O_2 and $\varepsilon = 4$ for CO_2 [108]. The next step consists in the calculation of $\gamma_{VJ\tau-V'J'\tau'}$ for various temperatures T using Eq. (3.2.1). For high T-values the temperature dependence of the line-broadening coefficients is very weak. The averaged γ-values for high temperatures were calculated in [108] as

$$\gamma^{av}(T) = \sum_{J\tau} \rho_{J\tau}(T)\gamma_{J\tau-J\tau}(T), \tag{3.8.2}$$

where $\rho_{J\tau}$ is the relative population of the level $J\tau$. The results of calculations by Eq. (3.8.2) were further approximated by the dependences

$$\begin{aligned} \gamma^{av}(T) &= \gamma^{av}(300)[300/T]^{n^{av}(T)}, \\ n^{av}(T) &= n^{av}(300)[300/T]^{\beta}, \end{aligned} \tag{3.8.3}$$

where the parameters $\gamma^{av}(300)$ = 136 cm^{-1} atm^{-1}, $n^{av}(300) = 0.84$, $\beta = 0.06$ for H_2O–CO_2, $\gamma^{av}(300)$ = 419 cm^{-1} atm^{-1}, $n^{av}(300) = 0.92$, $\beta = 0.07$ for H_2O–H_2O, $\gamma^{av}(300)$ = 95.2 cm^{-1} atm^{-1}, $n^{av}(300) = 0.86$, $\beta = 0.07$ for H_2O–N_2, and $\gamma^{av}(300)$ = 59.4 cm^{-1} atm^{-1}, $n^{av}(300) = 0.83$, $\beta = 0.06$ for H_2O–O_2 [108]. The validity of these parameters is limited to the rotational quantum numbers J, K_a corresponding to the energies below 2900 cm^{-1} while experimental energy levels for higher J and K_a are currently available. Table 3.31 shows a comparison of γ coefficients calculated with $\gamma(sur)$ model and with Eqs (3.8.1) and (3.8.2) for some lines.

Table 3.31. H_2O–N_2 and H_2O–H_2O γ coefficients (in 10^{-3} cm^{-1} atm^{-1}) calculated by various models*.

Line	H_2O–N_2			H_2O- H_2O		
	calc. [108]	$\gamma(sur)$	expt [69]	calc. [108]	$\gamma(sur)$	expt [21]
[1 1 1]→[2 0 2]	110	112#		440	510	475
[2 2 0]→[3 3 1]	93.7	94.4	92.7	431	422	460
[4 1 4→[5 0 5]	95.4	94.0#	90.6	447	460	475
[5 0 5]→[6 1 6]	84.7	83.3#	83.3	404	440	441
[5 2 3]→[6 3 4]	92.6	93.6		411	440	437
[6 1 6]→[6 2 5]	77.6	89.7	83.2	401	440	391
[9 4 5]→[8 5 4]	80.0	71.0		346	366	396

* For H_2O–N_2 system the parameters of $\gamma(sur)$ model are those of the last column of Table 3.20 (for the values marked by # the parameters of the second column are used); for H_2O–H_2O system the $\gamma(sur)$ parameters are those of Table. 3.16.

3.8.3. Exponential representation of Toth

For H_2O–H_2O, H_2O–air an empirical dependence on the rotational quantum numbers was proposed by Toth [22]:

$$\gamma = \exp\left[\sum_i a(i)x(i)\right], \tag{3.8.4}$$

where $x(i)$ depend on J, K_a of the upper and lower states and $a(i)$ are adjustable parameters. Twenty eight $x(i)$ elements were considered, with the maximal total power of J and K_a equal to seven. The results of Ref. [22] for the last seven transitions (experimental and calculated by Eq. (3.8.4)) are compared with $\gamma(sur)$-values of Eq. (3.3.4) in Table 3.32. The quality of line widths calculated with both models is very similar, and the γ-values depend weakly on the vibrational state. The temperature dependence $\gamma(T)$ can be obtained by Eq. (3.2.1) if the temperature exponents n are known.

Table 3.32. Calculated by various models and experimental H_2O–air and self-broadening coefficients for the $2\nu_3$ band (in cm^{-1} atm^{1}).

upper	lower	H_2O–air			H_2O–H_2O			
$J\,K_a\,K_c$	$J\,K_a\,K_c$	expt [22]	calc. [22]	$\gamma(sur)^a$	expt [22]	calc. [22]	$\gamma(sur)^b$	$\gamma(sur)^c$
6 5 2	5 4 1	0.6410	0.0647	0.0619	0.350	0.340	0.351	0.333
6 5 1	5 4 2	0.0610	0.0616	0.0619	0.330	0.336	0.351	0.333
6 6 1	5 5 0	0.0453	0.0454	0.0463	0.285	0.220	0.286	0.273
7 4 3	6 3 4	0.0769	0.0793	0.0777	0.415	0.394	0.415	0.403
7 5 3	6 4 2	0.0760	0.0699	0.0646	0.368	0.351	0.361	0.350
7 5 2	6 4 3	0.0626	0.0642	0.0646	0.305	0.341	0.361	0.350
8 6 3	7 5 2	0.0567	0.0552	0.0522	0.294	0.264	0.312	0.302

[a] Parameters obtained for the ν_2 band (Table 3.21); [b, c] parameters of the rotational and ν_2 bands (Table 3.16).

3.8.4. Polynomial representation

This model always mentioned in Sec. 3.7 for H_2O–H_2 system uses the polynomial form of Eq. (3.7.1). However, due to the limited radius of convergence, its asymptotic behaviour for high J- and K_a-values is incorrect. Figure 3.26 shows the H_2O–CO_2 line-broadening coefficients for $[J\,0\,J] \rightarrow [J\,1\,J-1]$ transitions calculated with $\gamma(sur)$ (Eq. (3.3.4), parameters of Table 3.21), polynomial (Eq. (3.7.1), parameters of Ref. [79]) and interpolation (Eq. (3.8.1)) models. The results of $\gamma(sur)$ and interpolation procedures are very close whereas those of the polynomial representation are quite different.

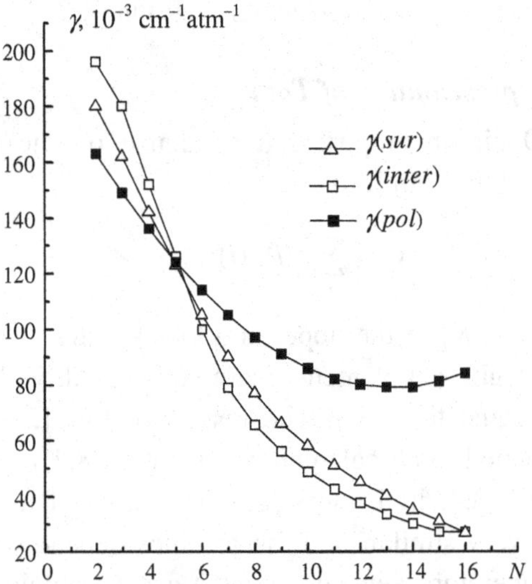

Fig. 3.26. H_2O–CO_2 line-broadening coefficients for $[J\,0\,J] \rightarrow [J\,1\,J-1]$ transitions calculated with various models: $\gamma(sur)$ of Eq. (3.3.4), $\gamma(inter)$ of Eq. (3.8.1) and $\gamma(pol)$ of Eq. (3.7.1).

Bibliography

1. Camy-Peyret C. and Flaud J.-M. (1975). PhD thesis. University Paris-VI, Paris (in French).
2. Bykov, A.D., Makushkin Yu.S. and Ulenikov O.N. (1989). *Vibration-rotational spectroscopy of water vapour molecule*. Nauka, Novosibirsk (in Russian).
3. Bykov A.D., Sinitsa L.N. and Starikov V.I. (1999). *Experimental and theoretical methods in the spectroscopy of water vapour*. Nauka, Novosibirsk (in Russian).
4. Gamache R.R., Hartmann J.-M. and Rosenmann L. (1994). Collisional broadening of water vapour lines – I. A survey of experimental results. *J Quant Spectrosc Radiat Transfer* 52: 481–499.
5. Gamache R.R. and Hartmann J.M. (2004). An intercomparison of measured pressure-broadening and pressure-shifting parameters of water vapour. *Can J Chem* 82: 1013–1027.
6. Starikov V.I. and Tyuterev Vl.G. (1997). Intramolecular ro-vibrational interactions and theoretical methods in the spectroscopy of non-rigid molecules. Spektr, Tomsk (in Russian).
7. Bykov A.D., Sinitsa L.N. and Starikov V.I. (2004). *Introduction to ro-vibrational spectroscopy of polyatomic molecules*. Spektr, Tomsk (in Russian).
8. Protasevich, A.E., Mikhailenko S.N. and Starikov V.I. (2002). Highly excited rotational states of H_2O, self-broadening and self-shifting, ν_2 band. *Atmos Oceanic Opt* 15: 790–796.
9. Aliev M.R. and Watson J.K.G. (1985). Higher-order effects in the vibration-rotation spectra of semirigid molecules. *Mol Spectrosc: Mod Res* 3: 1–69, Academic Press, New York.
10. Camy-Peyret C. and Flaud J.-M. (1985). Vibration-rotation dipole moment operator for asymmetric rotors. *Mol Spectrosc: Mod Res* 3: 70–117, Academic Press, New York.
11. Starikov V.I. (2001). Forth order rotational correction to the effective dipole moment of non-rigid asymmetric rotors. *J Mol Spectrosc* 206: 166–171.
12. Starikov V.I. and Mikhailenko S.N. (1992). Effective dipole moment for non-rigid molecules. *Atmos Oceanic Opt* 5: 129–137.
13. Starikov V.I. and Protasevich A.E. (2003). Effective polarizability operator for X_2Y-type molecules, Application to line-width and line-shift calculations of H_2O. *J Mol Struct* 646: 81–88.
14. Starikov V.I. and Protasevich A.E. (2004). Vibration-rotational polarizability operator for non-linear X_2Y type molecule. *Opt Spectrosc* 97: 121–128.
15. Mengel M. and Jensen P.A. (1995). Theoretical study of the Stark effect in triatomic molecules; application to H_2O. *J Mol Spectrosc* 169: 73–91.
16. Starikov V.I. (2001). Effective dipole moment for X_2Y asymmetric tops. *Opt Spectrosc* 91: 206–212.

17. Luo, Yi., Agren H., Vahtras O., Jorgensen P., Spirko V., and Hettema H. (1993). Frequency-dependent polarizabilities and first hyperpolarizabilities of H_2O. *J Chem Phys* 98: 7159–7164.
18. Murphy W.F. (1977). The Raleigh depolarization ratio and rotational Raman spectrum of water vapour and the polarizability components for the water molecule. *J Chem Phys* 67: 5877–5882.
19. Partridge H. and Schwenke D.W. (1997). The determination of an accurate isotope dependent potential energy surface for water from extensive *ab initio* calculations and experimental data. *J Chem Phys* 106: 4618–4639.
20. Barbier R.J. and Tennyson J. (2006). A high accuracy computed water line list. *Mon Not R Astron Soc* 368: 1087–1094.
21. Toth R.A., Brown L.R. and Plymate C. (1998). Self-broadened widths and frequency shifts of water vapour lines between 590 and 2400 cm^{-1}. *J Quant Spectrosc Radiat Trans* 59: 529–562.
22. Toth R.A. (2005). Measurements and analysis (using empirical functions for widths) of air- and self-broadening parameters of H_2O. *J Quant Spectrosc Radiat Trans* 94: 1–50.
23. Toth R.A. (2005). Measurements of positions, strengths and self-broadened widths of H_2O from 2900–8000 cm^{-1}: Line strength analysis of the 2nd triad bands. *J Quant Spectrosc Radiat Trans* 94: 51–107.
24. Grossman B.E. and Browell W.E. (1989). Spectroscopy of water vapour in the 720 nm wavelength region: line strengths, self-induced pressure shifts. *J Mol Spectrosc* 136: 264–294.
25. Jenouvrier A., Daumont L., Régalia-Jarlot L., Tyuterev V.G., Carleer M., Vandaele A.C., Mikhailenko S. and Fally S. (2007). Fourier transform measurements of water vapor line parameters in the 4200–6600 cm^{-1} region. *J Quant Spectrosc Radiat Trans* 105: 326–355.
26. Li H., Farooq A., Jeffries J.B. and Hanson R.K. (2008). Diode laser measurements of temperature-dependent collisional-narrowing and broadening parameters of Ar-perturbed H_2O transitions at 1391.7 and 1397.8 nm. *J Quant Spectrosc Radiat Trans* 109: 132–143.
27. Ibrahim N., Chelin P., Orphal J. and Baranov Y.I. (2008). Line parameters of H_2O around 0.8 μm studied by tunable diode laser spectroscopy. *J Quant Spectrosc Radiat Trans* 109: 2523–2536.
28. Golubiatnikov Yu.G. (2005). Shifting and broadening parameters of the water vapour 183-Gz line ($3_{1\,3}$–$2_{2\,0}$) by H_2O, O_2, N_2, CO_2, H_2, He, Ne, Ar, and Kr at room temperature. *J Mol Spectrosc* 230: 196–198.
29. Koshelev M.A., Tretyakov Yu.M., Golubiatnikov Yu.G., Parshin V.V., Markov V.N. and Koval I.A. (2007). Broadening and shifting of the 321-, 325- and 380-GHz lines of water vapour by pressure of atmospheric gases. *J Mol Spectrosc* 241: 101–108.
30. Cazzoli G., Puzzarini C., Buffa G. and Tarrini O. (2007). Experimental and theoretical investigation on pressure-broadening and pressure-shifting of the 22.2 GHz line of water. *J Quant Spectrosc Radiat Trans* 105: 438–449.
31. Cazzoli G., Puzzarini C., Buffa G. and Tarrini O. (2008). Pressure-broadening in the THz frequency region: The 1.113 THz line of water. *J Quant Spectrosc Radiat Trans* 109: 1563–1574.
32. Cazzoli G., Puzzarini C., Buffa G. and Tarrini O. (2008). Pressure-broadening of water lines in the THz frequency region: Improvements and confirmations for spectroscopic databases: Part I. *J Quant Spectrosc Radiat Trans* 109: 2820–2831.

33. Mandin J.-Y., Flaud J.-M. and Camy-Peyret C. (1981). New measurements and improved calculations of self-broadening coefficients of lines in the ν_2 band of $H_2{}^{16}O$. *J Quant Spectrosc Radiat Trans* 76: 483–494.
34. Langlois S., Birbeck T.P. and Hanson R.K. (1994). Diode Laser Measurements of H_2O Line Intensities and Self-broadening Coefficients in the 1.4-μm Region. *J Mol Spectrosc* 163: 27–42.
35. Rothman L.S., Gamache R.B., Tipping R.H., Rinsland C.P., Smith N.H., Benner D.C., Devi V.M., Flaud J.-M., Camy-Peyret C., Perrin A., Goldman A., Massie S.T., Brown L.R. and Toth R.A. (1992). The HITRAN molecular database: Editions of 1991 and 1992. *J Quant Spectrosc Radiat Trans* 48: 469–507; Rothman L.S., Barbe A., Benner D.C., Brown L.R., Camy-Peyret C., Carleer M.R., Chancea K., Clerbauxf C., Dana V., Devi V.M., Fayt A., Flaud J.-M., Gamache R.R., Goldman A., Jacquemart D., Jucks K.W., Lafferty W.J., Mandin J.-Y., Massie S.T., Nemtchinov V., Newnham D.A., Perrin A., Rinsland C.P., Schroeder J., Smith K.M., Smith M.A.H., Tang K., Toth R.A., Vander Auwera J., Varanasi P. and Yoshino K. (2003). The HITRAN molecular spectroscopic database: edition of 2000 including updates through 2001. *J Quant Spectrosc Radiat Trans* 82: 5–44; Rothman L.S., Jacquemart D., Barbe A., Benner C.D., Birk M., Brown L.R., Carleer M.R., Chackerian C., Jr., Chance K., Coudert L.H., Dana V., Devi V.M., Flaud J.-M., Gamache R.R., Goldman A., Hartmann J.-M., Jucks K.W., Maki A.G., Mandin J.-Y., Massie S.T., Orphal J., Perrin A., Rinsland C.P., Smith M.A.H., Tennyson J., Tolchenov R.N., Toth R.A., Vander Auwera J., Varanasi P. and Wagner G. (2005). The HITRAN 2004 molecular spectroscopic database (including updates through 2006). *J Quant Spectrosc Radiat Trans* 96: 139–204; Rothman L.S., Gordon I.E., Barbe A., Benner D.C., Bernath P.F., Birk M., Boudon V., Brown L.R., Campargue A., Champion J.-P., Chance K., Coudert L.H., Dana V., Devi V.M., Fally S., Flaud J.-M., Gamache R.R., Goldman A., Jacquemart D., Kleiner I., Lacome N., Lafferty W.J., Mandin J.-Y., Massie S.T., Mikhailenko S.N., Miller C.E., Moazzen-Ahmadi N., Naumenko O.V., Nikitin A.V., Orphal J., Perevalov V.I., Perrin A., Predoi-Cross A., Rinsland C.P., Rotger M., Šimečková M., Smith M.A.H., Sung K., Tashkun S.A., Tennyson J., Toth R.A., Vandaele A.C. and Vander Auwera J. (2009). The HITRAN 2008 molecular spectroscopic database. *J Quant Spectrosc Radiat Trans* 110: 533–572.
36. Protasevich A.E. (2003). Self-broadening and self-shifting of $H_2{}^{16}O$ lines in the region 1950–2750 cm^{-1}. *J Appl Spectrosc* 70: 574–578.
37. Labani B., Bonamy J., Robert D. and Hartmann J.M. (1987). Collisional broadening of rotation-vibration lines for asymmetric top molecules. III. Self-broadening case; application to H_2O. *J Chem Phys* 87: 2781–2789.
38. Starikov V.I. (2008). Calculation of the self-broadening coefficients of water vapor absorption lines using an exact trajectory model. *Opt Spectrosc*, 104: 513–523.
39. Starikov V.I. and Mikhailenko S.N. (1998). Analysis of experimental data for the first hexad {(040), (120), (200), (002), (021), (101)} of H_2O molecule interacting states. *J Mol Struct* 442: 39–53.
40. Starikov V.I. and Mikhailenko S.N. (1998). New analysis of experimental data for the second hexad {(050), (130), (210), (012), (031), (111)} of $H_2{}^{16}O$ molecule interacting states. *J Mol Struct* 449: 39–51.
41. Antony B.K., Gamache P.R., Szembek C.D., Niles D.L. and Gamache R.R. (2006). Modified complex Robert–Bonamy formalism calculations for strong to weak interacting systems. *Mol Phys* 104: 2791–2799.

42. Antony B.K., Neshyba S. and Gamache R.R. (2007). Self-broadening of water vapour transitions via the complex Robert–Bonamy theory. *J Quant Spectrosc Radiat Trans* 105: 148–163.
43. Antony B.K. and Gamache R.R. (2007). Self-broadened half-widths and self-induced line shifts for water vapor transitions in the 3.2–17.76 μm spectral region via complex Robert–Bonamy theory. *J Mol Spectrosc* 243, p. 113–123.
44. http://ara.lmd.polytechnique.fr/htdocs-public/products/GEISA/HTML-GEISA/
45. Hartmann J.M., Taine J., Bonamy J., Labani B. and Robert D. (1987). Collisional broadening of rotation-vibration lines for asymmetric top molecules. II. H_2O diode laser measurements in the 400–900K range; calculations in the 300–2000 K range. *J Chem Phys* 86: 144–157.
46. Townes C.H. and Schawlow A.L. (1955). *Microwave spectroscopy*. McGraw-Hill, New York.
47. Landau L.D. and Lifshitz E.M. (1965). *Quantum Mechanics*. Pergamon Press, New York.
48. Pickett H.M. (1980). Effects of velocity averaging on the shapes of absorption lines. *J Chem Phys* 73: 6090–6094.
49. Baldacchini G., Buffa G., Amato F.D., Pelagalli F. and Tarrini O. (1996). Variation in the sign of the pressure-induced line shifts in the ν_2 band of ammonia with temperature. *J Quant Spectrosc Radiat Trans* 55: 741–743.
50. Zou Q. and Varanasi P. (2003). Laboratory measurement of the spectroscopic line parameters of water vapor in the 610–2100 and 3000–4050 cm^{-1} regions at lower-tropospheric temperatures. *J Quant Spectrosc Radiat Trans* 82: 45–98.
51. Starikov V.I. (2008). Calculation of the self-broadened linewidth coefficients of D_2O absorption lines using exact trajectory model. *Atmos Oceanic Opt* 21: 757–760.
52. Hirschfelder J.O., Curtiss C.F. and Bird R.B. (1967). Molecular theory of gases and liquids, Wiley, New York.
53. Labani B., Bonamy J., Robert D., Hartmann J.M. and Taine J. (1986). Collisional broadening of rotation-vibration lines for asymmetric top molecules. I. Theoretical model for both distant and close collisions. *J Chem Phys* 84: 4256–4267.
54. Starikov V.I. and Protasevich A.E. (2005). Analytical representation for water vapor self-broadening coefficients. *Opt. Spectrosc* 98: 330–335.
55. Tsao C.J. and Curnutte B. (1962). Line-widths of pressure-broadened spectral lines. *J Quant Spectrosc Radiat Trans* 2: 41–91.
56. Starikov V.I. (2009). Analytical representation of half-width for molecular ro-vibrational lines in the case of dipole–dipole and dipole–quadrupole interactions. *Mol Phys* 107: 2227–2236.
57. Bauer A., Godon M., Kheddar M. and Hartmann J.M. (1989). Temperature and perturber dependences of water vapour line-broadening. Experiments at 183 GHz; calculations below 1000 GHz. *J Quant Spectrosc Radiat Trans* 41: 49–54.
58. Bauer A., Godon M., Kheddar M., Hartmann J.M., Bonamy J. and Robert D. (1987). Temperature and perturber dependences of water vapour 380 GHz line broadening. *J Quant Spectrosc Radiat Trans* 37: 531–539.
59. Bykov A., Lavrentieva N. and Sinitsa L. (2004). Semi-empiric approach to the calculation of H_2O and CO_2 line broadening and shifting. *Mol Phys* 102: 1653–1658.
60. Camy-Peyret C., Valentin A., Claveau C., Bykov A., Lavrentieva N., Saveliev V. and Sinitsa L. (2004). Half-width temperature dependence of nitrogen broadened lines in the ν_2 band of H_2O. *J Mol Spectrosc* 224: 164–175.

61. Zeninari V., Parvitte B., Courtois D., Lavrentieva N.N., Ponomarev Yu.N. and Durry G. (2004). Pressure-broadening and shift coefficients of H_2O due to perturbation by N_2, O_2, H_2 and He in the 1.39 μm region: experiment and calculations. *Mol Phys* 102: 1697–1706.
62. Tashkun S.A., Perevalov V.I., Teffo J.-L., Bykov A.D. and Lavrentieva N.N. (2003). CDSD-1000, the high-temperature carbon dioxide spectroscopic databank. *J Quant Spectrosc Radiat Trans* 82: 165–196.
63. ftp://ftp.iao.ru/pub/CDSD-1000
64. http://saga.atmos.iao.ru/
65. Bykov A.D., Lavrentieva N.N., Mishina T.P., Sinitsa L..N, Barber R.J., Tolchenov R.N. and Tennyson J. (2008). Water vapour line width and shift calculations with accurate vibration–rotation wave functions. *J Quant Spectrosc Radiat Trans* 109: 1834–1844.
66. Hodges J.T., Lisak D., Lavrentieva N., Bykov A, Sinitsa L., Tennyson J., Barber R.J. and Tolchenov R.N. (2008). Comparison between theoretical calculations and high-resolution measurements of pressure broadening for near-infrared water spectra. *J Mol Spectrosc* 249: 86–94.
67. Bykov A.D., Lavrentieva N.N., Petrova T.M., Sinitsa L.N., Solodov A.M., Barber R.J., Tennyson J. and Tolchenov R.N. (2008). Shift of the Centers of H_2O Absorption Lines in the Region of 1.06 μm. *Opt Spectrosc* 105: 25–31.
68. Lavrentieva N., Osipova A., Sinitsa L., Claveau C. and Valentin A. (2008). Shifting temperature dependence of nitrogen-broadened lines in the ν_2 band of H_2O. *Mol Phys* 106: 1261–1266.
69. Toth A.R. (2000). Air- and N_2-broadening parameters of water vapour: 604 to 2271 cm^{-1}. *J Mol Spectrosc* 201: 218–243.
70. Valentin A., Claveau C., Bykov A.D., Lavrentieva N.N., Saveliev V.N. and Sinitsa L.N. (1999). The Water-Vapour ν_2 Band Line-shift Coefficients Induced by Nitrogen Pressure. *J Mol Spectrosc* 198: 218–229.
71. Claveau C., Henry A., Hurtmans D. and Valentin A. (2001). Narrowing and broadening parameters of H_2O lines perturbed by He, Ne, Ar, Kr and nitrogen in the spectral range 1850–2140 cm^{-1}. *J Quant Spectrosc Radiat Trans* 68: 273–298.
72. Zeninari V., Parvitte B., Courtois, D, Pucher I., Durry G. and Ponomarev Yu.N. (2003). He- and H_2- induced broadening and shifting of H_2O lines in the region 1.39 μm. *Atmos Oceanic Opt* 16: 212–216.
73. Nagali V., Chou S.I., Baer D.S. and Hanson R.K. (1997). Diode-laser measurements of temperature-dependent half-widths of H_2O transitions in the 1.4 μm region. *J Quant Spectrosc Radiat Trans* 57: 795–809.
74. Durry G., Zeninari V., Parvitte B., Le Barbu T., Lefevre F., Ovarlez J. and Gamache R.R. (2005). Pressure-broadening coefficients and line strengths of H_2O near 1.39 μm: application to the in situ sensing of the middle atmosphere with balloon-borne diode lasers. *J Quant Spectrosc Radiat Trans* 94: 387–403.
75. Tretyakov M.Y., Parshin V.V., Koshelev M.A., Shanin V.N., Myasnikova S.E. and Krupnov A.F. (2003). Studies of 183 GHz water line: broadening and shifting by air, N_2 and O_2 and integral intensity measurements. *J Mol Spectrosc* 218: 239–245.
76. Seta T., Hoshina H., Kasai Y., Hosako I., Otani C., Loβow S., Urban J., Ekström M., Eriksson P. and Murtagh D. (2008). Pressure-broadening coefficients of the water vapour lines at 556.936 and 752.033 GHz. *J Quant Spectrosc Radiat Trans* 109: 144–150.
77. Markov V.N. and Krupnov A.F. (1995). Measurements of the pressure shift of the 1(10)–1(01) water line at 556 GHz produced by mixtures of gases. *J Mol Spectrosc* 172: 211–214.

78. Golubiatnikov Yu.G., Koshelev M.A. and Krupnov A.F. (2008). Pressure shift and broadening of 1_{10}–1_{01} water vapour lines by atmospheric gases. *J Quant Spectrosc Radiat Trans* 109: 1828–1833.
79. Dick M.J., Drouin B.J. and Pearson J.C. (2009). A collisional cooling investigation of the pressure broadening of the $1_{10} \leftarrow 1_{01}$ transition of water from 17 to 200 K. *J Quant Spectrosc Radiat Trans* 110: 619–627.
80. Gamache R.R., Neshyba S.P., Plateaux J.J., Barbe A., Regalia L. and Pollack J.B. (1995). CO_2-broadening of water-vapour lines. *J Mol Spectrosc* 170: 131–151.
81. Gamache R.R., Lynch R., Plateaux J.J. and Barbe A. (1997). Halfwidths and line shifts of water vapour broadening by CO_2: measurements and complex Robert–Bonamy formalism calculations. *J Quant Spectrosc Radiat Trans* 57: 485–496.
82. Poddar P., Bandyopadhyay A., Biswas D., Ray B. and Ghosh P.N. (2009). Measurement and analysis of rotational lines in the ($2\nu_1 + \nu_2 + \nu_3$) overtone band of H_2O perturbed by CO_2 using near infrared diode laser spectroscopy. *Chem Phys Lett* 469: 52–56.
83. Sagawa H., Mendrok J., Seta T., Hoshina H., Baron P., Suzuki K., Hosako I., Otani C., Hartogh P. and Kasai Y. (2009). Pressure-broadening coefficients of H_2O induced by CO_2 for Venus atmosphere. *J Quant Spectrosc Radiat Trans* 110: 2027–2036.
84. Lynch R., Gamache R.R. and Neshyba S.P. (1996). Fully complex implementation of the Robert–Bonamy formalism: half widths and line shifts of H_2O broadened by N_2. *J Chem Phys* 105: 5711–5721.
85. Jacquemart D., Gamache R. and Rothman L.S. (2005). Semi-empirical calculation of air-broadened half-widths and air pressure-induced frequency shifts of water-vapor absorption lines. *J Quant Spectrosc Radiat Trans* 96: 205–239.
86. Gamache R.R. and Hartmann J.-M. (2004). Collisional parameters of H_2O lines: effect of vibration. *J Quant Spectrosc Radiat Trans* 83: 119–147.
87. Wagner G., Birk M., Gamache R.R. and Hartmann J.-M. (2005). Collisional parameters of H_2O lines: effect of temperature. *J Quant Spectrosc Radiat Trans* 92: 211–230.
88. Mandin J.-Y., Chevillard J.-P., Flaud J.-M. and Camy-Peyret C. (1988). N_2-broadening coefficients of H_2O lines between 8500 and 9300 cm^{-1}. *J Mol Spectrosc* 132: 352–360.
89. Tolchenov R.N., Naumenko O., Zobov N., Shirin S., Polyansky O.L., Tennyson J., Carleer M., Coheur P.-F., Fally S., Jenouvrier A. and Vandaele A.C. (2005). Water vapour line assignments in the 9250–26000 cm^{-1} frequency range. *J Mol Spectrosc* 233: 68–76.
90. Thibault F. (1992). Spectral profiles and molecular collisions. PhD thesis, University Paris XI, Orsay (in French).
91. Smith E.W. (1981). Absorption and dispersion in the O_2 microwave spectrum at atmospheric pressures. *J Chem Phys* 74: 6658–6673.
92. Cherkasov, M.R. (1976). Bradening of overlapping-spectral lines by pressure.1. *Opt Spectrosc* 40: 7–13.
93. Lam K.S. (1977). Application of pressure-broadening theory to the calculation of atmospheric oxygen and water vapour microwave absorption. *J Quant Spectrosc Radiat Trans* 17: 351–358.
94 Cherkasov, M.R. (2000). Collisional interference of blended vibrational bands. *Atmos Oceanic Opt* 13: 329–337.
95. Bykov A.D., Lavrentieva N.N., Mishina T.P. and Sinitsa L.N. (2008). Influence of the interference between water vapor lines on the atmospheric transmission of near-IR radiation. *Opt Spectrosc* 104: 172–179.

96. Gamache R.R., Lynch R. and Brown L.R. (1996). Theoretical calculations of pressure-broadening coefficients for H_2O perturbed by hydrogen or helium gas. *J Quant Spectrosc Radiat Trans* 56: 471–487.
97. Brown L.R. and Plymate C. (1996). H_2 broadened $H_2{}^{16}O$ in four infrared bands between 55 and 4045 m^{-1}. *J Quant Spectrosc Radiat Trans* 56: 263–282.
98. Steyert D.W., Wang W.F., Sirota J.M., Donahue N.M. and Reuter D.C. (2004). Hydrogen and helium pressure broadening of water transitions in the 380–600 cm^{-1} region. *J Quant Spectrosc Radiat Trans* 83: 183–191.
99. Langlois S., Birbeek T.P. and Hanson R.K. (1994). Temperature-dependent collision-broadening parameters of H_2O lines in the 1.4-μm region using diode laser absorption spectroscopy. *J Mol Spectrosc* 167: 272–281.
100. Giesen T., Schieder R., Winnewisser G. and Yamada K.M.T. (1992). Precise measurements of pure broadening and shift for several H_2O lines in the ν_2 band by argon, nitrogen, oxygen and air. *J Mol Spectrosc* 153: 406–418.
101. Grossman B.E. and Browell W.E. (1989). Water-vapour line broadening and shifting by air, nitrogen, oxygen, and argon in the 720-nm wavelength region. *J Mol Spectrosc* 138: 562–596.
102. Lavrentieva N.N. and Solodov A.M. (1999). Shitting of water vapour lines by the pressure of argon and oxygen in $\nu_1 + \nu_2$ and $\nu_2 + \nu_3$ bands. *Atmos Oceanic Opt* 12: 1124–1128.
103. Gamache R.R. and Lynch R. (2000). Argon-induced halfwidths and line shifts of water vapour transitions. *J Quant Spectrosc Radiat Trans* 64: 439–456.
104. Solodov A.M. and Starikov V.I. (2008). Broadening and shift of lines of the $\nu_2 + \nu_3$ band of water vapour induced by helium pressure. *Opt Spectrosc* 105: 14–20.
105. Protasevich, A.E. and Starikov V.I. (2005). Calculation and simulation of absorption line broadening in water vapour by noble gas atoms. *Opt Spectrosc* 98: 528–534.
106. Schmücker N., Trojan Ch., Giesen T., Schieder R., Yamada K.M.T. and Winnewisser G. (1997). Pressure broadening and shift of some H_2O lines in the ν_2 band: Revisited. *J Mol Spectrosc* 184: 250–256.
107. Salimian S. and Hanson R.K. (1983). Absorption measurements of H_2O at high temperatures using a CO laser. *J Quant Spectrosc Radiat Trans* 30: 1–7.
108. Delaye C., Hartmann J.-M. and Taine J. (1989). Calculated tabulations of H_2O line broadening by H_2O, N_2, O_2 and CO_2 at high temperature. *Appl Opt* 28: 5080–5087.

Chapter 4

Pressure broadening and shifting of vibrotational lines of atmospheric gases

Water vapour is one of the strongest absorbers of radiation in the Earth's atmosphere, but some other gases also play an important role in the atmospheric radiative transfer processes: carbon dioxide CO_2, carbon monoxide CO, ozone O_3, methane CH_4. Minor atmospheric pollutants such as hydrogen sulphide H_2S, methyl fluoride CH_3F, methyl chloride CH_3Cl, ethylene C_2H_4, nitric oxide NO, nitrous oxide N_2O and nitrogen dioxide NO_2 are equally recognised for their implication in the fundamental chemical reactions occurring in the terrestrial atmosphere. As a consequence, a lot of spectroscopic studies are made for these molecules in order to obtain the line positions, intensities and widths. They are briefly summarised below, starting by asymmetric water-like molecules of X_2Y-type which have the same selection rules for vibrotational spectra.

4.1. Vibrotational lines of asymmetric X_2Y molecules

Between asymmetric tops of the X_2Y type most attention in the spectroscopic literature is paid to the H_2S, SO_2, O_3 and NO_2 molecules.

A vibrotational state (v_1 v_2 v_3) [J K_a K_c] of these molecules is characterised by three vibrational quantum numbers v_1, v_2, v_3 and three rotational quantum numbers J, K_a, K_c. In the NO_2 molecule, having an unpaired electron (total electronic spin $S = 1/2$), each vibrotational level is split by the spin–rotational interaction into two components ($J = N \pm S$) denoted by + or –, and the three rotational quantum numbers are N, K_a, K_c. For a given set of vibrotational quantum numbers the corresponding energies are obtained by a diagonalization of the effective Hamiltonian matrix of Eq. (3.1.1) in the basis of symmetry-adapted symmetric top rotational wave functions.

For H_2S, SO_2 and O_3 molecules the rotational Hamiltonian is usually written in a polynomial form (parametric dependence on the vibrational state is indicated by the superscript *V*)

$$H^{(V)} = \sum_{i+j=0} a_{ij}^{(V)} \vec{J}^{2i} J_z^{2j} + \sum_{i+j=0} b_{ij}^{(V)} \vec{J}^{2i} \{J_z^{2j}, J_{xy}^2\}, \tag{4.1.1}$$

where $a_{00} = E$, $a_{10} = (B + C)/2$, $a_{01} = A - (B + C)/2$; $b_{00} = (B - C)/2$, ... are defined by the energy E of the vibrational level and the rotational constants A, B, C; see Table 4.1 for molecular parameters.

Table 4.1. Equilibrium characteristics, normal frequencies, rotational constants and electro-optical parameters of the H_2S, SO_2, O_3 and SO_2 molecules*.

Molecule	r_0, Å	θ_0, °	ν_1, cm^{-1}	ν_2, cm^{-1}	ν_3, cm^{-1}	A, cm^{-1}	B, cm^{-1}
H_2S	1.3365	92.06	2614.56	1182.7	2625.0	10.36	9.0161
SO_2	1.451	118.5	1151.4	517.7	1361.8	2.027	0.344
O_3	1.2717	116.7	1103.1	700.9	1042.1	3.5537	0.4453
NO_2	1.197	134.15	1353.3	759.6	1671.1	8.0024	0.4337
Molecule	**C, cm^{-1}**	**μ, D**	**Q_{xx}, DÅ**	**Q_{yy}, DÅ**	**Q_{zz}, DÅ**	**Ω, DÅ^2**	**α, Å^3**
H_2S	4.7312	0.974					3.67
SO_2	0.294	1.627	−2.2	−2.2	4.4		3.78
O_3	0.3948	0.53	−0.7	2.1	−1.4		
NO_2	0.4104	0.32					3.02

* Here r_0 is the XY length, θ_0 is the angle $\angle XYX$. The molecule lies in the XOZ plane with the OX axis as the symmetry axis. The multipole values can be alternatively expressed in the atomic units: 1 a.u. = 2.54154 D for the dipole moment, 1 a.u. = $1.344911 \cdot 10^{-26}$ esu cm^2 for the quadrupole moment, and 1 a.u. = $0.711688 \cdot 10^{-34}$ esu cm^3 for the octupole moment. The parameters values are mainly taken from Refs [1–3]. Vibrational and rotational dependences of the H_2S effective dipole moment can be found in Ref. [4] and are not very significant. The *ab initio* calculated polarizability components $\alpha_{\alpha\beta}$ for H_2S are listed in Ref. [5] and the H_2S quadrupole moment components are given in Refs [6–8]. For the O_3 molecule the polarizability tensor components are α_{xx} = 2.6 Å^3, α_{yy} = 2.2 Å^3, and α_{zz} = 5.6 Å^3 [9], the *ab initio* calculated multipole moments can be found in Ref. [10]. The vibrational dependence of the O_3 dipole moment is given by $\mu(\nu_1, \nu_2, \nu_3) = \mu_e - 8.98 \cdot 10{-}3(\nu_1+1/2) - 7.09 \cdot 10{-}3(\nu_2+1/2) - 2.39 \cdot 10{-}3(\nu_3+1/2)$ (in D) with μ_e = 0.5425 D and $\mu(0, 0, 0)$ = 0.5333 D [11].

The rotational constants can be used for the estimation of the energy gaps between rotational states and, consequently, for the estimation of the rotational states density. In the first approximation, the rotational energies of NO_2 are calculated with the Hamiltonian of Eq. (4.1.1), where instead of $\vec{J}$ and J_α the operator $\vec{N}$ ($\vec{J} = \vec{N} + \vec{S}$) and its components N_α ($\alpha = x, y, z$) in the molecular frame are used. After that, the terms $\pm E_{SR}$ due to the spin–rotational coupling are added. The magnitude of splitting of vibrotational levels depends on the rotational quantum numbers N, K_a and (for large N- and small K_a-values) can reach 0.1 cm^{-1}.

The selection rules for vibrotational transitions in the X_2Y molecules are the same as for H_2O. For NO_2, the allowed transitions $\pm \leftrightarrow \pm$ ($\Delta S = \Delta\Sigma = 0$, $\Delta J = \Delta N$) are usually considered (Σ is the projection of $\vec{S}$ on the molecular z-axis); $\Delta\Sigma \neq 0$ give transitions of small intensity ($\Delta J \neq \Delta N$).

4.1.1. H_2S molecule

Hydrogen sulphide is a minor atmospheric pollutant produced mainly by natural sources and industrial processing of oil and coal containing sulphur. Since is inflammable and extremely toxic, its presence in the Earth's atmosphere constitutes an olfactory problem for human health.

The dipole moment of H_2S is approximately two times smaller than the dipole moment of H_2O. Since the main contribution in its vibrotational line-broadening coefficients is due to the dipole–dipole interaction, these coefficients appear to be two times smaller in the case of collisions with 'quadrupole' molecules (molecules with the quadrupole moment as the first non-zero multipole moment) and four times smaller in the self-broadening case.

Extensive experimental investigations of the collisional shift and broadening of H_2S absorption lines were performed in Berlin using a three-channel diode laser spectrometer. The broadening and shifting coefficients by O_2, H_2, D_2, N_2, and CO_2 pressure were determined for approximately 40 transitions ($J \leq 11$, $K_a \leq 4$) in the ν_2 band (1050–1325 cm^{-1}) [12, 13]. Self- and air-broadening coefficients for the ν_1, ν_2, ν_3, and $2\nu_2$ bands were obtained in Refs [14–16]. The broadening of H_2S lines by noble gases was studied in Refs [17, 18].

In the case of H_2S lines broadened by pressure of quadrupole molecules, the usual dependence of the broadening coefficient γ on the rotational quantum numbers and the quadrupole moment Q of the perturbing molecules was experimentally confirmed [12, 13]: γ-values increase with J decreasing and Q increasing. The Q-value is minimal for O_2 ($Q = 0.39 \cdot 10^{-26}$ esu cm^2) and maximal for CO_2 ($Q = 4.3 \cdot 10^{-26}$ esu cm^2). In the case of $H_2S–O_2$, the broadening coefficients $38.0 \leq \gamma \leq 73$ (in 10^{-3} cm^{-1} atm^{-1}) were obtained whereas for $H_2S–CO_2$ $81.4 \leq \gamma \leq 135.8$ (in 10^{-3} cm^{-1} atm^{-1}) were stated [12] (for non-overlapping lines). The comparison of N_2-broadening coefficients for the ν_1, ν_2 and ν_3 bands [13] showed their weak dependence on the vibrational quantum numbers.

The shift coefficients in the ν_2 band were found to have positive as well as negative values [12]. No dependence on the polarizability or the quadrupole moment of the perturbing molecule was observed, confirming thus a rather complex character of the line-shifting mechanism in H_2S.

As expected, in the self-broadening case [14–16] the broadening coefficients of H_2S are approximately four times smaller than those of the H_2O molecule. The comparison of these coefficients for the ν_1, ν_2, and $2\nu_2$ bands shows their weak dependence on the vibrational quantum numbers.

Some regularities in the behaviour of γ and δ coefficients of H_2S vibrotational lines were evidenced for collisions with noble gases [17, 18] (see Table 4.2). These coefficients were determined for the strongest lines with the restriction $\Delta K_a \leq 1$. Since for collisions with atoms the intermolecular potential is composed of the induction and dispersion interactions, γ and δ depend on the polarizability α_2 and the mass m_b of the perturber (see Appendix B). The smallest γ-values were obtained for broadening by Ne (α_2 = 0.397 Å^3, m_b = 20.18 a.m.u.), and the largest γ-values were found for broadening by Kr (α_2 = 2.48 Å, m_b = 83.80 a.m.u.).

Table 4.2. Broadening coefficients γ(in 10^{-3} cm^{-1} atm^{-1}) for H_2S ν_2 lines perturbed by noble gases [17]. Reproduced with permission.

ν, cm^{-1}	Band	$J\,K_a\,K_c \leftarrow$	$J\,K_a\,K_c$	H_2S–He	H_2S–Ne	H_2S–Ar	H_2S–Kr	H_2S–Xe
2714.5480	ν_3	4 3 1	3 1 2	58 ± 2	39 ± 4	69 ± 5	85 ± 3	71 ± 4
2693.2838	ν_1	4 4 0	3 3 1	45 ± 2	30.5 ± 1.9	60.4 ± 1.3	72.3 ± 1.9	55.3 ± 1.5
2740.9751	ν_1	5 3 3	4 0 4	49 ± 5	55 ± 3	85 ± 8	96 ± 9	81.9 ± 1.6
1293.2192	ν_2	5 4 1	4 3 2	37 ± 2	31 ± 7	54 ± 5	82 ± 6	88 ± 3
2713.8532	ν_1	5 4 1	4 3 2	43 ± 2	38 ± 2	59 ± 4	68 ± 5	67 ± 5
2740.0321	ν_1	6 4 2	5 3 3	40 ± 2	32.9 ± 1.3	50 ± 4	69 ± 4	87 ± 5
2717.9659	ν_1	6 5 2	5 4 1	37.8 ± 0.8	33 ± 4	47.2 ± 0.9	64 ± 5	61 ± 2
2717.3082	ν_3	6 5 1	5 5 0	39.3 ± 1.9	26.7 ± 1.2	50.3 ± 1.3	66.7 ± 1.5	61.3 ± 1.7
2740.9150	ν_1	7 6 2	6 5 1	30.9 ± 1.7	32.4 ± 1.2	50 ± 3	61 ± 6	64 ± 3
2716.0085	ν_1	8 4 5	7 3 4	40 ± 2	34.2 ± 1.9	51 ± 2	63 ± 3	62 ± 3
2715.0784	ν_1	8 3 5	7 4 4	43.7 ± 1.2	40 ± 2	53.6 ± 1.4	72 ± 2	63.4 ± 1.2
2741.7315	ν_3	8 4 4	7 4 3	47 ± 2	37 ± 3	56 ± 5	69 ± 6	65 ± 4
2741.7631	ν_1	8 6 3	7 5 2	35 ± 5	58 ± 4	70 ± 3	91 ± 3	92 ± 7
2762.5175	ν_1	8 7 1	7 6 2	35.6 ± 1.7	38.1 ± 1.5	48 ± 2	63.2 ± 1.7	63 ± 3
2740.1501	ν_3	8 7 2	7 7 1	59 ± 8	22 ± 5	57 ± 7	41 ± 7	63 ± 7
2741.5120	ν_3	8 7 1	7 7 0	31.8 ± 1.8	24 ± 2	46 ± 4	52.1 ± 1.1	54 ± 3
2714.2992	ν_1	9 3 7	8 2 6	35.9 ± 1.8	25.2 ± 1.9	44 ± 4	47 ± 7	53 ± 6
2714.2885	ν_1	9 2 7	8 3 6	37.1 ± 1.1	30 ± 2	42 ± 4	57 ± 4	53 ± 6
2761.4825	ν_1	9 7 3	8 6 2	40 ± 3	37.4 ± 1.9	57.1 ± 1.9	64 ± 5	75 ± 6
2739.3810	ν_1	10 5 6	9 4 5	52 ± 2	34.9 ± 1.4	45 ± 6	68 ± 4	59 ± 4
2761.4458	ν_1	12 6 7	11 5 6	38.3 ± 1.9	32 ± 3	52 ± 3	51 ± 5	56 ± 3
2742.4091	ν_1	15 0 15	14 1 14	30.2 ± 1.3	19.5 ± 0.9	26.8 ± 1.5	37 ± 2	35.6 ± 0.8
2762.8073	ν_1	18 1 18	17 0 17	26.6 ± 1	11 ± 2	13 ± 2	22 ± 3	28 ± 3

For several lines of the ν_2 band a collisional narrowing effect was clearly observed [18]. The narrowing coefficients were obtained for the H_2S–Ne [12 1 12]←

[11 0 11] (1295.09497 cm^{-1}) line: β_0 = (31±4)·10^{-3} cm^{-1} atm^{-1}, and for the H_2S–Kr [6 4 2]←[5 3 3] (1322.54907 cm^{-1}) line: β_0 = (21±2.3)·10^{-3} cm^{-1} atm^{-1}. No significant differences between the line-broadening and line-shifting coefficients obtained with Galatry, Rautian–Sobel'man and Voigt profile models were stated. Namely, $\gamma = 19.7\cdot10^{-3}$ cm^{-1} atm^{-1}, $\delta = -0.266\cdot10^{-3}$ cm^{-1} atm^{-1} and $\gamma = 56.7\cdot10^{-3}$ cm^{-1} atm^{-1}, $\delta = -1.6\cdot10^{-3}$ cm^{-1} atm^{-1} were deduced with the Voigt profile for the first and the second lines, respectively [18].

Theoretical line-width values for H_2S–H_2S, H_2S–N_2, and H_2S–O_2 systems at 300 K and H_2S–air system at 200 K were obtained [19] by the ATC theory for pure rotational transitions with $0 \le J \le 12$ and $0 \le K_a \le 12$. It was found that in the case of self-broadening the contribution from quadrupolar interactions gives about 20% of the total line width for low J-values. The computed H_2S–N_2-broadening coefficients agreed well with the experimental data [12] whereas too large differences between theory and experiment were observed for the H_2S–O_2 system (due to the small value of O_2 quadrupole moment and non-negligible role of other interactions in the intermolecular potential [19, 20]).

In Ref. [20], self- and H_2O-broadening coefficients in the ν_2 band of H_2S were calculated by the RB′ formalism and the broadening by quadrupole molecules was estimated by the ATC approach. The vibrotational wave functions for (000) and (010) vibrational states of H_2S were obtained by the method of Sec. 3.1.

4.1.1.1. Self-broadening case

The H_2S self-broadening coefficients were calculated in Ref. [20] for ν_2 lines with $J \le 15$, $K_a \le 13$, $\Delta K_a = |K_{a\,i} - K_{a\,f}| \le 3$ at 296 K (about 2000 transitions). For the lines with J, $K_a \le 10$ and $\Delta K_a \le 3$ the calculations were additionally performed for 400, 500, 700, 900 and 1300 K (about 1100 transitions for each temperature). Table 4.3 shows a very satisfactory agreement of some calculated and experimental [14] self-broadening coefficients (for about 50% of lines the disagreement is within experimental errors). An extended list of these coefficients for T = 296 K is given in Table 4.4 ($Q_{xx} = -Q_{zz} = -2.2$ DÅ and $Q_{yy} = 0$ were used).

In the same reference, both calculated and experimental broadening coefficients were also fitted to the analytical model $\gamma(sur)$ of Eq. (3.3.4) with the parameters of Eqs (3.3.6) and (3.5.1) listed in Table 4.5. The statistical analysis of this fitting is given in Table 4.6. It shows that the difference between the $\gamma(sur)$ model values and the input data is less than 20% for practically all lines.

Table 4.3. Experimental [14], RB′-calculated and $\gamma(sur)$-model values of self- and H_2O-broadening coefficients for the ν_2 band of H_2S (cm^{-1} atm^{-1}) [20]*. Reproduced with permission.

			H_2S–H_2S		H_2S–H_2O	
$J\,K_a\,K_c$ ←	$J\,K_a\,K_c$	expt	RB′	$\gamma(sur)$	RB′	$\gamma(sur)$
4 4 1	4 3 2	0.186(14)	0.165	0.175	0.252	0.261
2 1 1	2 2 0	0.181(14)	0.155	0.175	0.220	0.237
7 4 3	7 3 4	0.180(6)	0.156	0.161	0.252	0.266
2 2 1	2 1 2	0.180(12)	0.172	0.175	0.242	0.237
2 2 0	2 1 1	0.172(14)	0.156	0.175	0.219	0.237
4 4 0	4 3 1	0.171(1)	0.155	0.175	0.242	0.262
2 0 2	1 1 1	0.165(10)	0.163	0.162	0.210	0.210
4 1 3	4 0 4	0.161(8)	0.167	0.165	0.259	0.246
4 3 1	4 2 2	0.159(7)	0.161	0.177	0.231	0.264
4 2 3	4 1 4	0.158(23)	0.166	0.174	0.261	0.259
1 1 0	1 0 1	0.157(7)	0.152	0.159	0.192	0.200
7 5 2	7 4 3	0.155(9)	0.149	0.153	0.228	0.255
5 3 2	5 4 1	0.150(1)	0.155	0.169	0.226	0.264
5 5 0	5 4 1	0.150(8)	0.143	0.161	0.233	0.255
5 3 2	5 2 3	0.149(9)	0.167	0.172	0.246	0.268
7 3 4	7 2 5	0.148(17)	0.164	0.164	0.284	0.275
3 1 2	3 2 1	0.148(17)	0.163	0.175	0.227	0.249
3 2 1	3 3 0	0.141(7)	0.162	0.180	0.224	0.257
1 0 1	1 1 0	0.140(9)	0.151	0.159	0.190	0.203
8 6 2	8 5 3	0.121(17)	0.144	0.141	0.213	0.242

* Parameters for the model $\gamma(sur)$ of Eq. (3.3.4) are given in Table 4.5. The value Q = 0.71 a.u. [7] was used for the quadrupole moment of H_2S.

Table 4.4. RB′-calculated H_2S self-broadening coefficients for the ν_2 band (in cm^{-1} atm^{-1}).

(010) $J\,K_a\,K_c$ ←	(000) $J\,K_a\,K_c$	γ	(010) $J\,K_a\,K_c$ ←	(000) $J\,K_a\,K_c$	γ	(010) $J\,K_a\,K_c$ ←	(000) $J\,K_a\,K_c$	γ
0 0 0	1 1 1	0.153	3 1 2	3 0 3	0.167	4 3 1	4 2 2	0.161
1 0 1	2 1 2	0.156	3 2 2	4 3 1	0.166	4 1 3	4 2 2	0.172
1 1 0	1 0 1	0.146	3 3 0	2 2 1	0.172	4 4 1	5 3 2	0.163
1 0 1	1 1 0	0.146	3 0 3	2 1 2	0.167	4 4 1	4 3 2	0.167
1 1 1	2 2 0	0.161	3 2 1	4 3 2	0.167	4 4 1	5 5 0	0.156
1 1 0	2 2 1	0.163	3 1 3	4 0 4	0.166	4 4 0	5 5 1	0.155
1 1 1	2 0 2	0.161	3 2 1	3 3 0	0.161	4 3 2	4 2 3	0.173
2 1 2	2 2 1	0.171	3 3 0	4 4 1	0.168	5 1 4	6 2 5	0.170
2 2 1	3 1 2	0.168	3 2 2	3 3 1	0.171	5 3 2	5 2 3	0.170
2 1 1	2 2 0	0.152	3 1 2	2 2 1	0.168	5 5 1	5 4 2	0.151
2 0 2	2 1 1	0.158	4 1 3	4 0 4	0.169	5 4 2	6 5 1	0.154
2 0 2	1 1 1	0.161	4 3 1	5 2 4	0.172	5 1 5	6 0 6	0.167
2 0 2	3 1 3	0.166	4 3 1	4 4 0	0.159	5 1 5	6 2 4	0.174
2 2 0	3 1 3	0.165	4 2 2	5 3 3	0.171	5 3 2	6 4 3	0.165
2 2 0	1 1 1	0.162	4 1 4	3 0 3	0.166	5 1 5	4 0 4	0.165

2 2 1	2 1 2	0.170	4 2 2	4 3 1	0.161	5 0 5	5 1 4	0.170
2 2 0	2 1 1	0.153	4 1 4	3 2 1	0.167	5 2 3	4 3 2	0.173
2 1 2	3 0 3	0.167	4 0 4	4 1 3	0.170	5 3 2	5 4 1	0.157
2 1 2	1 0 1	0.155	4 2 3	5 3 2	0.174	5 3 3	5 4 2	0.167
2 2 0	3 3 1	0.166	4 1 4	5 0 5	0.165	5 2 4	6 3 3	0.178
2 1 2	3 2 1	0.166	4 2 3	3 3 0	0.172	5 2 4	5 3 3	0.174
2 2 1	1 1 0	0.163	4 1 3	3 2 2	0.172	5 3 2	4 4 1	0.165
2 1 1	2 0 2	0.158	4 0 4	3 1 3	0.165	5 4 2	5 3 3	0.165
2 2 1	3 3 0	0.171	4 2 3	4 3 2	0.175	5 3 3	4 2 2	0.170
2 1 1	3 2 2	0.162	4 2 2	3 3 1	0.169	5 3 2	4 2 3	0.173
3 0 3	3 1 2	0.168	4 3 2	3 2 1	0.166	5 3 3	4 4 0	0.169
3 0 3	4 1 4	0.166	4 1 4	4 2 3	0.169	5 5 1	6 4 2	0.153
3 2 2	2 1 1	0.162	4 3 2	5 4 1	0.166	5 5 0	4 4 1	0.156
3 2 2	4 1 3	0.171	4 1 4	5 2 3	0.176	5 1 5	4 2 2	0.172
3 3 0	3 2 1	0.161	4 1 3	5 2 4	0.174	5 2 4	6 1 5	0.170
3 1 3	3 2 2	0.169	4 0 4	5 1 5	0.165	5 4 2	6 3 3	0.162
3 3 1	4 4 0	0.168	4 3 2	5 2 3	0.172	5 4 2	4 3 1	0.158
3 1 3	4 2 2	0.173	4 2 2	3 1 3	0.172	5 2 3	5 3 2	0.170
3 3 0	4 2 3	0.173	4 3 1	3 2 2	0.165	5 4 1	6 3 4	0.166
3 1 3	2 0 2	0.166	4 3 2	4 4 1	0.169	5 5 0	5 4 1	0.147
3 3 1	2 2 0	0.165	4 2 2	4 1 3	0.172	5 3 2	6 2 5	0.173
3 1 3	2 2 0	0.164	4 2 3	4 1 4	0.168	5 0 5	4 1 4	0.165
...	...	...	...	...	...	...	...	...
8 6 2	8 5 3	0.147	8 5 4	8 4 5	0.147	9 2 8	10 1 9	0.137
8 6 3	7 5 2	0.145	8 1 7	8 2 6	0.158	9 7 3	10 8 2	0.131
8 6 2	7 7 1	0.129	8 2 6	7 1 7	0.166	9 2 8	8 3 5	0.155
8 2 7	8 1 8	0.153	8 3 6	8 2 7	0.156	9 2 7	9 1 8	0.146
8 3 6	8 4 5	0.156	8 4 5	8 5 4	0.150	9 7 2	8 8 1	0.119
9 2 8	8 1 7	0.150	9 3 7	8 4 4	0.158	10 4 7	11 5 6	0.142
9 2 7	10 3 8	0.140	9 3 6	10 4 7	0.140	10 3 8	11 2 9	0.126
9 4 6	8 3 5	0.150	9 9 0	10 10 1	0.079	10 7 3	10 8 2	0.136
9 8 1	8 7 2	0.116	10 1 10	10 2 9	0.136	10 1 9	11 2 10	0.124
9 5 4	9 4 5	0.150	10 6 4	10 5 5	0.148	10 7 3	11 8 4	0.141
9 5 4	10 4 7	0.151	10 2 8	10 1 9	0.132	10 4 6	11 5 7	0.135
9 5 5	10 4 6	0.137	10 2 9	11 1 10	0.124	10 7 4	11 8 3	0.142
9 4 5	9 3 6	0.146	10 6 4	9 5 5	0.149	10 4 7	9 5 4	0.153
9 3 7	10 2 8	0.139	10 6 5	9 5 4	0.148	10 4 7	10 5 6	0.139
9 3 6	10 2 9	0.143	10 6 5	11 5 6	0.132	10 0 10	11 11 1	0.133
9 2 8	10 3 7	0.150	10 6 4	11 5 7	0.149	10 3 8	10 2 9	0.132
9 0 9	8 1 8	0.151	10 5 6	11 6 5	0.149	10 3 8	9 2 7	0.141
9 5 5	10 6 4	0.154	10 6 5	10 5 6	0.136	10 0 10	10 1 9	0.136
9 2 7	8 3 6	0.153	10 5 6	10 6 5	0.140	10 7 4	10 8 3	0.126
9 6 3	8 7 2	0.138	10 1 10	9 0 9	0.143	10 4 7	10 3 8	0.134
9 8 2	10 7 3	0.135	10 2 9	9 1 8	0.138	10 8 2	10 7 3	0.139
9 3 6	8 2 7	0.156	10 2 8	9 1 9	0.144	10 8 3	10 7 4	0.125
9 4 5	8 3 6	0.157	10 2 9	10 1 10	0.133	10 8 2	11 7 5	0.133
9 5 5	8 6 2	0.149	10 6 4	11 7 5	0.146	10 8 3	11 7 4	0.143

9 6 4	8 7 1	0.133	10 6 5	11 7 4	0.151	10 0 10	9 1 9	0.143
9 5 4	8 6 3	0.146	10 2 8	11 3 9	0.126	10 4 7	11 3 8	0.130
9 6 4	10 7 3	0.147	10 5 5	10 6 4	0.153	10 4 6	11 3 9	0.136
9 8 2	8 7 1	0.120	10 5 5	11 6 6	0.140	10 4 7	9 3 6	0.141
9 5 4	10 6 5	0.147	10 1 10	11 0 11	0.133	10 3 7	10 2 8	0.135
9 0 9	10 1 10	0.144	10 6 5	10 7 4	0.134	10 3 7	11 2 10	0.132
9 6 3	10 7 4	0.145	10 2 9	11 3 8	0.137	10 3 8	10 4 7	0.137
9 2 7	8 1 8	0.155	10 5 6	11 4 7	0.133	10 8 2	10 9 1	0.117
9 8 1	10 7 4	0.125	10 2 8	10 3 7	0.137	10 3 7	11 4 8	0.130
9 3 6	8 4 5	0.150	10 1 10	9 2 7	0.147	10 3 8	9 4 5	0.146
9 8 2	10 9 1	0.108	10 1 9	9 2 8	0.138	10 8 2	11 9 3	0.124
9 4 6	9 3 7	0.145	10 6 4	10 7 3	0.149	10 8 3	11 9 2	0.124
9 8 1	10 9 2	0.109	10 5 5	10 4 6	0.142	10 8 3	10 9 2	0.111
9 9 0	10 8 3	0.110	10 5 5	11 4 8	0.143	10 9 1	10 8 2	0.122
9 6 3	8 5 4	0.150	10 2 9	10 3 8	0.135	10 9 2	11 8 3	0.131
9 9 0	8 8 1	0.089	10 7 4	11 6 5	0.138	10 3 8	11 4 7	0.140
9 0 9	9 1 8	0.148	10 5 5	9 4 6	0.148	10 4 6	9 3 7	0.148
9 6 4	8 5 3	0.150	10 5 6	10 4 7	0.137	10 9 1	11 8 4	0.119
9 6 4	9 5 5	0.137	10 2 8	9 3 7	0.141	10 4 6	10 3 7	0.139
9 9 1	8 8 0	0.089	10 2 9	9 3 6	0.144	10 3 7	9 4 6	0.142
9 2 8	9 1 9	0.145	10 7 3	11 6 6	0.145	10 9 2	10 8 3	0.114
9 6 4	10 5 5	0.139	10 7 4	10 6 5	0.131	10 9 2	11 10 1	0.101
9 3 7	9 4 6	0.147	10 5 6	9 4 5	0.144	10 3 7	10 4 6	0.140
9 4 6	10 3 7	0.138	10 1 9	10 2 8	0.135	10 9 1	11 10 2	0.101
9 3 6	9 4 5	0.148	10 3 7	9 2 8	0.147	10 9 1	10 10 0	0.089
9 9 1	10 8 2	0.116	10 1 10	11 2 9	0.138	10 2 8	11 1 11	0.130
9 3 7	10 4 6	0.150	10 1 9	10 0 10	0.134	10 9 2	10 10 1	0.088
9 6 3	9 5 4	0.150	10 4 6	9 5 5	0.142	10 10 1	11 11 0	0.074
9 6 3	10 5 6	0.150	10 4 6	10 5 5	0.146	10 10 0	11 9 3	0.105
9 9 1	10 10 0	0.079	10 7 3	10 6 4	0.149	10 10 1	11 9 2	0.111
10 10 1	10 9 2	0.096	10 10 0	10 9 1	0.097	10 10 0	11 11 1	0.074

Table 4.5. Parameters x_{kl} of the $\gamma(sur)$ model at 296 K and the temperature exponents n_k for determination of H_2S self-broadening coefficients (ν_2 band) [20]*. Reproduced with permission.

x_{11}	$0.275(5)\cdot 10^{-2}$	x_{40}	5.16(0.21)	n_1	−0.3918(56)
x_{20}	0.194(1)	x_{41}	−0.49(2)	n_2	−0.9200(31)
x_{21}	$-0.484(7)\cdot 10^{-2}$	x_{42}	$0.215(7)\cdot 10^{-1}$	n_3	−0.3687(127)
x_{30}	0.107(7)	x_{50}	$0.11(4)\cdot 10^{-1}$	n_4	−0.4394(134)
x_{31}	$0.306(4)\cdot 10^{-2}$				

* Parameters x_{11}, x_{20} and x_{21} are in cm^{-1} atm^{-1}, other parameters are dimensionless. At 296 K 2146 theoretical and experimental broadening coefficients were used. The temperature exponents n_k of Eq. (3.5.1) were deduced from 9670 data for 296 K ≤ T ≤ 1300 K.

Table 4.6. Quality of the $\gamma(sur)$ model fit to the calculated H_2S self-broadening coefficients at 296 K (ν_2 band) [20]*. Reproduced with permission.

Δ, %	N_L	N_L/N, %
0–10	1883	87.75
10–20	257	12
20–30	6	0.25

* $\Delta = |[\gamma(RB') - \gamma(sur)]/\gamma(RB')| \cdot 100\%$, N_L is the number of calculated $\gamma(sur)$ values with a given Δ, the total number of lines $N = 2146$, $J \le 15$, $K_a \le 13$, $\Delta K_a = |K_{a\,i} - K_{a\,f}| \le 3$.

An example of surface $\gamma(sur)$ for the H_2S self-broadening case is shown in Fig. 4.1. Figures 4.2 and 4.3 demonstrate the calculated dependences of self-broadening coefficients on the rotational quantum numbers for transitions $[J\ K_a\ K_c] \to [J{+}1\ K_a{+}1\ K_c{-}1]$: the first one shows the γ dependence on K_a for a given J and the second one shows the γ dependence on J for a given K_a.

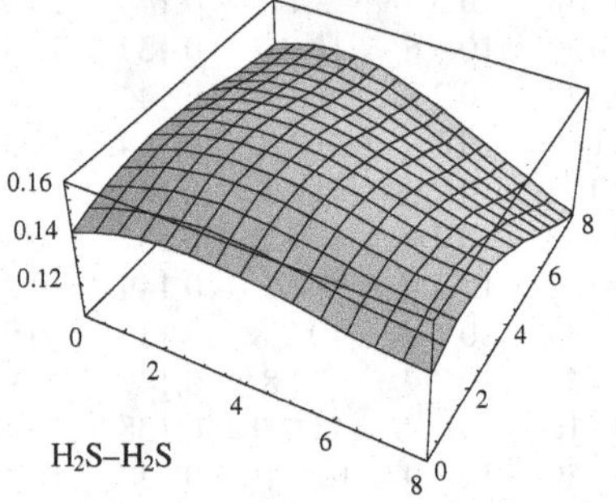

Fig. 4.1. Surface $\gamma(sur)$ for $J_i = 8$, $J_f = 8$ (ν_2 band).

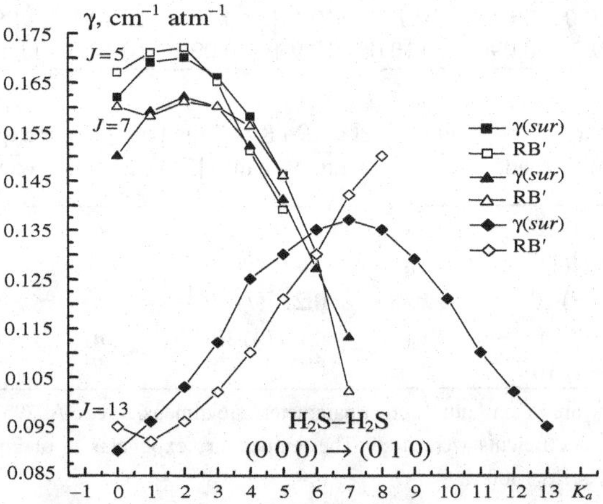

Fig. 4.2. Calculated dependence of H_2S self-broadening coefficients on K_a (for a given J) for some lines $[J\,K_a\,K_c] \to [J{+}1\,K_a{+}1\,K_c{-}1]$ of the ν_2 band [20]. Reproduced with permission.

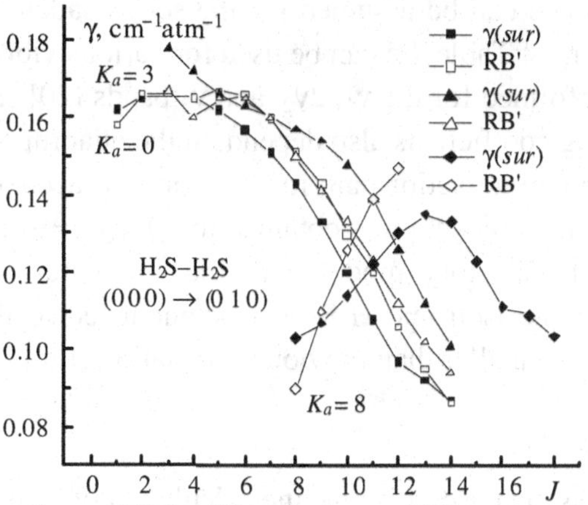

Fig. 4.3. Calculated dependence of H_2S self-broadening coefficients on J (for a given K_a) for some lines $[J\ K_a\ K_c] \rightarrow [J+1\ K_a+1\ K_c-1]$ of the ν_2 band [20]. Reproduced with permission.

The $\gamma(sur)$ model of Eq. (3.3.4) and the parameters of Table 4.5 allowed the calculation of the H_2S self-broadening coefficients in the ν_2 band for $J \leq 15$, $K_a \leq 13$ at 296 K $\leq T \leq$ 1300 K [20] which demonstrates a very weak rotational dependence for T > 700 K (Fig. 4.4).

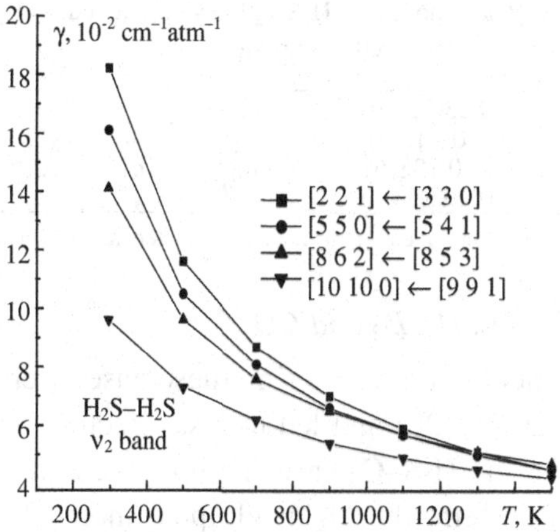

Fig. 4.4. Temperature dependence of H_2S self-broadening coefficients for some lines of the ν_2 band obtained with $\gamma(sur)$ model [20]. Reproduced with permission.

For the polar interacting molecules (large dipole moments) the main contribution to the self-broadening coefficients is given by the dipole–dipole interaction. Since the H_2S dipole moment weakly depends on the vibrational state [4], a weak vibrational dependence can be assumed for the self-broadening coefficients and the $\gamma(sur)$ parameters of Table 4.5 can be used for various vibrational bands.

Calculations performed for the ν_1, $2\nu_2$, and ν_3 bands [20] show that the calculated self-broadening coefficients also depend on the quadrupole moments so that the dipole–quadrupole interactions are to be taken into account. The best agreement with the experimental data was obtained for $Q_{xx} \cdot Q_{zz} \approx -6.0$ (D Å)^2 [6].

Using the exact trajectory model with the parameters ε/k_B = 309 K and σ = 3.59 Å [21] for the isotropic H_2S–H_2S potential led to the line-broadening coefficients about 5% smaller than previous calculations [20].

4.1.1.2. H_2O-broadening

The RB′ method was also used [20] for the calculation of γ coefficients of H_2S ν_2 lines perturbed by water vapour (J_i, $K_{a\,i} \leq 10$, $\Delta K_a \leq 3$). The calculated values given in Table 4.3 demonstrate larger values than for the case of self-broadening (due to the larger H_2O dipole moment). No comparison with experimental values was made due to the absence of measurements. The parameters of the $\gamma(sur)$ model (Eq. (3.3.4)) were also determined from 981 calculated RB′ values (J_i, $K_{a\,i} \leq 10$, $\Delta K_a \leq 3$) for T = 296 K (Table 4.7).

Table 4.7. Parameters of the $\gamma(sur)$ model for H_2S ν_2 lines broadened by H_2O at 296 K [20]*. Reproduced with permission.

x_{11}	$0.230(2) \cdot 10^{-2}$	x_{40}	4.63(16)
x_{20}	0.246(3)	x_{41}	−0.244 (2)
x_{30}	−0.138(6)	x_{50}	$0.10(1) \cdot 10^{-1}$
x_{31}	$0.293(3) \cdot 10^{-2}$		

* Parameters x_{11} and x_{20} are in cm^{-1} atm^{-1}, other parameters are dimensionless.

4.1.1.3. Broadening by N_2, O_2, H_2, D_2 and CO_2

The broadening coefficients of H_2S ν_2 lines by these gases were calculated in the framework of the ATC theory [20] and led to a satisfactory agreement with experimental data except for the H_2S–O_2 case. To improve agreement with experimental data for this last system, the O_2 quadrupole moment value was further kept free. The parameters of $\gamma(sur)$ model (Eq. (3.3.4)) for H_2S mixtures with the abovementioned gases were determined for two sets of input data. In the first set, only the experimental data [12] were included and resulted in the surface $\gamma_1(sur)$. In the second set, the ATC-computed values were added, leading to the surface $\gamma_2(sur)$ (Table 4.8). Figures 4.5–4.9 show a comparison of broadening coefficients calculated with both models $\gamma_1(sur)$ and $\gamma_2(sur)$, the ATC theory as well as experimentally determined values (the lines are numbered in decreasing order of experimental data).

Table 4.8. Parameters of the $\gamma_2(sur)$ model for H_2S ν_2 lines [20]*. Reproduced with permission.

Parameter	$H_2S–N_2$	$H_2S–O_2$	$H_2S–H_2$	$H_2S–D_2$	$H_2S–CO_2$
x_{20}	$9.64(17)\cdot10^{-2}$	$7.08(33)\cdot10^{-2}$	$8.48(29)\cdot10^{-2}$	$7.44(29)\cdot10^{-2}$	$15.19(35)\cdot10^{-2}$
x_{21}	$-1.38(17)\cdot10^{-3}$	$-0.76(2)\cdot10^{-3}$	$-0.81(15)\cdot10^{-3}$	$-0.62(2)\cdot10^{-3}$	$-2.72(1)\cdot10^{-3}$
x_{30}	0.122(37)	0.141(46)	0.180(37)	0.159(40)	0.271(80)
x_{31}	$-7.3(1.6)\cdot10^{-4}$	$-1.1(2)\cdot10^{-3}$	$-4.1(2)\cdot10^{-3}$	$-3.6(2)\cdot10^{-3}$	$-8.5(1)\cdot10^{-3}$
x_{40}	−1.32(8)	−0.95(9)	−0.70(4)	−0.93(11)	−0.17(1)
x_{41}	0.32(4)	0.34(5)	0.295(40)	0.34(6)	0.356(11)
x_{50}	$-1.56(19)\cdot10^{-2}$	$-1.12(1)\cdot10^{-2}$	$0.49(3)\cdot10^{-2}$	$0.37(4)\cdot10^{-2}$	$0.28(1)\cdot10^{-2}$
x_{51}	$9.6(7)\cdot10^{-4}$	$8.0(8)\cdot10^{-4}$			
$J_{i\,max}$	15	15	15	15	15
$K_{a\,i\,max}$	12	13	15	15	15
ΔK_a	≤ 3	≤ 3	≤ 3	≤ 3	≤ 3
N	2295	2173	2295	2295	2295

* Parameters x_{20} and x_{21} are in $cm^{-1}\,atm^{-1}$, other parameters are dimensionless; N denotes the total number of input data, $T = 296$ K.

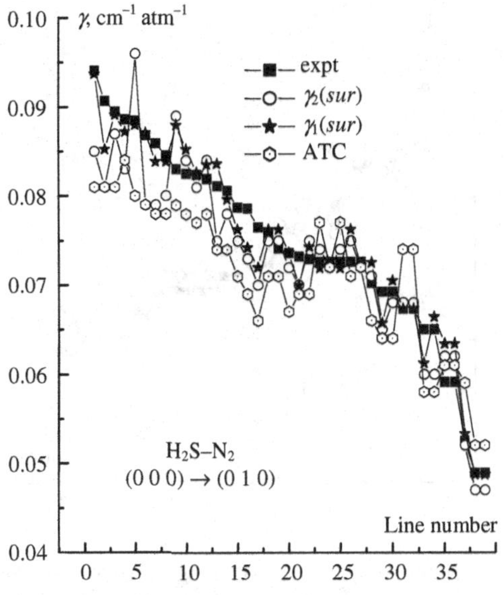

Fig. 4.5. Calculated and experimental [12] N_2-broadening coefficients for H_2S lines in the ν_2 band (lines are numbered in decreasing order of experimental data) [20]. Reproduced with permission.

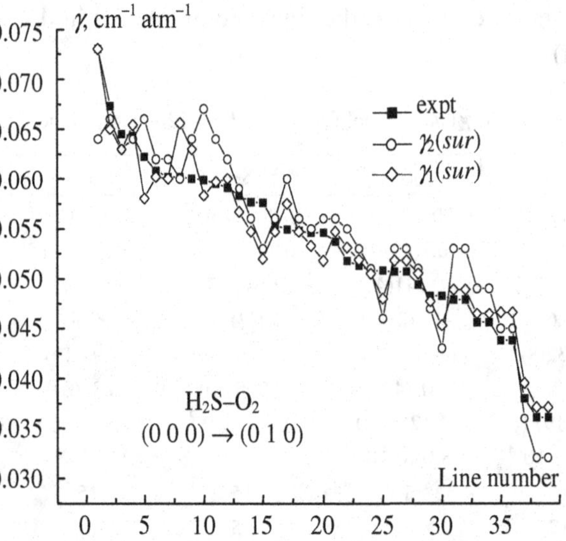

Fig. 4.6. Calculated and experimental [12] O_2-broadening coefficients for H_2S lines in the ν_2 band (lines are numbered in decreasing order of experimental data) [20]. Reproduced with permission.

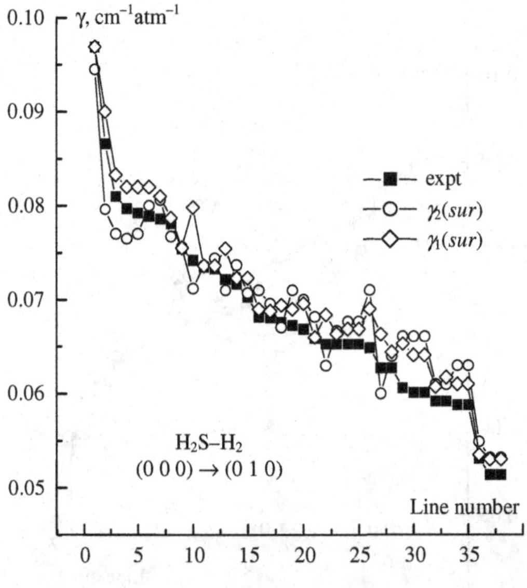

Fig. 4.7. Calculated and experimental [12] H_2-broadening coefficients for H_2S lines in the ν_2 band (lines are numbered in decreasing order of experimental data) [20]. Reproduced with permission.

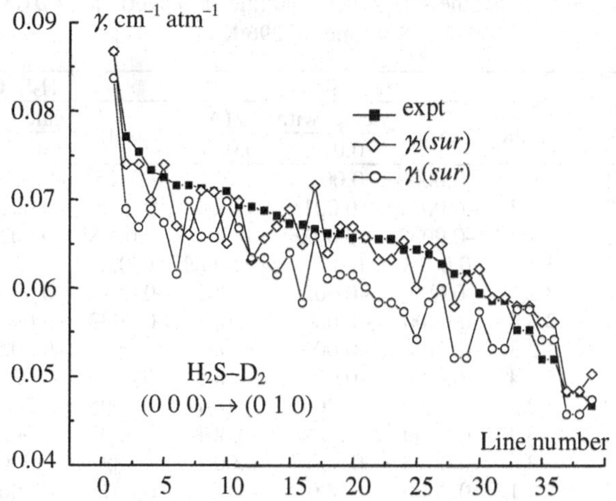

Fig. 4.8. Calculated and experimental [12] D_2-broadening coefficients for H_2S lines in the ν_2 band (lines are numbered in decreasing order of experimental data) [20]. Reproduced with permission.

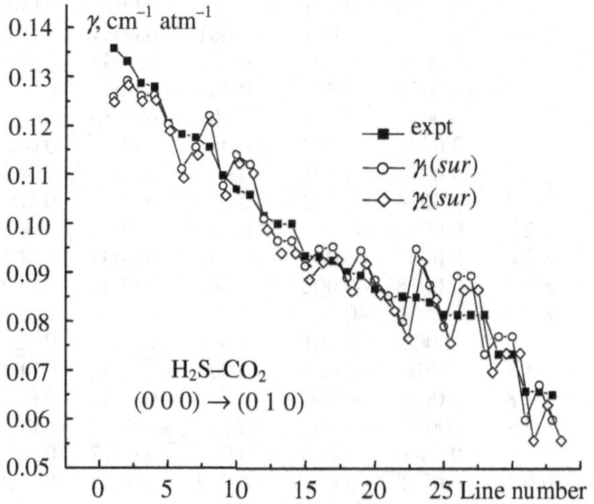

Fig. 4.9. Calculated and experimental [12] CO_2-broadening coefficients for H_2S lines in the ν_2 band (lines are numbered in decreasing order of experimental data) [20]. Reproduced with permission.

Despite the fact that the vibrational dependence of H_2S polarizability α is unknown, the difference $\Delta\alpha = \alpha(010) - \alpha(000)$ varying between 0.015 Å^3 and 0.020 Å^3 gives a reasonable agreement of the ATC-calculated values with the experimental data [12] for the pressure-induced shift coefficients δ in the ν_2 band (Table 4.9).

Table 4.9. Experimental [12] and the ATC-calculated line-shift coefficients δ (in cm^{-1} atm^{-1}) for H_2S ν_2 lines at 296 K.

ν, cm^{-1}	(010) $J\,K_a\,K_c$	← (000) $J\,K_a\,K_c$	H_2S–N_2 expt	H_2S–N_2 calc. with $\Delta\alpha$ (Å^3) 0.015	H_2S–N_2 calc. with $\Delta\alpha$ (Å^3) 0.020	H_2S–O_2 expt	H_2S–O_2 calc. with $\Delta\alpha$ (Å^3) 0.015	H_2S–O_2 calc. with $\Delta\alpha$ (Å^3) 0.020
1263.4153	3 3 1	2 0 2	–0.0000	0.001	0.000	0.0015	–0.000	–0.000
1149.7695	2 1 2	3 0 3	–0.0029	–0.001	–0.000	–0.0029	–0.002	–0.002
1149.4014	2 0 2	3 1 3	–0.0032	–0.002	–0.002	–0.0068	–0.002	–0.002
1144.2649	2 2 1	3 1 2	0.0013	0.000	0.000	0.0028	–0.001	–0.002
1253.4644	4 3 2	3 2 1	–0.0006	–0.002	–0.002	–0.0016	–0.000	–0.001
1266.9334	4 4 1	3 3 0	0.0000	–0.000	–0.001	–0.0033	–0.000	–0.001
1304.0056	5 3 2	4 2 3	–0.0000	–0.000	–0.000	0.0000	–0.002	–0.002
1122.8520	4 1 3	5 2 4	–0.0031	–0.002	–0.003	–0.0047	–0.002	–0.003
1117.4890	4 3 2	5 2 3	–0.0005	–0.000	–0.001	–0.0005	–0.000	–0.001
1322.5491	6 4 2	5 3 3	0.0000	–0.000	–0.001	–0.0007	–0.002	–0.002
1286.5657	6 4 3	5 3 2	0.0016	0.000	0.000	0.0000	–0.000	–0.001
1303.7024	6 5 2	5 4 1	0.0000	0.000	–0.000	–0.0018	–0.000	–0.001
1105.7183	5 2 3	6 3 4	–0.0017	–0.003	–0.003	–0.0043	–0.003	–0.003
1279.7048	7 2 5	6 3 4	–0.0013	–0.001	–0.001	–0.0014	–0.001	–0.002
1110.9890	6 1 6	7 0 7	–0.0047	–0.002	–0.002	–0.0047	–0.002	–0.003
1110.9888	6 0 6	7 1 7	–0.0047	–0.002	–0.002	–0.0047	–0.002	–0.003
1289.8918	8 3 6	7 2 5	–0.0008	–0.001	–0.001	–0.0013	–0.001	–0.002
1289.8357	8 2 6	7 3 5	–0.0015	–0.001	–0.001	–0.0021	–0.002	–0.002
1101.0584	7 1 7	8 0 8	–0.0037	–0.001	–0.002	–0.0047	–0.002	–0.003
1101.0584	7 0 7	8 1 8	–0.0037	–0.001	–0.002	–0.0047	–0.002	–0.003
1095.6958	7 2 6	8 1 7	–0.0017	–0.001	–0.002	–0.0036	–0.002	–0.002
1252.8633	8 2 6	8 1 7	–0.0022	–0.002	–0.002	–0.0031	–0.002	–0.003
1095.6985	7 1 6	8 2 7	–0.0017	–0.001	–0.002	–0.0036	–0.002	–0.002
1252.8745	8 3 6	8 2 7	–0.0014	–0.002	–0.002	–0.0042	–0.002	–0.003
1089.6241	7 3 5	8 2 6	–0.0016	–0.002	–0.002	–0.0044	–0.002	–0.003
1299.7693	9 3 7	8 2 6	–0.0037	–0.000	–0.001	–0.0037	–0.002	–0.002
1089.5537	7 2 5	8 3 6	–0.0018	–0.002	–0.002	–0.0036	–0.002	–0.003
1299.7585	9 2 7	8 3 6	–0.0030	–0.001	–0.001	–0.0024	–0.002	–0.002
1313.2622	9 3 6	8 4 5	0.0000	–0.001	–0.002	–0.0020	–0.002	–0.002
1263.0420	9 2 7	9 1 8	0.0010	–0.002	–0.002	–0.0037	–0.002	–0.003
1294.5083	10 2 9	9 1 8	–0.0016	–0.000	–0.001	–0.0017	–0.002	–0.003
1263.0442	9 3 7	9 2 8	0.0010	–0.002	–0.002	–0.0037	–0.002	–0.003
1294.5083	10 1 9	9 2 8	–0.0016	–0.000	–0.001	–0.0017	–0.002	–0.003
1286.8428	11 1 11	10 0 10	–0.0015	–0.002	–0.002	–0.0025	–0.003	–0.004
1273.0593	10 2 8	10 1 9	–0.0000	–0.002	–0.003	–0.0046	–0.003	–0.003
1273.0598	10 3 8	10 2 9	–0.0000	–0.002	–0.003	–0.0046	–0.003	–0.003
1295.0949	12 1 12	11 0 11	–0.0008	–0.002	–0.003	–0.0024	–0.003	–0.004
1295.0949	12 0 12	11 1 11	–0.0008	–0.002	–0.003	–0.0024	–0.003	–0.004

4.1.2. SO_2 molecule

Spectroscopic studies of SO_2 line-broadening coefficients in various frequency regions have been realised by different experimental groups (see, for example, Refs [16, 22]).

4.1.2.1. Self-broadening

The SO_2 self-broadening coefficients were measured [16] for 118 lines in the $\nu_1 + \nu_3$ band, for 10 lines in the ν_3 band, and for 10 lines in the $\nu_1 + \nu_2 + \nu_3 - \nu_2$ band in the region of 2464–2503 cm^{-1} ($2 \le J \le 48$, $0 \le K_a \le 16$). These coefficients vary from 0.24 to 0.504 $cm^{-1}\,atm^{-1}$. It was found that the dependence of the self-broadening coefficients on the rotational quantum number J is negligible but they decrease with K_a increasing. These experimental values can be used to determine the parameters of $\gamma(sur)$ model given by Eq. (3.3.4). These parameters are listed in Table 4.10 and the quality of fitting can be seen in Fig. 4.10 (the lines are numbered according to the list of experimental values).

Table 4.10. Parameters of the $\gamma(sur)$ model (Eq. (3.3.4)) and temperature exponents n_k (Eq. (3.5.1)) for SO_2 and O_3 line-broadening coefficients at 296 K (valid for lines with $\Delta K_a = 0$, 1 only).

Parameter	SO_2–SO_2	O_3–O_3	O_3–air	O_3–O_2
x_{20}	0.3824(79)	0.1098(7)	$0.9178(23)\cdot10^{-1}$	$0.8303(72)\cdot10^{-1}$
x_{21}	$0.266(44)\cdot10^{-2}$	$-0.1998(176)\cdot10^{-3}$	$-0.2788(45)\cdot10^{-3}$	$-0.3311(166)\cdot10^{-3}$
x_{22}	$-0.294(44)\cdot10^{-4}$	–	–	
x_{30}	$-0.591(31)\cdot10^{-1}$	$-0.4320(183)\cdot10^{-1}$	$0.6528(178)\cdot10^{-1}$	$0.8456(623)\cdot10^{-1}$
x_{31}	–	$0.6784(47)\cdot10^{-4}$	$-0.831(33)\cdot10^{-3}$	$-0.1267(145)\cdot10^{-2}$
x_{41}	$0.357(9)\cdot10^{-1}$	0.0	0.2258(61)	0.2480(66)
n_2		−0.6474(157)	−0.5409(67)	−0.6864(121)
n_4			0.5411(111)	
N	136	648	1140	385
ΔK_a	0	0, 1	0, 1	0, 1

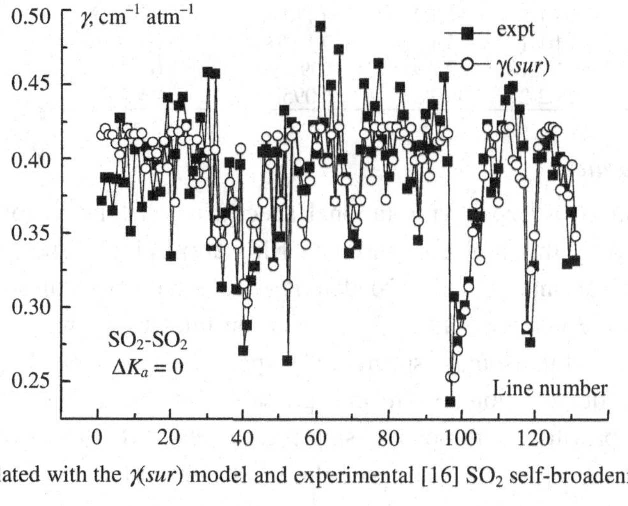

Fig. 4.10. Calculated with the $\gamma(sur)$ model and experimental [16] SO_2 self-broadening coefficients.

The self-broadening of SO_2 lines at terahertz frequencies was experimentally studied [23] for 84 lines ($4 \le J \le 59$, $0 \le K_a \le 10$) leading to values between 110 and 170 MHz KPa^{-1}. The microwave and the far-infrared regions were investigated in Refs [24, 25], respectively.

4.1.2.2. SO_2 line broadening by foreign gases

The SO_2-broadening coefficients by N_2, air, H_2, N_2,, He, Ne, Ar, Kr, and Xe in the ν_1 and ν_3 bands (1341–1357 cm^{-1}) were measured in Refs [22, 26].

The H_2- and He-broadening coefficients in the ν_1 band (near 1121 cm^{-1}) and ν_3 band (1341–1357 cm^{-1}) were respectively obtained for 21 ν_3 lines with $8 \le J \le 33$, $1 \le K_a \le 16$ and for 3 ν_1 lines with $8 \le J \le 36$, $7 \le K_a \le 9$ [22]. For the broadening by Xe, the measurements were made for 6 lines of the ν_3 band (1321–1366 cm^{-1}) with $4 \le J \le 61$, $0 \le K_a \le 6$ and for 15 lines in the ν_1 band (1072–1140 cm^{-1}) [22]. For the mixtures with noble gases the measurements were performed for 3 lines in the ν_3 band at 252–362 K which allowed an estimation of the temperature exponent: $0.72 < n < 0.99$ [22]. The SO_2-air-broadening coefficients for 7 lines were also obtained for the same temperature interval. Some room-temperature (296 K) γ- and n-values [22] are given in Table 4.11.

Table 4.11. SO_2–air line-broadening coefficients at 296 K and temperature exponents [22].

$J\,K_a\,K_c \leftarrow$	$J\,K_a\,K_c$	γ, cm^{-1} atm^{-1}	n
7 1 6	8 1 7	0.106(4)	0.79(19)
7 2 5	8 2 6	0.108(4)	0.84(14)
7 3 4	8 3 5	0.109(3)	0.74(20)
13 1 12	14 1 13	0.114(2)	0.72(23)
14 1 14	15 1 15	0.113(8)	0.81(19)
27 9 18	28 9 19	0.099(10)	0.93(39)
29 2 27	30 2 28	0.095(6)	0.99(31)

4.1.3. O_3 molecule

The investigation of ozone vibrotational spectra has a long history because of the role played by this molecule in the Earth's atmosphere. The precise knowledge of pure rotational O_3 lines broadening is required for studies of the Earth's upper atmosphere where about 80% of emission lines in the region 10–100 cm^{-1} are due to $^{16}O_3$ and its isotopic species. The spectroscopic parameters of ozone in the infrared frequency domain are indispensable for the retrieval of vertical O_3 concentration profiles by remote sensing techniques. A particular interest represents the spectral region of 5μm since the most intense band $\nu_1 + \nu_3$ of the triad

$2\nu_3/\nu_1 + \nu_3/2\nu_1$ is located at these wavelengths and the water vapour absorption is very weak. Many hundred articles have been published on this subject; their review up to 1998 is given in Ref. [27]. The spectroscopic data on this molecule are included in the HITRAN and GEISA databases. According to Ref. [27], the accuracy of line-broadening coefficients has to be better than 5% whereas a critical analysis of available experimental data shows that the actual precision of both experimental and theoretical values is at the 10–15% level. For the temperature exponents the situation is even worse.

A very complete study of rotational O_3 lines self-broadening and broadening by air and oxygen in the region 50–90 cm^{-1} was realised in Ref. [28] using a Fourier transform spectrometer. The broadening coefficient values were obtained for 101 transitions with $7 \le J \le 34$, and $3 \le K_a \le 11$. For many lines the temperature exponent n was also determined from 14 spectra recorded at 212–296 K and simultaneously fitted to obtain γ and n. Their averaged values are presented in the second column of Table 4.12. Detailed studies of some rotational O_3 lines perturbed by nitrogen, oxygen and air were also made in Refs [29–31], their temperature dependence for the range 195–300 K was investigated in Ref. [32]. CRB calculations for the temperature dependent air broadened rotational lines of ozone were made in Refs [31, 33].

Table 4.12. Averaged broadening/shifting coefficients (at 296 K) and temperature exponents for O_3 (vib)rotational lines; δ_0 (in cm^{-1} atm^{-1}) and δ' (in cm^{-1} atm^{-1} K^{-1}) are defined by Eq. (3.2.2).

Parameter	Rotational band [28] $7 \le J \le 34$ $3 \le K_a \le 11$	ν_2 band [34] $3 \le J \le 45$ $0 \le K_a \le 11, \Delta K_a = 1$	ν_1 band [34] $3 \le J \le 45$ $0 \le K_a \le 11, \Delta K_a = 1$
$\gamma_{O_3-O_3}$	0.0904(49)	0.097(9)	0.094(8)
γ_{O_3-air}	0.0752(46)	0.079(6)	0.075(5)
$\gamma_{O_3-O_2}$	0.0687(47)		
$n_{O_3-O_3}$	0.67(11)		
n_{O_3-air}	0.73(19)	0.53(8)	0.67(7)
$n_{O_3-O_2}$	0.67(1)		
δ_{0O_3-air}		−0.0008 (17)	−0.0007 (18)
δ'_{O_3-air}		$(-0.2 \pm 2.9)\cdot 10^{-5}$	$(-0.2 \pm 2.3)\cdot 10^{-5}$

Self-, air-broadening and shift coefficients of O_3 lines and their temperature dependence were measured for the ν_2 band [34] and the ν_1 band [35]. The room-temperature results [34] were reported for 370 transitions of the ν_2 band (630–800 cm^{-1}) with $0 \le J \le 45$, $1 \le K_a \le 12$. For 350 lines the temperature exponent of air-broadened line-width and air-induced line-shift coefficients were determined by the simultaneous fitting of 29 absorption spectra recorded at 210–302 K. The temperature dependence of the line-shift coefficients δ was approximated by Eq. (3.2.2). The averaged values of these broadening and shifting coefficients as well as the temperature exponents for the ν_1 and ν_2 bands are given in Table 4.12.

Nitrogen and oxygen-broadening coefficients of 112 ozone lines ($J \leq 40$, $K_a \leq 11$) at room temperature were measured and calculated by the RB method in Ref. [36]. For calculations, in particular, the atom–atom parameters for the O_3-N_2 and O_3-O_2 interactions were adjusted to match experimental values. On average, the calculated values agreed with the experimental results within 6%, with the maximum deviation being 14%. Simple empirical polynomial corrections of theoretical values as a function of the rotational quantum numbers allowed an improvement in agreement by up to 2% for most lines. Line-broadening coefficients of ozone lines by nitrogen were also calculated by Neshyba and Gamache [37], improving the previous theoretical estimations of O_3-N_2 and O_3-O_2 line widths by Hartmann *et al.* [38].

Broadening coefficients by N_2 and O_2 at low temperatures (227 K and 186 K) were measured [39] for the $\nu_1 + \nu_3$ band located at 2110 cm^{-1}. The temperature exponents were obtained for 35 lines, giving the averaged values $n(O_2) = 0.72(8)$, $n(N_2) = 0.71(8)$ (within two standard deviations).

Recently, Anthony *et al.* [40] applied the MCRB approach to the O_3-N_2 and O_3-O_2 line-broadening coefficients. They found that for these molecular systems characterised by 'intermediate' strength of radiator-perturber interactions the average difference between the CRB and MCRB line widths was 0.47% in the case of N_2 and 0.075% in the case of O_2; for the line shifts the differences were 0.986% and 0.999%, respectively (the CRB results giving a better agreement with experimental data).

Theoretical calculations for nitrogen and oxygen broadening of ozone lines in the region of 5μm ($\nu_1 + \nu_3$ band) were also made by the semi-classical RBE (Sec. 2.14) and the semi-empirical (Sec. 3.4) methods [41]. The semi-classical calculations were realised with the O_3 energy levels and wave functions obtained by a numerical diagonalization of the effective rotational Hamiltonian and freely available from the IAO ozone database [42]. For the semi-empirical approach the correction factor was chosen in the form of Eq. (3.4.13) and the second-order contributions in the long-range interaction potential were taken as

$$S_2 = S_2^{12e} + S_2^{22e} + S_2^{22p} + S_2^{02p} + S_2^{20p}, \qquad (4.1.2)$$

where the superscripts 1 and 2 stand for the tensorial order of the multipole interactions (02 — charge-quadrupole interaction, 12 — dipole–quadrupole interaction, 22 — quadrupole-quadrupole interaction) and the superscripts *e* and *p* distinguish the electrostatic and polarization parts. An example of calculated line widths for the *Q*-branch of the $\nu_1 + \nu_3$ band is presented in Fig. 4.11 together with the corresponding experimental and RB values of Ref. [36] (the specific combination of rotational quantum numbers for the horizontal axis is chosen to separate the line widths with the same *J* but different K_a values). As can be seen from this figure, the use of exact trajectories improves the agreement with experimental data for both perturbers.

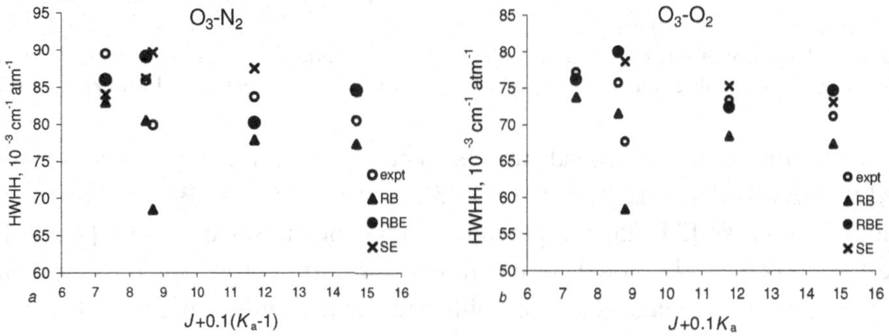

Fig. 4.11. Semi-classical RBE and semi-empirical SE line-broadening coefficients in comparison with experimental and semi-classical RB values [36] for the *Q*-branch of the $\nu_1 + \nu_3$ band of ozone perturbed by nitrogen (*a*) and oxygen (*b*) [41]. Reproduced with permission.

The adjustable parameters $c_1 = 2.7$ and $c_2 = 7.0$ of the SE correction factor were first determined by fitting to experimental data of all branches. For the O_3 molecule, however, the dipole moment is small (in comparison with that of H_2O), so that the correction factor well adapted for the water vapour lines had to be thoroughly tested for ozone. Separate fittings for transitions with various *J* (small $3 \le J \le 16$, middle $17 \le J \le 23$, high $24 \le J \le 41$) in various branches showed that these parameters are quite sensitive to the range of *J*-values. For example, if c_1, c_2 are fitted to experimental line widths with high *J*, their values lead to too small line-broadening coefficients for low *J* (crosses in Figure 4.12*a*). For the O_3–N_2 case the parameter c_2 obtained for high *J*-values differs from its value for low and middle *J* by a factor of two. In the case of O_3–O_2, C_2 is much larger for low *J*-values than for middle and high *J*-values, and the magnitude of both c_1 and c_2 is greater than for nitrogen broadening. This means that for the ozone molecule one pair of parameters c_1, c_2 (which corresponds to the dominant contribution $l = 1$ for H_2O) is not sufficient to correctly describe all the line widths.

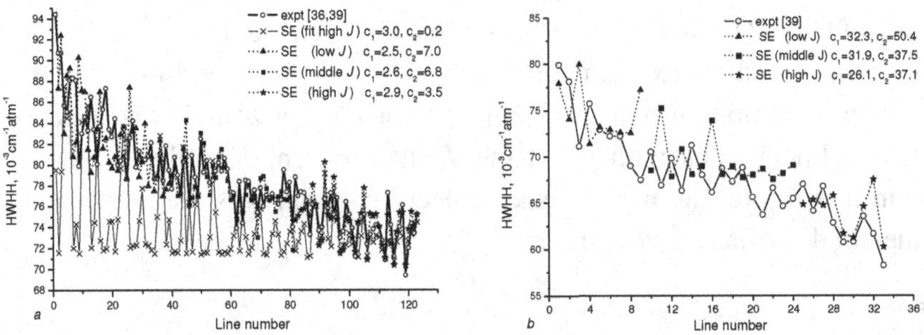

Fig. 4.12. Influence of separate small, middle, high *J*-value fittings on the c_1, c_2 parameters (P- and *R*-branches used simultaneously) for O_3–N_2 (*a*) and O_3–O_2 (*b*) [41]. Reproduced with permission.

While the ozone line broadening has been quite well studied experimentally and theoretically, the data on its line shifts are very sparse. Besides the already mentioned works [34, 35] the previous studies of the same authors [43, 44] as well as Refs [45–47] should be mentioned for currently available experimental results. No experimental data are published for the purely rotational line shifts of O_3 because of their too small values. QFT and ATC calculations were realised for nitrogen-, oxygen-, and air-broadening by Gamache *et al.* [48, 49]; for the rotational line shifts induced by the same perturbers the CRB approach was employed by Drouin *et al.* [31].

The line-shift mechanism has a more complicated character than the line broadening, so that many factors unimportant for the line widths become crucial for the line shifts: very pronounced dependences on the vibrational quantum numbers, type of perturber, isotopic species. The semi-empirical approach of Ref. [41] was also used in Ref. [50] to estimate ozone line shifts by N_2 and O_2 pressure in the $\nu_1 + \nu_3$, $2\nu_1$, and $2\nu_3$ bands. The calculated shifts agree with measurements within 0.001 cm^{-1} atm^{-1} for 98% of O_3–N_2 lines and for 87% of O_3–O_2 lines. The polarizability tensor components in the excited vibrational states were also obtained by least-squares fitting to some experimental O_3–N_2 line shifts: α = 2.89 $Å^3$ and α_{zz} = 5.14 $Å^3$ for the (101) state, α = 2.85 $Å^3$ and α_{zz} = 5.01 $Å^3$ for the (200) state, and α = 2.88 $Å^3$ and α_{zz} = 4.68 $Å^3$ for the (002) state; they were further used in calculation of O_2-induced line shifts. In the same reference, the temperature exponents n were also determined using the relation

$$\delta(T) = \delta(T_0)(T/T_0)^{-n} \tag{4.1.3}$$

with the reference temperature T_0 = 297 K. The calculated shifts are given in Tables 4.13–4.15 for the $\nu_1 + \nu_3$, $2\nu_1$ and $2\nu_3$ bands, respectively.

Table 4.13. Semi-empirical and experimental [45] O_3 line-shift coefficients (in 10^{-3} cm^{-1} atm^{-1}) and temperature exponents n for the $\nu_1+\nu_3$ band (lines used for fitting are marked with an asterisk) [50]. Reproduced with permission.

J'	K_a'	K_c'	J	K_a	K_c	O_3–N_2			O_3–O_2	
						δ(SE)	δ(expt)	n(SE)	δ(SE)	δ(expt)
42	2	41	43	2	42	−3.28	−4.2	0.78	−4.54	−5.3
38*	5	34	39	5	35	−3.25	−4.0	0.94	−4.69	−4.8
37	4	33	38	4	34	−2.60	−4.0	1.10	−3.73	−4.4
34	6	29	35	6	30	−3.37	−3.6	1.10	−4.71	−4.9
32	5	28	33	5	29	−2.92	−3.3	1.06	−4.22	−4.3
31	6	25	32	6	26	−3.27	−3.6	1.10	−4.53	−4.3
31	5	26	32	5	27	−2.80	−2.7	1.12	−3.99	−3.7
31	1	30	32	1	31	−2.85	−3.8	0.83	−4.06	−4.5
30	5	26	31	5	27	−2.86	−3.7	1.09	−4.15	−4.3
31	0	31	32	0	32	−3.52	−4.4	0.74	−4.76	−4.7
30	1	30	31	1	31	−3.50	−4.2	0.74	−4.76	−4.5
28	6	23	29	6	24	−3.14	−3.5	1.10	−4.37	−3.9
29	0	29	30	0	30	−3.48	−3.2	0.75	−4.72	−4.7
27	6	21	28	6	22	−3.11	−3.0	1.10	−4.33	−3.8
28	2	27	29	2	28	−2.89	−3.5	0.88	−4.12	−4.2
28	1	28	29	1	29	−3.46	−3.1	0.76	−4.71	−4.2
27	1	26	28	1	27	−2.66	−3.3	0.87	−3.82	−4.0
27*	0	27	28	0	28	−3.43	−3.3	0.75	−4.66	−3.8
25	5	20	26	5	21	−2.88	−3.0	1.15	−4.13	−3.4
25	3	22	26	3	23	−2.17	−2.9	1.08	−3.26	−3.5
24	6	19	25	6	20	−3.01	−3.0	1.10	−4.27	−3.3
25	1	24	26	1	25	−2.58	−3.1	0.90	−3.72	−4.0
24	5	20	25	5	21	−2.80	−3.0	1.13	−4.03	−3.5
24	4	21	25	4	22	−2.62	−3.2	1.05	−3.87	−4.0
23	6	17	24	6	18	−2.98	−2.4	1.11	−4.26	−3.5
22*	8	15	23	8	16	−3.24	−3.2	1.04	−4.68	−3.3
23	5	18	24	5	19	−2.78	−2.8	1.13	−4.01	−3.6
22	7	16	23	7	17	−3.14	−2.8	1.07	−4.52	−3.6
23	4	19	24	4	20	−2.38	−2.9	1.13	−3.53	−3.4
22	5	18	23	5	19	−2.73	−3.2	1.12	−3.95	−3.9
21	7	14	22	7	15	−3.10	−3.1	1.06	−4.49	−3.4
23	0	23	24	0	24	−3.27	−3.1	0.79	−4.56	−4.0
20	8	13	21	8	14	−3.14	−3.0	1.02	−4.45	−3.4
20	7	14	21	7	15	−3.06	−3.1	1.05	−4.41	−3.4
19	7	12	20	7	13	−3.02	−2.4	1.04	−3.37	−3.5
14	3	12	15	3	13	−2.25	−2.1	1.01	−3.45	−2.6
10	4	7	11	4	8	−2.28	−1.8	0.99	−3.39	−2.4
8	4	5	9	4	6	−2.25	−2.2	0.94	−3.36	−2.5
3*	2	1	2	2	0	−2.39	−2.0	0.81	−4.66	−1.2
27	0	27	26	0	26	−3.43	—	0.75	−4.53	−3.0
37	6	31	36	6	30	−3.26	—	1.09	−4.01	−3.3
37	2	35	36	2	34	−2.74	−3.2	1.03	−4.54	−3.9

Table 4.14. Semi-empirical and experimental [46] line-shift coefficients (in 10^{-3} cm^{-1} atm^{-1}) for the $2\nu_1$ band of O_3–N_2 (lines used for fitting are marked with an asterisk) [50]. Reproduced with permission.

J'	K_a'	K_c'	J	K_a	K_c	SE	expt
25*	6	20	26	7	19	−2.23	−1.9
28	5	23	29	6	24	−2.37	−1.7
14	7	7	15	8	8	−1.71	−1.8
28	4	24	29	5	25	−2.92	−1.5
24	4	20	25	5	21	−2.57	−1.7
29	3	27	30	4	26	−1.12	−1.2
13*	3	11	14	4	10	−2.11	−2.8
43*	1	43	44	0	44	−2.56	−2.5
42	0	42	43	1	43	−2.56	−3.4
18	2	16	19	3	17	−2.60	−2.4
32	0	32	33	1	33	−2.34	−2.2
37	4	34	38	3	35	−0.89	−1.7
25	1	25	26	0	26	−2.13	−1.2
25	2	24	26	1	25	−1.32	−2.7
17	1	17	18	0	18	−1.85	−2.1
12	0	12	13	1	13	−2.52	−1.7
15	2	14	16	1	15	0.04	−2.3

Table 4.15. Semi-empirical and experimental [46] O_3–N_2 line-shift coefficients (in 10^{-3} cm^{-1} atm^{-1}) for the $2\nu_3$ band (lines used for fitting are marked with an asterisk) [50]. Reproduced with permission.

J'	K_a'	K_c'	J	K_a	K_c	SE	expt
8*	8	0	9	9	1	−3.95	−4.8
21*	8	14	21	9	13	−6.31	−7.0
26	6	20	26	7	19	−6.20	−6.2
18	6	12	18	7	11	−5.68	−5.2

For the case of self-, N_2-, and O_2-broadening of O_3 lines the model $\gamma(sur)$ (Eqs (3.3.4) and (3.3.6)) can also be used. The necessary parameters are given in Table 4.10. They were deduced from the experimental data [28, 31, 34, 39] including rotational as well as vibrational ν_2 and $\nu_1 + \nu_3$ bands. Since for 90% of measurements [28] the difference between $\gamma(sur)$ and experimental values is less than 5%, the quality of fitting can be considered to be very high. (For any experimental datum this difference is inferior to 10%.) The comparison of $\gamma(sur)$ and corresponding experimental values for the self-broadened rotational O_3 lines and O_3-O_2 lines of the band $\nu_1 + \nu_3$ can be seen in Figs 4.13, 4.14, respectively (the lines are numbered in increasing order of experimental values). For the O_3-O_2 vibrotational lines the $\gamma(sur)$ model correctly reproduces the experimental values in the temperature range 186–296 K.

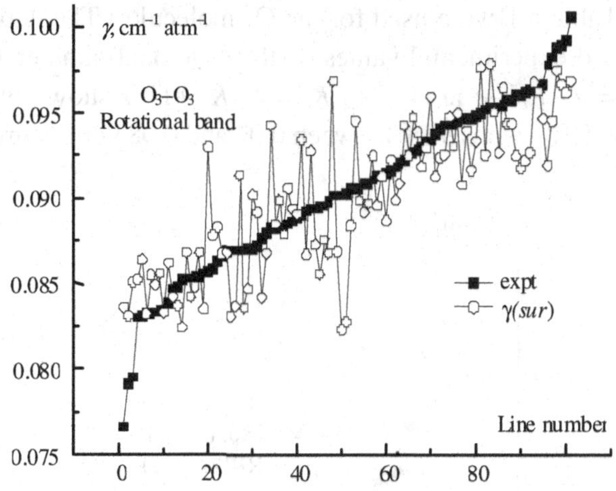

Fig. 4.13. Experimental [28] and calculated by the $\gamma(sur)$ model O_3 self-broadening coefficients at 296 K (lines are numbered in increasing order of experimental data).

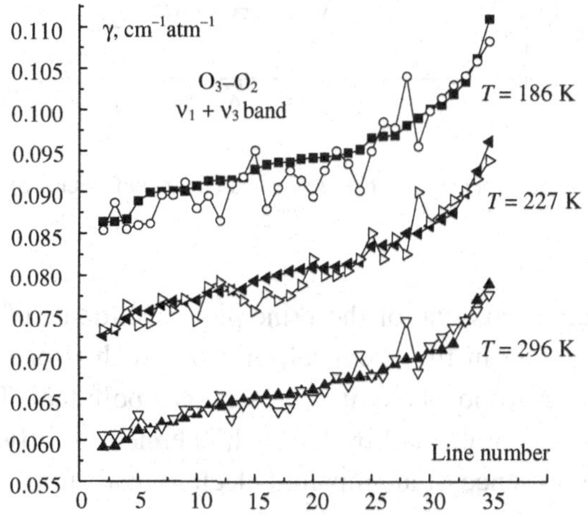

Fig. 4.14. Experimental [28] (solid symbols) and calculated with the $\gamma(sur)$ model (empty symbols) O_3–O_2 line-broadening coefficients at various temperatures (lines are numbered in increasing order of experimental data).

The O_3 self-broadening coefficients calculated in the framework of the RB′ approach for 101 lines of the rotational band show that for 93 lines the difference with experimental values [28] $\Delta = |\gamma(\text{expt}) - \gamma(\text{calc})|/\gamma(\text{expt})$ is smaller that 5% while for the other eight lines it is between 5 and 10%. (For these calculations the dipole moment, the quadrupole moment components and the polarizability tensor components of Table 4.1 were used for the O_3 molecule.) The dependence of the RB′-calculated and experimental values on the rotational number K_a for the transitions $[J\ K_a\ K_c = J - K_a] \rightarrow [J{+}1\ K_a{+}1\ K_c = J - K_a + 1]$ is shown in Fig. 4.15. For $J \geq 13$, $K_a > 9$ the RB′ values vary between 0.07 and 0.083 cm^{-1} atm^{-1}.

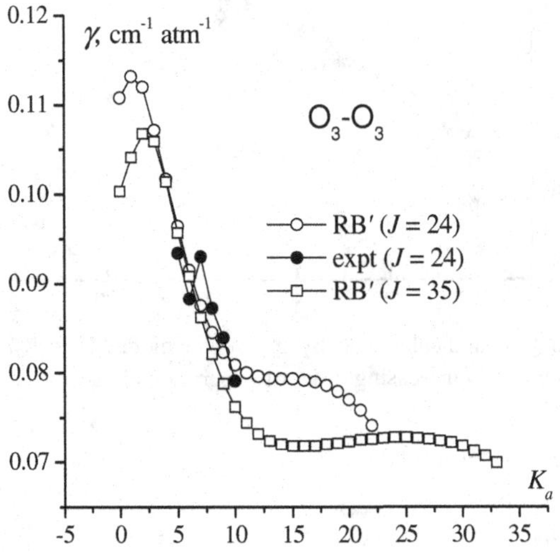

Fig. 4.15. RB′-calculated and experimental [28] self-broadening coefficients of O_3 rotational lines.

4.1.4. NO_2 molecule

Nitrogen dioxide represents one of the principle constituents of the stratosphere and plays a decisive role in the chemical reactions with ozone in the upper atmosphere. In the lower troposphere, it is present as a pollutant. The investigation of its spectral lines is complicated by the high chemical reactivity of this molecule. Moreover, the presence of an unpaired electron leads to the splitting of each vibrotational state into two components. There are a few papers only [51–55] which report some measurements of NO_2 line shape parameters. Semi-classical calculations of NO_2 widths were made by Tejwani and Yeung [56, 57].

Experimental values of NO_2 self-, N_2-, and O_2-broadening coefficients in the $\nu_1 + \nu_3$ band (2633–3511 cm^{-1}) were obtained from Fourier transform spectra of mainly unresolved doublets [52]. These coefficients were determined for the lines with $N'' \leq 55$ in the self-broadening case and with $N'' \leq 40$ for the N_2- and O_2-broadening cases (N'' is the rotational quantum number of the lower transition level). A slight dependence of γ on N'' was stated, whereas no significant dependence was observed on K_a. In a very few cases, however, the spin–rotational splitting was quite important and the doublet components were observable separately. For these cases it was noted that in general the ++ component is 10% broader than the — component. The influence of the spin–rotational splitting error on the broadening coefficient value was additionally estimated: an error of 0.0001 cm^{-1} typically induces a systematic error of about 5% in the self-broadening coefficient deduced from a pure NO_2 spectrum at 50 Torr. Simple semi-empirical relations were also given for estimation of the broadening coefficients (in 10^{-3} cm^{-1} atm^{-1}) at 296 K and $N'' \leq 40$, for example, for $\nu_1 + \nu_3$ NO_2 lines:

$$\gamma_{NO_2-N_2} = 86.0 - 0.765N'' + 0.0063N''^2, \tag{4.1.4}$$

$$\gamma_{NO_2-O_2} = 76.9 - 0.709N'' + 0.0046N''^2, \tag{4.1.5}$$

$$\gamma_{NO_2-air} = 84.1 - 0.753N'' + 0.0059N''^2. \tag{4.1.6}$$

When $N'' < 40$ and $K_a'' < 10$, the uncertainly of γ_{NO_2-air} is about 10%. Since no important vibrational dependence was stated for the existing experimental data, these formulae can be used for other vibrational bands. The Ne-, Ar-, Kr-, and Xe-broadening coefficients of NO_2 lines in the ν_3, $\nu_2 + \nu_3 - \nu_2$ and $2\nu_2$ bands were obtained in Ref. [51].

4.2. Vibrotational lines of ethylene

Ethylene C_2H_4 (symmetry group D_{2h}) is one of the most important pollutants in the Earth's atmosphere [58] produced by automobile traffic, industrial processes, plant life and forest fires; the latter can be even detected from space owing to the presence of C_2H_4 [59]. It is also detected in the atmospheres of Titan (where it is produced by methane photo-dissociation) [60], Jupiter [61] and Saturn [62].

Due to the high density of C_2H_4 absorption lines, the measurements of line widths require the use of a high-resolution spectrometer and quite low pressures in order to minimise the line overlapping. No broadening coefficients for this molecule had been reported up to the year 2000, except the results of Brannon and Varanasi [63] for the transition $12_{1\,12} \rightarrow 12_{0\,12}$ in the ν_7 band perturbed by He, H_2, and N_2 pressure at 152, 202, 252, and 295 K. For the same band ν_7 a series of measurements for self-, H_2- and N_2-broadening coefficients (the most important for atmospheric applications) at room and low temperatures was realised with the diode-laser spectrometer of Namur (resolution of about $5 \cdot 10^{-4}$ cm^{-1}) [64–69].

In Ref. [64], the authors measured N_2-broadening coefficients for 35 ethylene lines ($3 \le J \le 23$, $0 \le K_a \le 4$, $1 \le K_c \le 22$) located in the spectral range 880–1024 cm^{-1} (ν_7 band) using Voigt and Rautian–Sobel'man profile models (Table 4.16). They observed that these coefficients globally decrease with increasing J and that for the transitions with the same J-value the measured line widths generally increase with increasing K_a. Approximate RB calculations (considering the C_2H_4 molecule as a prolate symmetric top) were also realised with a model potential including anisotropic dispersion contribution and electrostatic terms up to hexadecapole moments:

$$V_{aniso}(r,\theta) = -4\varepsilon_{12}\gamma_1\left(\frac{\sigma_{12}}{r}\right)^6 P_2(\cos\theta) + V_{Q_1Q_2} + V_{Q_1\Phi_2} + V_{\Phi_1Q_2} + V_{\Phi_1\Phi_2}, \qquad (4.2.1)$$

where ε_{12} and σ_{12} are the Lennard–Jones parameters for the C_2H_4-N_2 pair (ε_{11} = 199.2 K, σ_{11} = 4.523 Å and ε_{22} = 92.91 K, σ_{22} = 3.816 Å), γ_1 = 0.164 Å^3 is the polarizability anisotropy of C_2H_4, and only the Q_{zz} (C_2H_4) = −3.70 DÅ, Q_{zz} (N_2) = −1.40 DÅ, and Φ_{zzzz} (C_2H_4) = 7.85 DÅ^3, Φ_{zzzz} (N_2) = 3.40 DÅ^3 components of the quadrupole and hexadecapole moments are taken into account. The calculated line-broadening coefficients were found to be in satisfactory agreement with the measurements for $J \le 12$, but for higher J-values they decreased more rapidly, leading to a significant underestimation of line broadening at $J \ge 19$. The authors ascribed this disagreement to the insufficient dispersion contribution and to the absence of induction and short-range repulsive interactions in their model potential.

The room-temperature H_2-broadening of 34 ethylene lines ($2 \le J \le 23$, $0 \le K_a \le 4$, $2 \le K_c \le 22$) in the ν_7 band (919–1023 cm^{-1}) with Voigt and Rautian–Sobel'man profiles was experimentally studied in Ref. [65] (see Table 4.16). For their semi-classical calculations the authors used the same model potential of Eq. (4.2.1) but introduced in the anisotropic dispersion contribution an adjustable parameter A_2 (A_2 = 0.55) instead of the anisotropy polarizability γ_1 to match better the experimental data. For the hydrogen molecule ε_{22} = 32.0 K, σ_{22} = 2.944 Å, Q_{zz} = 0.6522 DÅ, and Φ_{zzzz} = 0.1264 DÅ^3 were used. Despite the adjustable parameter, the calculated γ-values were too low in comparison with the measurements even for small J-values; this discrepancy has again been attributed to the insufficient model potential. However, a global decrease of the broadening coefficients with J increasing and their general increase with K_a increasing were roughly predicted by the calculations.

Table 4.16. Experimental C_2H_4–N_2 line-broadening coefficients γ(in 10^{-3} cm^{-1} atm^{-1}) in the ν_7 band at 297 K [64, 65]. Reproduced with permission.

ν, cm^{-1}	J K_a K_c ← J K_a K_c		C_2H_4–N_2		C_2H_4–H_2	
			Voigt	Rautian-Sobel'man	Voigt	Rautian-Sobel'man
880.8869	21 2 20	22 3 20	86.1(3.5)	88.6(4.1)		
919.8311	9 1 8	10 2 8	91.0(2.3)	93.2(2.4)	118.7(2.3)	119.8(2.8)
923.4693	15 3 12	15 4 12	91.1(2.2)	92.5(2.8)	116.7(4.3)	118.2(3.9)
923.5326	7 1 6	8 2 6	93.7(2.8)	96.5(2.3)	117.8(3.1)	119.8(3.4)
926.6493	7 0 7	8 1 7	94.5(2.7)	97.0(5.3)	117.7(3.7)	119.7(4.2)
926.7510	13 2 12	13 3 10	90.0(2.6)	94.0(2.6)		
927.0739	4 1 4	5 2 4	97.4(2.5)	99.6(2.9)	119.7(4.8)	121.7(4.0)
927.5198	12 2 11	12 3 9	91.0(2.8)	94.0(2.8)	118.7(2.9)	120.4(3.5)
931.3365	2 1 2	3 2 2	102.8(3.6)	107.4(3.0)	121.5(4.4)	123.2(4.0)
931.3790	9 1 9	9 2 7	96.3(3.2)	98.7(4.2)		
931.5019	11 2 9	11 3 9	91.4(3.0)	93.8(3.5)	119.9(3.1)	121.4(3.0)
931.8042	2 1 1	3 2 1	100.7(3.3)	103.6(3.8)		
931.8834	5 0 5	6 1 5	96.0(2.8)	98.0(3.6)	119.9(2.8)	121.6(2.5)
936.1210	4 1 4	4 2 2	100.1(2.7)	102.2(3.4)	121.6(2.7)	123.0(2.7)
939.4905	13 3 10	14 2 12	91.1(3.4)	93.0(4.3)	120.4(3.0)	121.4(3.2)
943.3529	13 1 12	13 2 12			115.2(3.2)	116.5(3.0)
947.0301	8 0 8	8 1 8	92.9(4.1)	96.3(5.4)	107.2(3.2)	108.4(3.3)
947.3806	9 0 9	9 1 9	86.5(2.4)	88.0(3.1)	107.0(2.8)	108.1(2.8)
947.7002	10 0 10	10 1 10	84.0(2.1)	84.8(2.4)	105.1(3.3)	106.2(3.6)
948.2275	12 0 12	12 1 12	79.6(1.7)	80.8(2.6)	99.9(3.3)	101.7(3.1)
951.3717	6 1 6	6 0 6	92.4(2.4)	93.0(3.3)	111.4(2.7)	112.8(2.4)
951.7394	5 1 5	5 0 5			108.7(5.6)	111.3(5.3)
959.3072	11 1 10	10 2 8	91.6(2.6)	93.8(2.7)	120.2(2.7)	120.7(3.0)
959.4075	6 2 5	6 1 5	96.9(3.6)	98.6(5.3)		
959.5877	14 5 9	15 4 11	85.98(2.3)	88.9(3.4)		
959.9277	5 2 4	5 1 4	97.0(2.2)	100.0(2.6)		
960.0282	22 3 20	21 4 18	83.3(3.4)	89.2(2.8)	112.9(3.7)	114.8(4.1)
960.0769	9 4 5	10 3 7	95.0(3.6)	97.1(4.4)	121.2(4.1)	122.4(4.3)
962.9599	20 3 17	19 4 15	86.3(2.8)	89.7(2.6)	118.3(3.5)	119.5(3.4)
963.1025	16 2 14	15 3 12	87.3(2.6)	91.1(3.7)	118.1(3.3)	119.3(2.9)
966.9867	3 2 2	2 1 2			124.0(3.5)	124.8(3.7)
967.2054	11 3 9	11 2 9			121.6(4.0)	122.7(3.8)
970.7542	8 1 7	7 0 7			119.2(2.8)	120.1(3.0)
970.8737	11 3 8	11 2 10			120.5(3.2)	121.1(3.5)
982.5852	23 4 19	23 3 21	85.1(2.8)	83.3(2.1)	115.7(4.9)	117.2(4.6)
982.7607	7 3 5	6 2 5	95.8(2.9)	97.1(2.8)	121.9(2.8)	122.6(3.1)
1011.5135	20 1 19	19 0 19	84.0(2.9)	84.6(3.1)	105.4(4.0)	107.1(3.6)
1011.8617	20 3 18	19 2 18	85.6(2.8)	86.1(3.5)	113.6(3.5)	115.0(3.5)
1015.3474	20 4 17	19 3 17	86.6(2.3)	87.7(3.5)	112.4(4.5)	113.7(3.7)
1023.1018	20 5 16	19 4 16			116.8(2.7)	118.1(2.6)
1023.4364	24 3 22	23 2 22	81.5(2.0)	83.2(3.2)	109.7(3.5)	110.4(3.6)
1023.7362	21 5 16	20 4 16	86.0(2.7)	87.7(3.5)		

The self-broadening coefficients for a few transitions ($3 \le J \le 17$, $1 \le K_a \le 4$, $1 \le K_c \le 14$) in the ν_7 band (919–982 cm^{-1}) were experimentally obtained in Ref. [66] at 298 and 174 K. Each spectral line was fitted to Voigt, Rautian–Sobel'man and speed-dependent Rautian–Sobel'man profiles. The latter model gives larger γ-values than the previous models. For the semi-classical RB calculations the ethylene molecule was again treated as a prolate symmetric top, and the interaction potential only included the electrostatic interactions (quadrupole-quadrupole, quadrupole-hexadecapole, and hexadecapole- hexadecapole) because of a quite large value of C_2H_4 quadrupole moment. Since the line-broadening coefficients calculated with Q_{zz} = 3.70 DÅ were about 20% larger that the experimental results, the authors adjusted this value to Q_{zz} = 3.30 DÅ, in order to improve the agreement with experiment. Comparison of experimental values for room and low temperatures allowed a determination of the temperature exponents according to Eq. (3.2.1); their theoretical values were also calculated. It was found that n is practically constant for $J < 17$, with the experimental value of 0.75 and theoretical value of 0.78. The measured and calculated C_2H_4-C_2H_4 line-broadening coefficients of Ref. [66] are reproduced in Table 4.17.

The low-temperature (173.2 K) measurements of 35 hydrogen-broadening coefficients in the ν_7 band of C_2H_4 ($2 \le J \le 19$, $0 \le K_a \le 4$, $1 \le K_c \le 18$) were realised by the same authors in Ref. [67], using Voigt and Rautian–Sobel'man profiles. Room- and low-temperature RB line-broadening coefficients were also calculated with a new model potential including pairwise atom–atom interactions. For these calculations, however, the ethylene molecule was again considered as linear, so that the γ dependence on K_c was completely ignored. The line widths calculated for 296.5 K showed satisfactory agreement with the experimental results whereas for 173.2 K they were significantly smaller than the measurements. The theoretical temperature exponents demonstrated a behaviour almost independent of K_a, decreasing progressively from 0.72 to 0.58 for the studied range of J; the experimental n-values were almost constant, with the average values of 0.73 and 0.71 for the Voigt and Rautian–Sobel'man profiles, respectively. Analogous studies were also made for the N_2-broadening case [68].

Infrared heterodyne spectroscopy with a resolution of about $1.7 \cdot 10^{-4}$ cm^{-1} was used in Ref. [69] to study self- and nitrogen-broadening coefficients of the ν_{10} $19_{10,9} \to 20_{11,10}$ line at 296 K; γ_{self} = 0.1244(92) cm^{-1} atm^{-1} and $\gamma_{C2H4-N2}$ = 0.0757(99) cm^{-1} atm^{-1} were obtained.

Table 4.17. Experimental and RB-calculated C_2H_4 self-broadening coefficients (in 10^{-3} cm^{-1} atm^{-1}) for the ν_7 band at 298 K and 174 K [66]. Reproduced with permission.

ν, cm^{-1}	$J\,K_a\,K_c \leftarrow$	$J\,K_a\,K_c$	expt Voigt	expt Rautian-Sobel'man	expt SDR	calc.
			T = 298 K			
919.83110	9 1 8	10 2 8	120.9(4.0)	121.8(3.7)	122.7(4.3)	121.4
923.46929	15 3 12	15 4 12	117.8(3.3)	118.7(3.5)	119.6(3.8)	117.4
923.53259	7 1 6	8 2 6	124.6(3.9)	125.1(3.4)	125.8(3.4)	124.6
926.64931	7 0 7	8 1 7	122.2(3.4)	126.2(4.1)	128.2(4.6)	121.4
927.07387	4 1 4	5 2 4	118.0(4.0)	119.7(4.2)	121.1(4.4)	132.3
931.33652	2 1 2	3 2 2	131.5(3.8)	132.8(4.3)	134.4(4.1)	138.6
931.37896	9 1 9	9 2 7	121.8(4.3)	123.4(3.4)	124.7(3.2)	122.6
931.80418	2 1 1	3 2 1	130.1(4.7)	132.7(5.1)	135.2(4.7)	138.6
931.88343	5 0 5	6 1 5	123.8(4.7)	125.9(3.5)	127.3(3.5)	124.6
936.12095	4 1 4	4 2 2	130.9(3.0)	132.9(3.6)	134.3(4.0)	134.9
951.73937	5 1 5	5 0 5	118.6(4.3)	122.7(4.3)	124.8(4.8)	126.6
966.98673	3 2 2	2 1 2	130.5(3.7)	133.0(4.7)	135.3(4.2)	138.6
967.15689	18 2 16	17 3 14	120.3(3.8)	122.0(3.1)	123.1(3.3)	108.2
967.20539	11 3 9	11 2 9	116.3(3.1)	119.2(4.1)	120.8(3.9)	122.6
970.75418	8 1 7	7 0 7	120.9(3.1)	122.9(3.6)	124.4(3.6)	121.4
970.87375	11 3 8	11 2 10	121.5(3.2)	124.1(4.5)	125.6(5.0)	122.6
982.39175	7 3 4	6 2 4	123.5(2.9)	125.4(3.0)	127.0(3.0)	132.3
			T = 174 K			
919.83110	9 1 8	10 2 8	186.6(8.3)	187.6(9.3)	189.1(9.4)	194.2
923.46929	15 3 12	15 4 12	182.9(9.3)	185.1(10.1)	186.8(10.4)	185.6
923.53259	7 1 6	8 2 6	194.4(7.9)	196.2(8.7)	197.9(9.1)	198.8
926.64931	7 0 7	8 1 7	193.6(6.1)	195.2(7.1)	197.2(7.4)	194.9
927.07387	4 1 4	5 2 4	190.1(5.9)	191.9(7.1)	193.5(7.5)	208.1
931.33652	2 1 2	3 2 2	219.8(6.3)	221.0(6.9)	222.6(7.1)	217.6
931.37896	9 1 9	9 2 7	189.8(6.0)	192.1(6.2)	194.0(6.8)	196.0
931.80418	2 1 1	3 2 1	207.3(12.2)	209.4(10.5)	212.2(10.4)	217.6
931.88343	5 0 5	6 1 5	201.3(6.2)	203.5(7.0)	205.5(7.2)	197.6
936.12095	4 1 4	4 2 2	197.7(6.9)	199.2(7.5)	200.8(7.7)	212.1
951.73937	5 1 5	5 0 5	196.1(7.8)	198.4(8.5)	200.4(8.8)	199.8
966.98673	3 2 2	2 1 2	189.4(4.3)	190.9(4.5)	192.4(4.5)	217.6
967.15689	18 2 16	17 3 14	171.6(7.2)	173.4(7.8)	174.8(8.0)	167.8
967.20539	11 3 9	11 2 9	183.6(5.0)	184.8(5.6)	186.1(5.8)	195.2
970.75418	8 1 7	7 0 7	186.5(5.2)	187.2(5.8)	188.6(6.3)	194.9
970.87375	11 3 8	11 2 10	186.8(6.4)	189.3(7.7)	191.1(8.1)	195.2
982.39175	7 3 4	6 2 4	189.3(5.2)	191.3(5.8)	192.4(6.5)	208.0
982.76068	7 3 5	6 2 5	200.4(7.0)	202.2(7.1)	204.0(7.2)	208.0

The important for atmospheric applications case of nitrogen-broadened line widths of ethylene was recently addressed [70] in the framework of the RBE approach extended to the case of asymmetric tops. The intermolecular interaction potential was traditionally composed of atom–atom and electrostatic (quadrupole-quadrupole) terms and thoroughly tested for the convergence of all radial components with respect to the accounted values of l_1, l_2, l. The general dependence of the line-broadening coefficients in the different branches of the ν_7 band is visualized in Fig. 4.16a (in order to separate the lines with the identical J-values but different K_a-values the usual abscissa J is incremented by $0.2K_a$). In this figure, the calculated RBE values are compared with experimental data and RB results of Refs [64, 68]. For middle J-values in the P-branch the RBE values are clearly better than the RB calculations [68]. In the Q-branch, the RBE results are closer to the experimental results for some values of J but are more distant for other rotational quantum numbers. In the R-branch, the agreement of the RBE calculations with the measurements is nearly perfect (with the single exception for J = 19), contrary to the RB results which visibly overestimate or underestimated the line broadening.

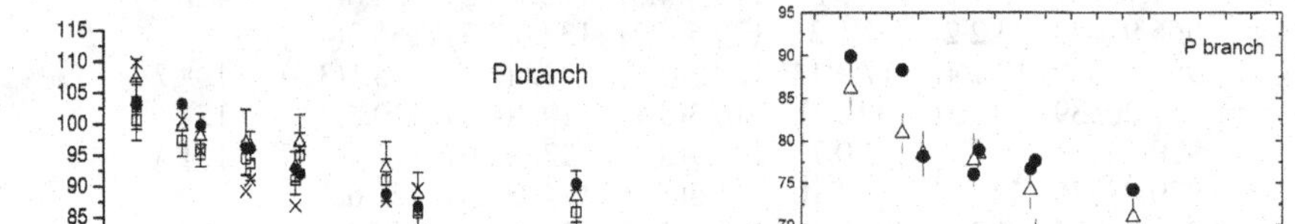

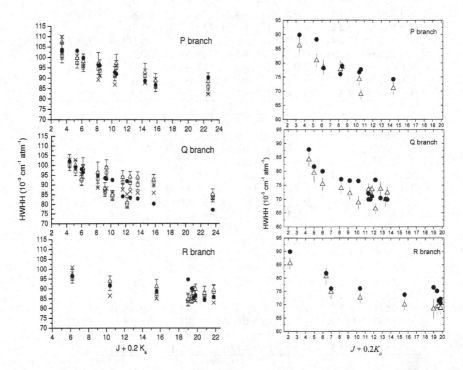

Fig. 4.16. Dependence of C_2H_4–N_2 (left) and C_2H_4–Ar (right) line-broadening coefficients on the rotational quantum numbers for the ν_7 band. Experimental results [64, 66, 71]: open squares — Voigt profile, open triangles — Rautian–Sobel'man profile. Theoretical results: crosses — RB results [68], full circles — RBE results [70, 71]. Reproduced with permission.

The case of C_2H_4 line broadening by noble gases is of minor interest for the terrestrial and planetary atmospheres. It is, however, extremely important from the theoretical viewpoint since it allows a precise analysis of atom–atom interactions and therefore a deeper insight into the mechanism of collisional line broadening. Room-temperature measurements and RBE calculations for 32 C_2H_4–Ar lines ($3 \le J \le 19$, $0 \le K_a \le 4$, $2 \le K_c \le 19$) in the ν_7 band (919–1023 cm^{-1}) were realised in Ref. [71]. Fittings of experimental line shapes with Rautian–Sobel'man profile model yielded slightly larger collisional line widths than those obtained using the Voigt profile. For the RBE calculations a rough atom–atom potential model resulted in quite a good agreement with the experimental data (see Fig. 4.16). The C_2H_4–Ar broadening at low-temperature was considered in Ref. [72]. Room-temperature Ne- and Kr-broadening coefficients were experimentally determined in Ref. [73].

4.3. Vibrotational lines of symmetric tops

Among the symmetric top rotors the ammonia NH_3 and CH_3X (X = D, F, Br, Cl) molecules represent the main interest for atmospheric applications. Ammonia is an important pollutant in the Earth's atmosphere detected in mixtures with other pollutants in industrial areas. Its interest for astrophysics is due to the discovery of large quantities of gaseous ammonia in the atmospheres of Jupiter and Saturn. In Jupiter's atmosphere, it is the fourth most abundant constituent after hydrogen, helium and methane. Spectroscopic studies of the CH_3D molecule are needed for the correct interpretation of atmospheric spectra of Jupiter, Saturn, and Titan. The atmospheres of Neptune and Uranus are also known to contain CH_3D. Methyl chloride CH_3Cl is an important source of chlorine atoms involved in chemical reactions with ozone in the Earth's atmosphere.

The number of vibrational modes $3n - 6$ in the abovementioned molecules is determined by the number of atoms n. For NH_3 there are four vibrational modes denoted $\nu_1(A_1')$, $\nu_2(A_2'')$, $\nu_{3t}(E')$, and $\nu_{4t}(E')$ ($t = a, b$); two last modes are doubly degenerated. The symmetries of D_{3h} group are given in parentheses for the plane configuration. This choice is due to the fact that the intramolecular potential function has two minima, so that there are two equilibrium configurations constantly transforming one into another, and the reference molecular configuration is taken to be the planar configuration of D_{3h} symmetry group. All energy levels are split and the components are labelled by symbols + and – (or by symbols s and a) which reflect the symmetry (parity) of the wave function with respect to the inversion motion (tunnelling of the N atom through the plane formed by the three H atoms).

The vibrational states of NH_3 are denoted as $(v_1 v_2 v_3^{l_3} v_4^{l_4})$, where v_k are vibrational quantum numbers and $l_t = \pm v_t, \pm(v_t - 2), \pm 1$ or 0. The rotational wave functions of rigid symmetric top molecule $|JK\rangle$ are defined by two rotational quantum numbers J and $K = 0, \pm 1, \pm 2 \ldots \pm J$. The rotational energies $E(J, K)$ in the first approximation are given by

$$E(J,K) = BJ(J+1) + (B-C)K^2 - D_J[J(J+1)]^2 - D_{JK}J(J+1)K^2 - D_K K^4 + \cdots, \qquad (4.3.1)$$

where B, C are rotational and D_J, D_K, D_{JK},... are centrifugal distortion constants. For the energy levels with $|K| = 3p$ ($p = 1, 2, \ldots$) an additional term is included in this formula to account for the K–type splitting (C_3 symmetry axis). This splitting is maximal for the energy levels with $|K| = 3$, and the additional term $\Delta E_3(J)$ reads

$$\Delta E_3(J) = (-1)^J [J(J+1)][J(J+1)-2][J(J+1)-6][\eta_3^0 + \eta_3^1 J(J+1) + \cdots], \qquad (4.3.2)$$

with the known constants η_3^0 and η_3^1. In general, the vibrotational energy levels are obtained by the diagonalization of the Hamiltonian matrix of Eq. (3.1.1) (which is built for a given polyad of interacting inversion and vibration states) in the appropriate basis of the rotational wave functions $|JK\rangle$. The selection rules for the rotational quantum number J are the same as for X_2Y molecules: $\Delta J = 0$ for the Q-branch, $\Delta J = +1$ for the R-branch, and $\Delta J = -1$ for the P-branch. More complete selection rules are given in Table 4.18.

Table 4.18. Selection rules for vibration-rotation-inversion (VRI) transitions in X_3Y molecules with inversion symmetry [74].

Transitions between	Parallel bands (μ_z changing)	Perpendicular bands (μ_x,μ_y changing)	Mixed bands (μ_x, μ_y, μ_z changing)	
	Transitions between states of same symmetry and opposite parity			
VRI states	$-\Delta K + \Delta l_3 + \Delta l_4 = 3p, p = 0, \pm 1, \pm 2 \ldots$			
Vibrational states	$A_1' \leftrightarrow A_1'$ $A_2' \leftrightarrow A_2'$	$A_1' \leftrightarrow E'$ $A_2'' \leftrightarrow E'$	$E' \leftrightarrow E'$	
Inversion states	$a \leftrightarrow s$	$s \leftrightarrow s$ $a \leftrightarrow a$	$a \leftrightarrow s$ $s \leftrightarrow a$	$s \leftrightarrow s$ $a \leftrightarrow a$
Rotational states	$\Delta J = 0, \pm 1, \Delta K = 0$ for $K \neq 0$ $\Delta J = \pm 1, \Delta K = 0$ for $K = 0$	$\Delta J = 0, \pm 1, \Delta K = \pm 1$	$\Delta J = 0, \pm 1, \Delta K = 0$ for $K \neq 0$ $\Delta J = \pm 1, \Delta K = 0$ for $K = 0$	$\Delta J = 0, \pm 1, \Delta K = \pm 1$

It can be seen from this table that the parity changing (with respect to the inversion) correlates with ΔK. In particular, in the ν_1 and ν_2 bands with $\Delta K = 0$ transitions between different parity components ($\pm \leftrightarrow \mp$) are possible; in the ν_3 and ν_4 bands with $\Delta K = \pm 1$ the transitions occur between the levels of the same parity ($\pm \leftrightarrow \pm$).

In the notation of rotational transitions $_RQ_0[J_i\ K_i] \rightarrow [J_f\ K_f]$, the letter Q means that $\Delta J = 0$, the index R stands for $\Delta K = K_f - K_i = 1$ and the index 0 denotes that $K_i = 0$. Sometimes an alternative notation $^{\Delta K}\Delta J(J, K)_s$ is used with the index s reminding the rule $s \rightarrow s$ for the inversion components.

4.3.1. NH_3 molecule

Broadening and shifting coefficients of ammonia spectral lines in various frequency regions have been extensively studied experimentally and theoretically for collisions with various partners (He, Ar, H_2, O_2, N_2, CO_2, and so on) [75–96]. The most interesting for atmospheric applications are the $2\nu_2$ and ν_4 vibrational bands located in the region 5–6 μm. It is in this region that ammonia absorption lines in the atmosphere of Jupiter were detected [78]. At the same time ammonia is a good candidate for studying line interference. Indeed, for the overlapping lines the intensity transfer provokes a considerable modification in the line shapes and strongly influences the values of spectral parameters extracted by fitting to assumed model profiles. Various studies of NH_3 inversion-rotation spectra have proved the necessity of accounting for the line mixing.

For the NH_3-H_2 system the room-temperature collisional line widths at various pressures were measured [78] for 66 transitions in the ν_4 band (1470–1600 cm^{-1}) using the Voigt profile model. Theoretical line-broadening coefficients calculated by the semi-classical RB method with a model potential composed of electrostatic, induction and dispersion contributions were also reported and compared to the measurements. For the collision-induced transitions with $\Delta K = 0$ only, the calculated broadening coefficients were underestimated but showed correct behaviour as functions of J and K. The consideration of a dispersion term including transitions with $\Delta K = \pm 3$ allowed an improvement in the line-width magnitude, but was not able to correctly predict the K-dependence of these line widths. Analogous theoretical results were obtained for the broadening coefficients of inversion-rotation transitions and for the Q-branch of the ν_1 parallel band. The molecular parameters of the NH_3-H_2 interaction potential are given in Table 4.19.

Table 4.19. Molecular parameters for the interaction potentials of some symmetric tops*.

	μ, D	Q, DÅ	Ω, D Å^2	α, Å^3	A_{II}, Å^4	$A_\perp$, Å^4	γ, Å^3	ε, K	σ, Å
NH_3(GS) [78]	1.4719	–2.32	1.20	2.18	–0.0633	0.793	0.054	294.3	3.018
NH_3 (v_4) [78]	1.4554								
CH_3F [97]	1.8585	–0.4	–1.27	2.61	1.55	2.49	0.0294	195.9	3.427
CH_3Cl [98]	1.8959	1.23		4.55			0.113	368.4	3.584
CH_3D [99]	0.0	0.0	3.10						
PH_3	0.574	2.60	0.0						

* Two values for the NH_3 dipole moment correspond to the initial (ground state — GS) and final (v_4) states; its vibrational dependence is given by $\mu(v_1, v_2, v_3, v_4) = 1.5610 + 7.2\cdot10^{-3}(v_1 + 1/2) - 2.271\cdot10^{-1}(v_2 + 1/2) + 3.75\cdot10^{-2}(v_3 + 1) - 1.65\cdot10^{-2}(v_4 + 1)$ [86]. A_{II} and $A_\perp$ are the higher polarizabilities, and γ is the polarizability anisotropy. The Lennard–Jones parameters ε and σ for the isotropic potential are determined from the second virial coefficients (detailed discussion of three ε, σ, Q sets can be found in Ref. [78]). In addition, we have $\Phi(CH_3F) = 0.979$ D Å^3 [97], $\Phi(CH_3D) = 6.55$ D Å^3 [99].

Table 4.20 gives the NH_3-H_2-broadening coefficients of Ref. [78]. In this table, the calculation (*a*) was realised with all terms of the anisotropic potential (dependent on μ, Q, Ω, α, γ, $A_\perp$ and A_{II}), using the analytical expressions of their contributions to $S_2\,[r_c(b)]$ obtained by Leavitt [100]. In the calculation (*b*), the coefficient $4\varepsilon_{12}(\sigma_{12})^6$ was used instead of $3/2\bar{u}\,\alpha_1\alpha_2$ for the dispersion interaction but did not give a satisfactory agreement with experimental data. An additional term $V_{\alpha_2 A_1'}$ was therefore introduced in the potential (calculation (*c*)) and allowed some improvement of theoretical results. Finally, the last calculation (*c′*) accounted for the virtual transitions with $\Delta K = -3$ for the small rotational quantum numbers $K = 0, 1, 2$ leading to much better results for small *K*-values but overestimated results for high *J*-values.

The same vibrational band v_4 was studied by the Fourier transform technique in Ref. [79]. The root-temperature high-resolution spectra of pure ammonia and its mixtures with Ar, He, and H_2 were recorded at various pressures. The line mixing was investigated on the inversion doublets, so that for this particular case of two interfering lines the collisional absorption coefficient was derived from the general Eq. (2.2.25) for the spectral function:

$$K(\nu) = \frac{P_{NH_3}}{\pi}\,\mathrm{Im}\left\{\frac{[S_1(\nu-\nu_2')+S_2(\nu-\nu_1')]-i\left(S_1P\gamma_2+S_2P\gamma_1+2P\xi\sqrt{S_1S_2}\right)}{[(\nu-\nu_2')(\nu-\nu_1')-P^2\gamma_2\gamma_1+P^2\xi^2]-i[P\gamma_1(\nu-\nu_2')+P\gamma_2(\nu-\nu_1')]}\right\}, \tag{4.3.3}$$

which gives in the first order

$$K^1(\nu) = \frac{P_{NH_3}}{\pi}\,\mathrm{Im}\left\{\frac{S_1(1+iPY_1)}{(\nu-\nu_1')-iP\gamma_1}+\frac{S_2(1+iPY_2)}{(\nu-\nu_2')-iP\gamma_2}\right\}. \tag{4.3.4}$$

Table 4.20. NH_3–H_2-broadening coefficients (in 10^{-3} cm^{-1} atm^{-1}) for the ν_4 band [78]. Reproduced with permission.

Line	expt	calc. (*a*)	calc. (*b*)	calc. (*c*)	calc. (*c′*)
${}^PP(2,1)_s$ ${}^PP(2,1)_a$	96.0(4.6) 91.7(3.3)	52.4	63.4	87.9	106.1
${}^PP(2,2)_s$ ${}^PP(2,2)_a$	100.5(5.2) 100.5(3.2)	53.1	64.7	84.1	106.8
${}^PP(3,1)_s$ ${}^PP(3,1)_a$	89.4(3.5) 90.1(3.0)	48.0	58.2	83.8	102.1
${}^PP(3,2)_a$	93.9(4.7)	51.8	62.2	82.4	103.7
${}^PP(3,3)_s$ ${}^PP(3,3)_a$	98.0(3.4) 98.0(5.4)	49.8	61.5	90.8	102.4
${}^PP(4,1)_s$ ${}^PP(4,1)_a$	80.8(3.1) 79.5(3.3)	43.6	52.9	78.1	96.4
${}^PP(4,2)_a$	85.8(4.2)	46.9	56.4	78.2	98.5
${}^PP(4,3)_s$ ${}^PP(4,3)_a$	86.8(3.9) 85.6(4.7)	50.3	60.5	88.7	99.0
${}^PP(4,4)_s$ ${}^PP(4,4)_a$	91.9(5.0) 92.2(5.0)	47.5	59.3	96.6	96.6
${}^PP(5,1)_s$ ${}^PP(5,1)_a$	73.4(3.0) 73.7(3.0)	39.6	47.9	72.3	90.6
${}^PP(5,2)_s$ ${}^PP(5,2)_a$	76.2(3.3) 77.0(3.8)	42.1	50.7	72.8	92.6
${}^PP(5,3)_s$ ${}^PP(5,3)_a$	79.8(3.1) 78.5(3.8)	46.0	55.2	84.3	93.9
${}^PP(5,4)_s$ ${}^PP(5,4)_a$	86.3(4.2) 86.3(3.9)	48.7	58.8	93.6	93.6
${}^PP(5,5)_s$ ${}^PP(5,5)_a$	87.1(4.4) 87.4(4.4)	45.8	57.7	90.5	90.5
${}^PP(6,1)_s$ ${}^PP(6,1)_a$	64.6(2.8) 63.5(2.0)	36.0	43.3	66.6	85.1
${}^PP(6,2)_s$	72.4(3.1)	38.0	45.6	67.0	86.8
${}^PP(6,3)_s$ ${}^PP(6,3)_a$	74.4(2.9) 75.4(3.1)	41.4	49.7	78.9	88.3
${}^PP(6,4)_s$ ${}^PP(6,4)_a$	81.5(3.5) 80.5(3.8)	45.0	54.2	88.9	88.9

Line	expt	calc. (*a*)	calc. (*b*)	calc. (*c*)	calc. (*c′*)
${}^PP(6,5)_s$ ${}^PP(6,5)_a$	85.8(4.2) 85.1(3.4)	47.1	57.3	88.3	88.3
${}^PP(6,6)_s$ ${}^PP(6,6)_a$	87.9(4.8) 87.9(4.5)	44.7	56.6	84.7	84.7
${}^PP(7,1)_s$	63.8(3.0)	32.7	39.1	61.5	80.0
${}^PP(7,2)_a$	64.3(3.8)	34.3	41.0	61.6	81.4
${}^PP(7,3)_s$ ${}^PP(7,3)_a$	67.6(3.5) 66.1(2.5)	37.3	44.6	73.6	82.7
${}^PP(7,4)_s$ ${}^PP(7,4)_a$	74.2(4.0) 76.0(4.0)	40.9	48.9	83.8	83.8
${}^PP(7,5)_s$ ${}^PP(7,5)_a$	78.0(3.8) 77.5(4.0)	44.0	53.1	84.3	84.3
${}^PP(7,6)_s$ ${}^PP(7,6)_a$	78.2(3.7) 78.7(4.1)	45.6	55.9	83.3	83.3
${}^PP(7,7)_s$ ${}^PP(7,7)_a$	81.3(3.3) 82.3(3.8)	43.8	55.8	79.0	79.0
${}^PP(8,1)_a$	55.9(2.6)	29.8	35.5	57.3	75.8
${}^PP(8,2)_s$	59.1(3.6)	31.2	37.1	57.0	76.8
${}^PP(8,3)_s$ ${}^PP(8,3)_a$	63.7(2.3) 65.3(3.0)	33.7	40.1	67.9	77.6
${}^PP(8,4)_a$	70.9(3.6)	37.0	44.1	78.2	78.2
${}^PP(8,5)_s$	74.2(3.2)	40.3	48.3	79.5	79.5
${}^PP(8,6)_a$	74.9(3.7)	43.1	52.2	79.9	79.9
${}^PP(8,7)_a$	75.4(4.5)	44.4	54.8	78.7	78.7
${}^PP(8.8)_s$ ${}^PP(8.8)_a$	79.5(4.3) 76.5(3.7)	43.0	55.1	74.6	74.6
${}^RP(2,0)_a$	107.9(3.7)	55.0	65.7	88.4	107.9
${}^RP(3,0)_s$	88.1(3.5)	49.0	59.0	82.5	102.0
${}^RP(4,0)_a$	72.4(2.5)	44.1	53.3	77.2	96.3
${}^RP(5,0)_s$	70.9(3.4)	39.9	48.1	71.7	90.4
${}^RP(7,0)_s$	59.0(3.2)	32.8	39.2	61.4	79.9
${}^RP(8,0)_a$	58.0(2.4)	30.0	35.6	57.4	75.8
${}^RP(9,0)_s$	56.3(2.6)	27.7	32.6	54.0	71.8

In Eqs (4.3.3) and (4.3.4), S_1 and S_2 are the line intensities, $\nu'_1 = \nu_1 + P\delta_1$ and $\nu'_2 = \nu_2 + P\delta_2$ are the wave numbers including the collisional shift; P is the total pressure and P_{NH_3} is the partial pressure; the interference parameters Y_1, Y_2 and ξ are related to the off-diagonal elements of the relaxation matrix. Three approaches were used for spectra analysis. In the first approach, all the spectra recorded at 15 pressures from 30 to 1200 mbar were simultaneously fitted to Eq. (4.3.2) to extract the parameters S_1, S_2, ν_1, ν_2, δ_1, δ_2, γ_1, γ_2, ξ. In the second and third approaches, the spectra were fitted one by one (pressure by pressure) using Eqs (4.3.3) and (4.3.4). The deduced parameters were therefore ν'_1, ν'_2, S_1, S_2, P_{NH_3}, P, γ_1, γ_2 and $P\xi$(or PY_1, PY_2). Their analysis showed that Eqs (4.3.3) and (4.3.4) lead to the identical results if the lines are not strongly mixed.

Table 4.21 shows the experimental and theoretical values of self-, H_2-, Ar-, and He-broadening coefficients of NH_3 vibrotational lines [79] (the line shifts were not given in Ref. [79] because of their too large uncertainties). The broadening coefficients were computed by the RB approach with the anisotropic potential composed of long-range electrostatic interactions:

$$V_{aniso} = V_{\mu_1\mu_2} + V_{\mu_1 Q_2} + V_{Q_1\mu_2} + V_{Q_1Q_2} + V_{\mu_1\Omega_2} + V_{\Omega_1\mu_2} + V_{\Omega_1 Q_2} + V_{Q_1\Omega_2} + V_{\Omega_1\Omega_2}. \quad (4.3.5)$$

The isotropic potential was taken in the Lennard–Jones 12-6 form and the necessary parameters ε and σ were derived from the best fit of the second virial coefficients to the experimental values (see Table 4.19). For the case of broadening by He it was noted that for almost all lines the ratio $\gamma_{NH_3-H_2}/\gamma_{NH_3-He}$ was about 2.7 and the mean value of $\gamma_{NH_3-Ar}/\gamma_{NH_3-He}$ was about 1.55. A simple semi-empirical expression was proposed for the off-diagonal relaxation matrix elements responsible for the intensity transfer between overlapping lines:

$$\langle k|\Lambda|l\rangle = 0.5A[K(il \to ik) + K(fl \to fk)], \quad (4.3.6)$$

where $K(il \to ik)$ is the rotational cross-section for the transition between the lower levels from which the optical transition $l \to k$ starts. The parameter A was supposed constant for all lines within a given branch and obtained by fitting to experimental data. A satisfactory agreement between calculated and experimental values of the line-mixing parameters ξ was found for the self-broadening case, where the intermolecular interactions are strongly dominated by the dipole–dipole term; for the other collisional partners the agreement was quite poor. The most striking and original result of Ref. [79] is that the coupling between the symmetric and anti-symmetric components was found to give up to 80% of the self-broadened NH_3 width.

Table 4.21. NH_3 line-broadening coefficients (in cm^{-1} atm^{-1}) for the ν_4 band at 296 K [79]. Reproduced with permission.

$^{\Delta K}\Delta J(J, K)$	NH_3–NH_3		NH_3–H_2		NH_3–Ar		NH_3–He	
	expt	calc.	expt	calc.	expt	calc.	expt	calc.
$^PP(2,2)_s$	0.581(19)	0.619		0.0840	0.0605(23)	0.0548	0.0355(38)	0.0364
$^PP(2,2)_a$	0.584(12)	0.619		0.0840	0.0501(15)	0.0548	0.0355(38)	0.0364
$^PP(3,1)_s$	0.336(3)	0.309	0.0919(33)	0.0838	0.0524(10)	0.0504	0.0349(43)	0.0328
$^PP(3,1)_a$	0.347(4)	0.309	0.0929(37)	0.0838	0.0533(13)	0.0504	0.0355(33)	0.0328
$^PP(3,3)_s$	0.620(3)	0.674	0.0950(54)	0.0908	0.0553(24)	0.0538	0.0361(48)	0.0342
$^PP(3,3)_a$	0.616(5)	0.674	0.0967(49)	0.0908	0.0566(25)	0.0538	0.0365(39)	0.0342
$^PP(4,3)_s$	0.523(4)	0.562	0.0898(47)	0.0887	0.0493(17)	0.0507	0.0338(33)	0.0339
$^PP(4,3)_a$	0.522(9)	0.562	0.0926(46)	0.0877	0.0509(18)	0.0507	0.0355(42)	0.0339
$^PP(4,4)_s$	0.607(8)	0.698	0.0893(46)	0.0966	0.0526(29)	0.0529	0.0339(52)	0.0326
$^PP(4,4)_a$	0.619(4)	0.698	0.0901(77)	0.0966	0.0529(20)	0.0529	0.0329(33)	0.0326
$^PP(5,3)_s$	0.483(4)	0.499	0.0817(40)	0.0843	0.0463(19)	0.0465	0.0317(26)	0.0317
$^PP(5,3)_a$	0.485(3)	0.499	0.0849(76)	0.0843	0.0535(22)	0.0465	0.0345(30)	0.0317
$^PP(5,4)_s$	0.552(9)	0.606	0.0913(37)	0.0935	0.0494(14)	0.0496	0.0342(39)	0.0324
$^PP(5,4)_a$	0.544(6)	0.606	0.0916(77)	0.0935	0.0490(17)	0.0496	0.0335(36)	0.0324
$^PP(5,5)_s$	0.598(7)	0.708	0.0819(44)	0.0905	0.0480(23)	0.0525	0.0302(35)	0.0315
$^PP(5,5)_a$	0.598(11)	0.708	0.0874(48)	0.0905	0.0521(18)	0.0525	0.0322(43)	0.0315
$^PP(6,4)_s$	0.496(8)	0.536	0.0796(37)	0.0889	0.0438(18)	0.0454	0.0301(34)	0.0303
$^PP(6,4)_a$	0.486(10)	0.536	0.0796(38)	0.0889	0.0438(16)	0.0454	0.0307(43)	0.0303
$^PP(6,5)_s$	0.555(8)	0.627	0.0896(52)	0.0883	0.0479(19)	0.0490	0.0328(47)	0.0311
$^PP(6,5)_a$	0.553(6)	0.627	0.0886(58)	0.0883	0.0475(13)	0.0490	0.0321(39)	0.0311
$^PP(6,6)_s$	0.595(11)	0.713	0.0807(52)	0.0847	0.0479(22)	0.0522	0.0300(40)	0.0308
$^PP(6,6)_a$	0.609(12)	0.713	0.0840(52)	0.0847	0.0508(24)	0.0522	0.0312(40)	0.0308
$^PP(7,6)_s$	0.575(15)	0.638	0.0835(54)	0.0833	0.0444(15)	0.0488	0.0288(31)	0.0300
$^PP(7,6)_a$	0.553(7)	0.638	0.0820(46)	0.0833	0.0441(15)	0.0488	0.0288(36)	0.0300
$^PP(7,7)_s$	0.592(7)	0.716	0.0785(48)	0.0790	0.0458(22)	0.0522	0.0285(34)	0.0303
$^PP(7,7)_a$	0.587(7)	0.716	0.0791(36)	0.0790	0.0469(18)	0.0522	0.0286(42)	0.0303
$^PP(8,7)_s$	0.533(22)	0.645	0.0763(48)	0.0787	0.0425(17)	0.0489	0.0267(33)	0.0293
$^PP(8,7)_a$	0.590(11)	0.645	0.0756(49)	0.0787	0.0444(28)	0.0489	0.0275(27)	0.0293
$^PP(8,8)_s$	0.576(10)	0.719	0.0760(46)	0.0745	0.0446(20)	0.0520	0.0274(33)	0.0299
$^PP(8,8)_a$	0.582(15)	0.719	0.0756(45)	0.0745	0.0446(20)	0.0520	0.0269(37)	0.0299

The experimental NH_3 self-broadening coefficients from Refs [79, 85] were used in Ref. [101] to determine the parameters of the analytical models $\gamma(sur)$ given by Eqs (3.3.3) and (3.3.4). Both models yield practically the same quality of fitting. The parameters of $\gamma(sur)$ model (3.3.4) are given in Table 4.22 and the comparison between theoretical $\gamma(sur)$ and experimental data is presented in Fig. 4.17 (lines are numbered in increasing order of experimental values).

Room-temperature experimental and theoretical Ar- and He-broadening coefficients for 62 absorption lines of NH_3 in the *P*-branch of the ν_4 band (1450–1600 cm^{-1}) were reported in Ref. [84]. RB calculations were performed with the model potential of Eq. (2.8.14) for $J \leq 9$ by varying the potential parameters A_1, R_1, A_2, and R_2 (first and second lines of Table 4.23). The theoretical results were practically identical for $a \leftrightarrow a$ and $s \leftrightarrow s$ transitions. The agreement between theory and experiment was satisfactory for $J \leq 6$ and the calculations correctly predicted the *J*- and *K*-dependences of the broadening coefficients. The same potential parameters were also used for calculation of broadening coefficients for the ν_1 band characterised by different selection rules. Since the agreement between calculated and experimental data was quite satisfactory (except two lines), the vibrational dependence of the potential parameters is probably small.

Table 4.22. Parameters of the $\gamma(sur)$ model (3.3.4) for NH_3 self-broadening coefficients in the ν_2 and ν_4 bands at 296 K (*N* is the number of lines used in fitting).

	ν_2, ${}^QR$ lines	ν_4, ${}^PP + {}^RP$ lines	all lines
x_{20}	0.830(41)	0.682(38)	0.613(208)
x_{21}	−0.0114(17)	−0.02382(19)	
x_{30}	−0.1154(25)	−0.3094(448)	−0.1816(161)
x_{31}		0.01182(33)	$0.350(68)\cdot 10^{-2}$
x_{40}		2.352(380)	3.329(422)
x_{41}	0.9673(441)	0.1367(537)	0.2085(387)
x_{42}	−0.01896(16)		0.0109(182)
x_{50}		0.0960(162)	−0.1601(186)
x_{51}			0.0119(15)
N	62	38	100

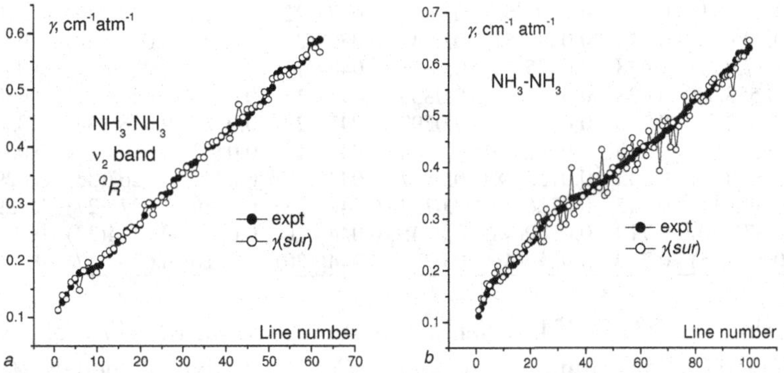

Fig. 4.17. Experimental [79, 85] and calculated by the $\gamma(sur)$ model NH_3 self-broadening coefficients for the ν_2 (*a*) and the ν_4 (*b*) bands [101]. Reproduced with permission.

Table 4.2.3. Parameters of the model potential of Eq. (2.8.14) for the NH_3–He and NH_3–Ar systems. Reproduced with permission from Refs [83, 84].

	A_1	A_2	R_1	R_2
NH_3–He [84]	0.18	0.20	0.32	0.20
NH_3–Ar (ν_4) [84]	0.20	0.20	0.50	0.40
NH_3–Ar ($2\nu_2$) [83]	0.20	0.20	0.40	0.30

Recent results for shifting coefficients and line-mixing parameters for the NH_3 molecule at room temperature are given in Refs [83, 86–91]. In Ref. [83], 66 vibrotational lines in the ν_4 band and 10 lines in the $2\nu_2$ band of NH_3 perturbed by H_2 and Ar were studied. In Ref. [86], the ν_4 band lines (1470–1600 cm^{-1}) of NH_3 perturbed by CO_2 and He were considered. Analyses of the spectra were performed taking into account the interference effects and the Rozenkranz approximation (Eq. (1.4.1)) for the absorption coefficient. The collisional parameters for a given temperature were obtained by a non-linear least-squares multi-pressure fitting procedure. The RB formalism was used to calculate the pressure-induced shifting coefficients. The main contribution to these coefficients is given by the variation of the isotropic potential between the initial and final vibrational states. For NH_3-H_2 shift calculations [83], Eq. (2.6.3) (with v and b replaced by v_c' and r_c) was used in the first order term S_1. The vibrational dependence of NH_3 dipolar moment is given by the formula from Table 4.19. Since the vibrational dependence of the NH_3 polarizability is unknown, it was considered that $\Delta\alpha = \alpha_f - \alpha_i = 0.05$ Å^3 for the ν_4 band and $\Delta\alpha = 0$ for the $2\nu_2$ band. Since the shifts in the $2\nu_2$ band are positive, only the positive contribution in S_1 term, arising from the vibrational dependence of the ammonia dipole moment, was taken into account. The calculated and experimental coefficients δ for NH_3-H_2 and other NH_3-A systems are given in Table 4.24.

In the case of perturbation by Ar, the model potential of Eq. (2.8.14) was chosen for shift calculations [83]. Its parameters deduced from the best fitting of the line-broadening coefficients are given in Table 4.23. The vibrational dephasing contribution was obtained in the form

$$S_1 = -\frac{3}{2}\frac{\Delta C_6}{C_6}\frac{\pi\varepsilon\sigma}{\hbar v_c'}\left(\frac{\sigma}{r_c}\right)^5 + \frac{63}{64}\frac{\Delta C_6}{C_6}\frac{\pi\varepsilon\sigma}{\hbar v_c'}\left(\frac{\sigma}{r_c}\right)^{11} y \tag{4.3.7}$$

where $y = (\Delta C_{12}/C_{12})/(\Delta C_6/C_6)$ and $\Delta C_n = C_{n\,f} - C_{n\,i}$, $n = 6, 12$. The coefficients $\Delta C_6/C_6$ and y were considered as adjustable parameters. It was found that $\Delta C_6/C_6 = 0.014$ and $y = 2.0$ in the ν_4 band and $\Delta C_6/C_6 = 0.03$ and $y = 5.0$ in the $2\nu_2$ band. The same vibrational dephasing contribution of Eq. (4.3.7) was used in Ref. [86] for NH_3-He shift calculation. In this case $\Delta C_6/C_6 = 0.01$ and $y = 1.5$ were obtained for the ν_4 band. For NH_3-CO_2 shift calculations only the long-range part of the isotropic potential was taken into account and it was found that in the ν_4 band $\Delta C'_6/C'_6 = 0.02$ (C'_6 stands for the dispersion force constant only).

Table 4.24. Line-shift coefficients δ (in 10^{-3} cm^{-1} atm^{-1}) in the ν_4 band of NH_3 perturbed by H_2, Ar, He and CO_2 at 296 K [83, 86]. Reproduced with permission.

${}^{\Delta K}\Delta J(J, K)$	NH_3–H_2 [83]		NH_3–Ar [83]		NH_3–He [86]		NH_3–CO_2 [86]	
	expt	calc.	expt	calc.	expt	calc.	expt	calc.
				ν_4				
${}^{P}P(2,1)_s$	−4.16(1.03)	−5.25	−6.05(0.93)	−5.79	−0.70(0.32)	−1.47	−8.74(0.85)	−10.16
${}^{P}P(2,1)_a$	−4.47(1.05)	−5.25	−7.22(0.92)	−5.79	−1.34(0.29)	−1.47	−4.50(0.64)	−10.16
${}^{P}P(2,2)_s$	−4.15(0.82)	−3.68	−8.77(1.22)	−5.68	−1.70(0.41)	−1.49	−9.66(0.90)	−4.90
${}^{P}P(2,2)_a$	−2.62(0.61)	−3.68	−5.89(1.52)	−5.68	−0.95(0.18)	−1.49	−9.28(0.88)	−4.90
${}^{P}P(3,1)_s$	−3.42(0.42)	−4.93	−4.70(0.83)	−5.80	−1.14(0.56)	−1.42	−3.16(0.56)	−9.94
${}^{P}P(3,1)_a$	−3.32(0.42)	−4.93	−4.68(1.11)	−5.80	−0.91(0.09)	−1.42	−5.21(1.38)	−9.94
${}^{P}P(3,2)_s$	−4.63(0.79)	−4.86	−5.64(0.92)	−5.79	−1.78(0.48)	−1.43	−8.31(1.94)	−7.03
${}^{P}P(3,2)_a$	−4.14(1.71)	−4.86	−7.35(1.03)	−5.79	−0.83(0.38)	−1.43		
${}^{P}P(3,3)_s$	−4.61(2.00)	−3.72	−6.35(1.10)	−5.64	−1.34(0.18)	−1.47	−2.97(0.11)	−5.06
${}^{P}P(3,3)_a$	−4.85(0.67)	−3.72	−5.17(1.36)	−5.64	−1.07(0.22)	−1.47	−10.79(0.28)	−5.06
${}^{P}P(4,1)_s$	−5.48(0.69)	−4.84	−6.63(1.19)	−5.68	−1.09(0.35)	−1.37	−2.06(1.52)	−6.94
${}^{P}P(4,1)_a$	−4.42(0.66)	−4.84	−3.03(0.45)	−5.68	−1.11(0.08)	−1.37	−5.03(0.83)	−6.94
${}^{P}P(4,2)_a$	−6.25(0.29)	−4.88	−6.09(0.93)	−5.75	−2.04(0.23)	−1.30	−3.48(0.46)	−4.58
${}^{P}P(4,3)_s$	−3.92(0.72)	−4.02	−5.27(0.48)	−5.73	−0.84(0.51)	−1.40	−4.06(1.70)	−3.33
${}^{P}P(4,3)_a$	−3.20(0.93)	−4.02	−4.21(0.93)	−5.73	−1.15(0.37)	−1.40	−5.01(1.15)	−3.33
${}^{P}P(4,4)_s$	−3.04(0.52)	−3.69	−4.03(0.71)	−5.62	−1.01(0.28)	−1.46	−4.26(0.96)	−4.58
${}^{P}P(4,4)_a$	−4.78(0.92)	−3.69	−4.13(0.91)	−5.62	−1.31(0.31)	−1.46	−11.59(1.12)	−4.58
${}^{P}P(5,1)_s$	−2.95(0.17)	−4.87	−4.98(0.75)	−5.46	−0.95(0.14)	−1.31	−4.20(0.58)	−7.80
${}^{P}P(5,1)_a$	−4.33(0.48)	−4.87	−3.51(1.06)	−5.46	−1.10(0.25)	−1.31	−4.13(0.29)	−7.80
${}^{P}P(5,2)_s$	−5.60(0.85)	−5.03	−4.41(0.89)	−5.57	−1.49(0.30)	−1.19	−2.57(0.04)	−5.35
${}^{P}P(5,2)_a$	−5.11(0.91)	−5.03	−5.75(0.92)	−5.57	−1.23(0.20)	−1.19	−2.41(0.63)	−5.35
${}^{P}P(5,3)_s$	−4.30(0.94)	−4.09	−5.43(0.61)	−5.68	−1.57(0.27)	−1.31	−2.05(0.25)	−3.03
${}^{P}P(5,3)_a$	−4.02(0.46)	−4.09	−5.02(1.23)	−5.68	−0.69(0.09)	−1.31	−7.00(0.51)	−3.03
${}^{P}P(5,4)_s$	−2.63(0.15)	−3.52	−2.82(0.98)	−5.67	−0.80(0.27)	−1.40	−5.55(1.47)	−2.76
${}^{P}P(5,4)_a$	−4.24(1.36)	−3.52	−4.10(0.57)	−5.67	−1.18(0.31)	−1.40	−2.39(0.62)	−2.76
${}^{P}P(5,5)_s$	−2.66(0.09)	−4.06	−5.37(0.70)	−5.60	−1.12(0.32)	−1.44	−4.59(0.58)	−4.35
${}^{P}P(5,5)_a$	−5.52(0.87)	−4.06	−4.32(1.55)	−5.60	−1.87(0.33)	−1.44	−8.66(0.75)	−4.35
${}^{P}P(6,1)_s$	−2.79(0.27)	−4.99	−3.41(0.72)	−5.13	−0.81(0.18)	−1.25	−4.27(0.67)	−9.72
${}^{P}P(6,1)_a$	−3.27(0.85)	−4.99	−5.94(0.86)	−5.13	−0.98(0.15)	−1.25	−6.24(0.42)	−9.72
${}^{P}P(6,3)_s$	−5.21(0.50)	−4.39	−5.05(1.09)	−5.48	−1.54(0.33)	−1.19	−2.70(0.32)	−4.48
${}^{P}P(6,3)_a$	−3.56(0.26)	−4.39	−3.16(0.41)	−5.48	−1.27(0.18)	−1.19	−4.27(0.70)	−4.48
${}^{P}P(6,4)_s$	−3.24(0.63)	−3.52	−4.21(0.53)	−5.58	−1.11(0.35)	−1.33		
${}^{P}P(6,4)_a$	−2.40(0.35)	−3.52	−4.84(0.92)	−5.58	−1.65(0.34)	−1.33	−7.26(2.16)	−3.08
${}^{P}P(6,5)_s$	−2.29(0.27)	−3.58	−5.86(1.14)	−5.62	−1.62(0.25)	−1.39	−6.05(1.03)	−3.10
${}^{P}P(6,5)_a$	−4.22(0.38)	−3.58	−5.42(0.54)	−5.62	−0.88(0.17)	−1.39	−7.65(0.87)	−3.10
${}^{P}P(6,6)_s$	+2.96(0.53)	−4.43	−2.23(0.66)	−5.58	−1.66(0.18)	−1.46	+10.95(0.60)	−4.33
${}^{P}P(6,6)_a$	−6.62(0.35)	−4.43	−3.77(1.81)	−5.58	−1.45(0.17)	−1.46	−11.15(2.03)	−4.33
${}^{P}P(7,1)_s$	−3.09(0.53)	−5.18	−5.80(0.95)	−4.70	−1.00(0.55)	−1.16	−4.65(1.26)	−10.67
${}^{P}P(7,2)_a$	+1.48(0.83)	−5.37	−3.99(1.07)	−4.90	−0.95(0.45)	−0.95		
${}^{P}P(7,3)_s$	−4.50(0.79)	−4.69	−2.03(0.65)	−5.15	−0.97(0.28)	−1.11	−4.36(1.76)	−6.16
${}^{P}P(7,3)_a$	−4.85(0.52)	−4.69	−5.70(1.33)	−5.15	−1.28(0.14)	−1.11	−5.61(0.37)	−6.16
${}^{P}P(7,4)_s$	+1.44(0.66)	−3.81	−3.90(1.62)	−5.36	+0.36(0.11)	−1.23	+8.05(1.31)	−4.07

$^PP(7,4)_a$	−3.71(1.68)	−3.81	−3.82(1.61)	−5.36	+1.40(0.61)	−1.23	+7.61(0.85)	−4.07
$^PP(7,5)_s$	+2.42(0.36)	−3.42	−5.04(0.39)	−5.49	+1.34(0.45)	−1.35	+3.18(1.71)	−3.26
$^PP(7,5)_a$	−4.09(0.55)	−3.42	−5.41(0.66)	−5.49	−0.97(0.32)	−1.35	−4.89(0.72)	−3.26
$^PP(7,6)_s$	−2.74(0.56)	−3.70	−3.46(0.75)	−5.57	−1.04(0.35)	−1.44	+5.58(0.92)	−3.41
$^PP(7,6)_a$	−3.13(0.41)	−3.70	−3.86(0.53)	−5.57	−0.86(0.38)	−1.44	−7.45(1.24)	−3.41
$^PP(7,7)_s$	+1.32(0.52)	−4.74	−4.69(0.28)	−5.55	−1.54(0.64)	−1.49	−3.00(0.98)	−4.35
$^PP(7,7)_a$	−6.13(0.60)	−4.74	−4.57(0.60)	−5.55	−0.85(0.40)	−1.49	−4.28(0.71)	−4.35
$^PP(8,1)_a$	−2.91(1.00)	−5.44	−2.90(0.65)	−4.19	−1.12(0.55)	−1.11	−6.09(0.84)	−11.57
$^PP(8,3)_s$	−2.43(0.63)	−4.95	−4.65(1.02)	−4.76	−0.91(0.18)	−1.02	−5.34(1.21)	−7.63
$^PP(8,3)_a$	−5.66(1.57)	−4.95	−4.95(1.35)	−4.76	−1.27(0.31)	−1.02		
$^PP(8,4)_a$	−4.64(0.98)	−4.18	−4.04(1.00)	−5.06	+1.43(0.27)	−1.13	−4.32(0.02)	−5.11
$^PP(8,5)_s$	−4.44(1.50)	−3.64	−7.57(0.91)	−5.36	−1.17(0.52)	−1.27	−7.12(0.97)	−3.81
$^PP(8,6)_a$	−2.23(0.96)	−3.42	−5.95(1.06)	−5.42	−0.80(0.34)	−1.37	−6.84(0.47)	−3.36
$^PP(8,7)_s$	−2.39(0.99)	−3.87	−3.73(1.19)	−5.59	−0.81(0.42)	−1.45	−3.26(0.58)	−3.57
$^PP(8,7)_a$	−3.21(1.66)	−3.87	−4.54(1.50)	−5.59	−1.28(0.18)	−1.45		
$^PP(8,8)_s$	+4.53(0.83)	−5.04	−3.73(1.22)	−5.55	+0.99(0.45)	−1.48	+3.19(0.82)	−4.34
$^PP(8,8)_a$	−4.37(0.66)	−5.04	−2.44(1.05)	−5.55	−1.10(0.27)	−1.48	−5.11(1.84)	−4.34
$^RP(2,0)_a$	−8.50(0.49)	−5.89	−8.85(1.16)	−5.64	−2.64(0.75)	−1.50	−9.72(0.64)	−11.61
$^RP(2,0)_s$	−4.86(1.27)	−8.08	−3.26(0.90)	−5.80	−0.58(0.12)	−1.43	−8.24(0.43)	−11.92
$^RP(4,0)_a$	−4.39(0.77)	−8.77	−5.06(0.93)	−5.72	−0.98(0.32)	−1.37	−5.79(0.18)	−7.61
$^RP(5,0)_s$	−6.61(0.75)	−8.97	−5.99(0.50)	−5.46	−0.50(0.50)	−1.31	−4.01(0.93)	−8.00
$^RP(7,0)_s$	−4.17(1.69)	−8.82	−3.88(0.71)	−4.67				
$^RP(8,0)_a$	−2.66(0.41)	−8.74	−3.67(0.42)	−4.25	−0.79(0.07)	−1.11	−9.23(0.28)	−13.75
$^RP(9,0)_s$	−6.76(0.80)	−8.59	−5.36(1.49)	−3.82	−1.38(0.78)	−1.03		
		$2\nu_2$						
$^QP(4,3)_a$	13.92(6.85)	10.81	20.27(2.00)	14.42				
$^QQ(3,3)_a$	12.31(0.92)	11.73	14.96(2.00)	14.60				
$^QQ(4,3)_a$	20.89(4.64)	11.80	18.29(5.16)	14.91				
$^QQ(4,4)_a$	19.40(0.72)	12.48	18.45(2.87)	15.07				
$^QQ(5,5)_a$	15.94(0.89)	13.13	14.64(3.50)	15.65				
$^QQ(6,6)_a$	17.20(2.09)	13.74	20.42(1.18)	16.04				
$^QQ(7,7)_a$	17.81(4.38)	14.26	27.11(3.28)	16.28				
$^QQ(8,8)_a$	21.68(4.40)	14.67	16.70(3.76)	16.41				
$^QQ(9,9)_a$	15.78(1.22)	15.00	6.28(0.42)	16.46				
$^QQ(10,9)_a$	11.39(0.49)	15.55	10.57(1.98)	18.57				

The most important for the Earth's atmosphere cases of NH_3 line broadening by N_2, O_2, air and CO_2 in various bands were considered in Refs [87–89, 91–96]. Table 4.25 contains empirical [92–94] N_2-, O_2-, air- and CO_2-broadening coefficients for the ν_4 band (theoretical values of Ref. [95] marked by an asterisk are given if experimental values are not available). Fabian *et al.* [82] measured collisional line widths in the ν_2 band of NH_3 perturbed by N_2, O_2, and air. Line broadening and shift coefficients by N_2, O_2 and air pressure for five lines in the *R*-branch of the ν_2 band (10 μm region) were experimentally obtained in Ref. [87]. The temperature dependence of H_2-, N_2-, O_2- and air-broadening coefficients in the ν_2 band was studied by Nemtchinov *et al.* [96] (measurements at 200, 255 and 296 K). The $2\nu_2$ and ν_4 bands of NH_3 in mixture with O_2 were investigated in Ref. [89].

Table 4.25. Experimental [92–94] and calculated (noted by an asterisk) [95] line-broadening coefficients (in 10^{-3} cm^{-1} atm^{-1}) for the ν_4 band of NH_3 perturbed by O_2, N_2, CO_2 and air.

Line	ν, cm^{-1}	NH_3–N_2 [92]	NH_3–CO_2 [93]	NH_3–O_2 [94]	NH_3–air [94]
$^RP(2,0)_a$	1586.8714	105.9*	207.9(10.1)	66.7(2.0)	110.3(3.3)
$^PP(2,1)_a$	1591.1050	103.9(5.1)	197.7(10.1)	62.4(2.5)	100.6(4.6)
$^PP(2,2)_a$	1595.0801	109.6*	203.5*	64.1(3.0)	102.7(4.8)
$^RP(3,0)_s$	1567.9930	98.7(5.1)	159.7(10.1)	60.8(1.8)	97.3(2.8)
$^PP(3,1)_s$	1571.8315	98.9(5.1)	144.5(7.6)	59.6(1.8)	97.3(2.3)
$^PP(3,1)_a$	1572.4882	101.4(5.1)	159.4*	58.6(1.8)	98.9(2.3)
$^PP(3,2)_s$	1575.8518	101.5*	178.7*	66.9(2.5)	99.4(3.8)
$^PP(3,3)_s$	1579.3615	106.5(5.1)	167.3(7.6)	67.7(1.8)	103.4(3.0)
$^PP(3,3)_a$	1579.6363	109.0(5.1)	167.3(7.6)	65.9(2.0)	103.4(3.0)
$^RP(4,0)_s$	1552.1574	93.8(5.1)	124.2(5.1)	56.8(1.8)	91.5(2.8)
$^PP(4,1)_s$	1553.6294	97.4*	121.7(5.1)	56.5(1.8)	89.5(2.5)
$^PP(4,1)_a$	1554.7340	97.4*	119.1(5.1)	55.3(2.8)	89.5(4.6)
$^PP(4,2)_s$	1558.2494	101.6*	152.1*	58.6(3.3)	92.3(4.6)
$^PP(4,3)_s$	1560.8895	104.9*	162.2(7.6)	63.4(1.8)	98.6(3.0)
$^PP(4,3)_a$	1561.3835	104.9*	162.2(7.6)	62.6(1.8)	98.1(3.0)
$^PP(4,4)_s$	1563.8239	103.9(5.1)	167.3(7.6)	63.4(1.8)	98.9(3.0)
$^PP(4,4)_a$	1564.0824	103.9(5.1)	167.3(7.6)	62.9(1.8)	99.1(3.0)
$^RP(5,0)_s$	1533.7968	90.5*	91.3(5.1)	50.0(1.5)	78.1(2.3)
$^PP(5,1)_s$	1536.2088	88.7(5.1)	108.1*	51.5(1.8)	80.4(2.3)
$^PP(5,1)_a$	1538.0102	90.6*	106.5(5.1)	51.5(1.8)	80.9(3.3)
$^PP(5,2)_s$	1539.7596	96.3(5.1)	132.0*	52.0(2.0)	87.7(3.5)
$^PP(5,2)_a$	1541.0041	95.2*	132.1*	51.5(1.5)	87.7(3.5)
$^PP(5,3)_s$	1542.9798	96.3(5.1)	144.5(7.6)	55.3(2.3)	90.0(1.5)
$^PP(5,3)_a$	1543.8550	96.3(5.1)	157.6*	56.0(2.3)	89.7(4.6)
$^PP(5,4)_s$	1545.8043	103.7*	136.9(7.6)	55.8(2.3)	90.8(4.6)
$^PP(5,4)_a$	1546.3315	103.7*	136.9(7.6)	56.3(1.8)	91.5(2.8)
$^PP(5,5)_s$	1548.1839	101.4(5.1)	169.9(7.6)	58.3(1.8)	91.8(2.8)
$^PP(5,5)_a$	1548.4290	98.9(5.1)	169.9(7.6)	57.3(1.8)	92.5(2.8)
$^PP(6,1)_s$	1519.6627	81.7*	91.7*	48.2(2.0)	77.6(2.0)
$^PP(6,1)_a$	1522.3847	81.8*	106.5(5.1)	47.4(1.5)	78.8(1.8)
$^PP(6,3)_s$	1525.7625	91.3*	124.2(5.1)	50.9(1.5)	79.3(2.3)
$^PP(6,3)_a$	1527.0615	91.3*	143.2*	52.0(2.0)	79.6(3.0)
$^PP(6,4)_s$	1528.3774	97.0*	165.9*	53.7(2.8)	85.9(4.3)
$^PP(6,4)_a$	1529.2897	97.0*	136.9(7.6)	54.5(2.3)	84.9(4.3)
$^PP(6,5)_s$	1530.6141	93.8(4.7)	139.4(7.6)	54.8(1.5)	90.7(2.8)
$^PP(6,5)_a$	1531.1591	93.8(4.7)	136.9(7.6)	54.2(1.5)	89.0(2.5)
$^PP(6,6)_s$	1532.4503	98.9(4.9)	152.1(7.6)	55.3(1.5)	91.3(2.8)
$^PP(6,6)_a$	1532.6830	96.3(4.8)	152.1(7.6)	54.3(1.5)	91.8(2.8)
$^RP(7,0)_s$	1502.7553	65.9(3.3)	91.3(5.1)	42.3(1.8)	69.7(2.8)
$^PP(7,1)_s$	1504.0218	72.2*	98.9(5.1)	46.9(1.3)	69.5(2.0)
$^PP(7,3)_s$	1509.1379	81.7*	121.7(5.1)	47.9(1.5)	73.3(2.3)

$^{P}P\,(7,4)_s$	1511.5985	88.3*	153.9*	50.9(2.5)	73.3(3.5)
$^{P}P\,(7,4)_a$	1513.0437	88.3*	124.2(12.7)	48.9(2.5)	74.8(3.5)
$^{P}P\,(7,5)_s$	1513.6480	88.7(4.4)	131.8(7.6)	52.7(1.5)	78.6(2.3)
$^{P}P\,(7,5)_a$	1514.6055	91.3(4.6)	173.3*	53.2(2.0)	78.1(3.0)
$^{P}P\,(7,6)_s$	1515.3243	91.3(4.6)	157.2(10.1)	53.7(2.0)	89.5(3.5)
$^{P}P\,(7,6)_a$	1515.8819	91.3(4.6)	157.2(10.1)	53.2(1.5)	92.0(2.8)
$^{P}P\,(7,7)_s$	1516.6309	96.3(4.8)	172.4(7.6)	53.5*	96.0*
$^{P}P\,(7,7)_a$	1516.8521	98.9(4.8)	172.4(7.6)	54.0(2.5)	89.2(4.6)
$^{R}P\,(8,0)_a$	1493.8635	71.0(3.6)	71.0(5.1)	40.8(1.5)	65.0(2.5)
$^{P}P\,(8,1)_a$	1494.2421	68.4(3.4)	101.4(10.1)	45.1(1.3)	67.7(2.3)
$^{P}P\,(8,3)_s$	1493.5269	73.5(3.7)	116.6(5.1)	46.6(1.3)	70.5(2.0)
$^{P}P\,(8,3)_a$	1496.1740	72.0*	120.7*	44.9(2.3)	73.0(3.5)
$^{P}P\,(8,4)_a$	1497.6749	78.6*	143.7*	45.6(2.3)	74.3(3.8)
$^{P}P\,(8,5)_s$	1497.3332	86.2(4.3)	162.2*	47.2(2.3)	74.8(3.5)
$^{P}P\,(8,6)_a$	1499.8028	94.3*	178.9*	47.4(2.3)	77.3(3.8)
$^{P}P\,(8,7)_a$	1500.5103	86.2(4.3)	194.7*	48.4(1.5)	80.1(2.5)
$^{P}P\,(8,8)_s$	1500.7334	98.9(4.9)	210.0*	52.0*	95.5*
$^{P}P\,(8,8)_a$	1500.9439	83.7(4.2)	210.0*	49.9(2.0)	82.6(3.3)

Calculations of N_2-, O_2-, air- and CO_2-broadening coefficients for NH_3 lines ($J \leq 8$) in the *P*-branch of the ν_4 band were performed in Ref. [95] using the Korff and Leavitt model (Eq. (2.8.4)) accounting for the inversion motion of NH_3. The anisotropic intermolecular potential included 18 terms for electrostatic (up to octupole-hexadecapole), induction and dispersion interactions. The dispersion contributions were calculated from the ionization energies u_1 and u_2 of the molecules through the term $\bar{u} = u_1 u_2 /(u_1 + u_2)$. The trajectory model included the influence of the isotropic potential in the Lennard–Jones form. These calculations demonstrated the dominant character of electrostatic interactions for the NH_3-N_2 and NH_3-CO_2 systems and the importance of the dipole-hexadecapole and dispersion terms. Different NH_3 parameters were introduced for symmetric and anti-symmetric levels in order to compare the broadening coefficients for *a*-*a* and *s*-*s* transitions. However, the calculations showed that for both kinds of transitions the differences are always inferior to 0.3%. It was also noted that the negative sign of the NH_3 quadrupole moment plays an important role for the NH_3-O_2 broadening but that it is less important in the case of NH_3-N_2.

4.3.2. PH_3 molecule

The knowledge of spectral parameters of phosphine (pyramidal molecule of C_{3v} group) is mainly required for astronomical applications. It is used as a tracer of vertical circulation and non-equilibrium chemistry in the atmospheres of Jovian planets (Jupiter, Saturn). Its interest from a spectroscopic point of view is due to the presence of numerous Coriolis and anharmonic interactions.

The collisional broadening of PH_3 spectral lines by N_2, H_2, PH_3, He and Ar in the parallel ν_2 and bending ν_4 bands (near 1000–1100 cm^{-1}) was studied in Refs [102–108] (the case of hydrogen-broadening [105, 106] is of particular interest for the atmospheres of the giant planets).

In Ref. [102], N_2-broadening coefficients were determined for 40 lines in the ^QR-branch of the ν_2 band and 21 lines in the ^PP-, ^RP-, ^SP-, and ^PQ- branches of the ν_4 band ($1 \leq J \leq 16$, $1 \leq K \leq 11$) at 297 K. The obtained γ-values (see Table 4.26) ranged between 0.0575 and 0.0842 cm^{-1} atm^{-1}. For the same system low-temperature measurements (173 K) were realised for 41 lines ($1 \leq J \leq 13$, $0 \leq K \leq 10$) in the ^QR-branch of the ν_2 band and ^PP, ^RP, and ^SP branches of the ν_4 band [103]. Three profile models were used for experimental line shape fitting: Voigt, Rautian–Sobel'man and SDR profiles. It was found that the last two models give larger values of the broadening coefficients than the Voigt model. Semi-classical calculations considering PH_3 as a linear molecule were also performed with a model potential composed of electrostatic and atom–atom contributions. The temperature dependence of the line widths was also determined both experimentally and theoretically.

The case of self-broadened PH_3 spectral lines in the region 995–1093 cm^{-1} at 298 K was discussed in Ref. [104]. It was found that for the lines with the quantum numbers $1 \leq J \leq 14$ and $0 \leq K \leq 1$ the self-broadening coefficients γ range from 0.072 to 0.108 cm^{-1} atm^{-1} (Table 4.26). For nine lines self-broadening coefficients were also measured at 173.4 K, which allowed the determination of the temperature exponents for the temperature interval 173–298 K: $0.72 \leq n \leq 1.03$.

H_2-broadening coefficients for 41 transitions in the ^QR-branch of the ν_2 band and in the ^PP-, ^RP- and ^PQ-branches of the ν_4 band with $2 \leq J \leq 16$, $0 \leq K \leq 11$ (995–1106 cm^{-1}) were measured in Ref. [105] at 297 K. As can be seen from Table 4.26, they are approximately 6% lower than the self-broadening coefficients [104].

The temperature dependence of H_2- and He-broadening coefficients for the bands $2\nu_2$, $\nu_2 + \nu_4$, $2\nu_4$ (1950–2150 cm^{-1}) and for the ν_2 band (1016–1093 cm^{-1}) was studied in Refs [106–107]. At 173 K [106], for the lines of the ν_2 band with $2 \leq J \leq 11$ and $0 \leq K \leq 9$ H_2-broadening coefficients vary between 0.126 and 0.150 cm^{-1} atm^{-1}. For the lines with $J \leq 13$ the averaged temperature exponent n given by Eq. (3.2.1) is equal to 0.73 for H_2-broadening and 0.30 for He-broadening [107]. He- and Ar-broadening was considered in Ref. [108].

Table 4.26. Experimental N_2-, H_2-, and self-broadening coefficients (in 10^{-3} cm^{-1} atm^{-1}) for the ν_2 and ν_4 bands of PH_3 at 297 K (asterisks mark the lines with undistinguishable components).

Line	ν, cm^{-1}	Band	PH_3–N_2 [102]	PH_3–PH_3 [104]	PH_3–H_2 [105]
^{Q}R (1, 0)	1008.8789	ν_2	83.9(2.3)		
^{Q}R (1, 1)	1009.0800	ν_2	83.1(2.4)		
^{Q}R (2, 0)	1016.7311	ν_2	80.0(1.7)	108.7(3.5)	107.2(3.6)
^{Q}R(2,1)	1016.9293	ν_2		107.6(3.0)	104.9(2.2)
^{P}P (3, 1)	1092.7072	ν_4	82.9(2.5)	110.6(3.3)	108.0(3.7)
^{Q}R (3, 3)*	1026.0232	ν_2	78.7(2.4)		
^{R}P (4, 1)	1089.2008	ν_4	83.1(2.8)	111.3(2.9)	106.7(3.2)
^{Q}R (4, 3)*	1033.1920	ν_2	76.2(2.7)		
^{R}P (5, 1)	1081.4295	ν_4	84.2(3.6)		
^{S}P (5, 1)	1081.4565	ν_4	84.2(3.1)		
^{P}P (5, 2)	1073.5258	ν_4	77.5(2.2)		102.4(2.2)
^{R}P (6, 0)	1073.3856	ν_4	77.2(2.1)		
^{R}P (6, 1)	1074.0873	ν_4	76.8(2.2)	110.5(3.0)	101.5(2.4)
^{Q}R (7, 0)	1051.4496	ν_2	73.3(2.2)	107.4(3.7)	95.5(2.3)
^{Q}R(7,1)	1051.6224	ν_2		105.8(2.5)	98.5(3.1)
^{Q}R(7,2)	1052.1428	ν_2		103.8(3.3)	95.9(3.4)
^{Q}R(7,5)	1055.8929	ν_2		95.5(3.7)	96.2(2.3)
^{R}P (7, 0)	1066.6132	ν_4	76.2(2.3)	108.5(3.3)	99.3(2.3)
^{P}P (7, 3, A_1)	1054.6746	ν_4	80.7(1.8)		
^{P}P (7, 3, A_2)	1054.7362	ν_4	80.3(1.8)		
^{R}P (7, 3, A_1)	1069.4686	ν_4	82.6(4.0)		
^{R}P (7, 3, A_2)	1069.4253	ν_4	81.1(3.0)		
^{P}P(7,4)	1051.8799	ν_4		106.3(3.0)	98.4(2.9)
^{Q}R (7, 4)	1054.2621	ν_2	73.8(2.3)		
^{Q}R (8, 2)	1058.3011	ν_2	72.1(2.4)		
^{R}P (8, 3, A_1)	1062.3679	ν_4	80.5(2.6)		
^{R}P (8, 3, A_2)	1062.4475	ν_4	81.3(2.7)		
^{P}P (8, 4)	1044.0292	ν_4	72.2(1.9)		
^{P}P(8,5)	1041.0058	ν_4		107.3(3.2)	
^{Q}R (8, 7)	1066.2652	ν_2	70.4(2.0)	97.8(2.6)	90.4(2.1)
^{R}P (9, 1)	1054.1826	ν_4	73.2(2.4)		
^{R}P (9, 2)	1054.8536	ν_4	74.4(1.9)		
^{Q}R (9, 4)	1066.2191	ν_2	69.3(2.0)	101.4(2.5)	95.4(2.1)
^{Q}R (9, 6)*	1069.6077	ν_2	69.1(1.5)		
^{Q}R (9, 9)*	1077.8521	ν_2	67.5(2.3)	93.7(2.7)	78.0(1.8)
^{Q}R (10, 0)	1069.4045	ν_2	68.6(3.0)		
^{R}P(10,1)	1048.1455	ν_4		101.4(2.7)	98.0(2.3)
^{Q}R (10, 1)	1069.5589	ν_2	69.0(1.6)		
^{Q}R (10, 2)	1070.0241	ν_2	68.7(1.9)	100.4(2.6)	94.1(2.1)
^{Q}R (10, 5)	1073.3566	ν_2	67.4(1.6)		
^{Q}R (10, 7)	1077.3456	ν_2	67.7(1.7)	96.9(2.8)	89.8(2.4)
^{Q}R (10, 10)	1086.6031	ν_2	65.7(1.9)		
^{R}P (11, 3)	1043.3918	ν_4	72.8(2.2)		
^{Q}R (11, 4)	1077.4376	ν_2	67.8(1.8)	97.6(2.9)	92.7(2.0)
^{Q}R (11, 5)	1078.8235	ν_2	66.6(2.4)		
^{R}P (11, 6)	1043.8355	ν_2	68.4(1.7)		

^{Q}R (11, 7)	1082.6384	ν_2	64.4(2.5)		
^{Q}R (11, 8)	1085.1194	ν_2	63.8(2.8)	96.6(2.4)	87.9(2.2)
^{Q}R (12, 2)	1081.0750	ν_2	65.5(2.1)	94.0(3.2)	92.6(2.2)
^{Q}R (12, 3,A_2)	1081.8240	ν_2	67.1(1.9)		
^{Q}R (12, 4)	1082.8143	ν_2	66.5(3.5)		
^{Q}R (12, 6)*	1085.7949	ν_2	63.8(1.6)		
^{Q}R (12, 9)*	1092.9228	ν_2	61.7(1.6)	92.3(2.5)	85.2(2.3)
^{P}Q (12, 9)*	1105.2097	ν_4	65.5(3.1)		87.6(2.1)
^{Q}R (13, 0)	1085.8367	ν_2	63.6(1.5)		
^{Q}R (13, 1)	1085.9736	ν_2	64.2(1.6)		
^{Q}R (13, 2)	1086.3865	ν_2	64.0(1.6)		
^{Q}R (13, 5)	1089.3284	ν_2	62.1(2.6)	93.9(2.3)	
^{Q}R (13, 7)	1092.8169	ν_2	62.0(1.5)	92.6(3.0)	
^{Q}R (13, 10)	1100.7499	ν_2	58.6(1.6)		
^{Q}R (14, 3, A_1)	1092.1806	ν_2	66.1(2.7)		
^{Q}R (14, 3, A_2)	1092.2904	ν_2	62.4(3.2)		
^{Q}R (14, 4)	1093.1723	ν_2	62.1(1.6)	91.1(3.0)	
^{Q}R (14, 10)	1105.2756	ν_2	60.9(2.5)		
^{P}Q (14, 11)	1101.0797	ν_4	57.5(2.2)		
^{Q}R (15, 6)*	1100.7972	ν_2	59.8(1.7)		
^{Q}R (15, 8)	1104.5939	ν_2	58.2(1.6)		
^{Q}R (16, 0)	1101.1202	ν_2	61.6(1.8)		

According to Table 4.26 the N_2-broadening coefficients globally decrease with increasing J for both ν_2 and ν_4 bands; for a given J and $K \approx J$ they have also a tendency to decrease. The global behaviour of the self-broadening coefficients with increasing J is quite similar, but significantly low values are observed for the lines $^{Q}R(9,9)$ and $^{P}P(11,11)$ with $K = J$. The H_2-broadening coefficients are roughly constant, except for the lines with K approaching or equal to J for which noticeably lower values are observed.

4.3.3. CH_3A-molecules (A = Cl, F, D)

These molecules are involved in the processes responsible for the depletion of ozone layer. Their line-broadening coefficients for mixtures with various gases and at various temperatures were therefore studied both experimentally and theoretically.

4.3.3.1. Methyl chloride CH_3Cl

For the CH_3Cl molecule particular attention was paid to the vibrational bands ν_2, ν_3, ν_5 and ν_6. The most important for the terrestrial atmosphere case of perturbation by N_2 and O_2 was studied in Refs [98, 109–110] (see also references cited therein). In Ref. [109], N_2-broadening coefficients of 29 vibrotational lines in the P- and R-branches of the ν_3 band (684–764 cm^{-1}) were measured using the Voigt profile model. For the lines with $3 \le J \le 46$ and $3 \le K \le 9$ it was found that $0.095 < \gamma < 0.123$ cm^{-1} atm^{-1} at 296.5 K. The same band was studied in Ref. [98] at 203.2 K. The coefficients γ were obtained for 87 lines ($2 \le J \le 22$, $0 \le K \le 6$) in the ^QP- and ^QR-branches with Voigt and Rautian–Sobel'man model profiles. For the lines with $J \le 22$ the room-temperature N_2-broadening coefficients were also re-measured. Their theoretical estimation was made by the RB approach using the mean thermal velocity approximation and the interaction potential composed of 13 terms. These coefficients are given in Table 4.27 (for the RB calculations the hyper-polarizabilities of CH_3F were arbitrarily used). The temperature dependence of N_2-broadening coefficients was considered from experimental and theoretical viewpoints. It was found that the temperature exponent n is practically independent of the quantum number K but that it depends strongly on the quantum number m: for m increasing from 1 to 4 n increases from 0.82 to 0.84, then for m = 17–19 it decreases to 0.57 and for m = 24 increases again to 0.64.

Table 4.27. Experimental (Voigt profile) and RB-calculated CH_3Cl–N_2 broadening coefficients (in 10^{-3} cm^{-1} atm^{-1}) for the ν_3 band ($\Delta K = 0$) at 296.5 K [98]. Reproduced with permission.

Line	ν, cm^{-1}	expt	calc.	Line	ν_0, cm^{-1}	expt	calc.
R(3,1)	736.3074	127.7(4.2)	126.2	*P*(11,2)	722.6491	117.6(3.2)	123.4
R(3,2)	736.2930	124.8(3.5)	120.2	*P*(11,3)	722.6261	115.7(3.5)	122.3
R(3,3)	736.2690	112.9(4.2)	107.5	*P*(11,7)	722.4430	118.2(3.8)	111.9
R(4,3)	737.1171	122.6(3.9)	116.8	*P*(12,6)	721.5324	113.8(4.5)	115.4
P(5,0)	728.9314	124.5(3.6)	126.9	*P*(17,8)	716.4449	113.0(4.7)	106.3
P(5,1)	728.3268	125.6(3.7)	127.7	*P*(17,9)	716.3692	108.2(3.8)	105.0
P(5,2)	728.3127	119.6(4.2)	123.9	*P*(21,4)	712.5553	108.4(2.9)	102.5
P(5,3)	728.2893	119.5(3.7)	116.8	*P*(21,6)	712.4663	105.8(3.9)	101.7
P(5,4)	728.2566	117.5(5.0)	103.6	*P*(21,7)	712.4087	107.0(2.9)	101.2
R(8,4)	740.3977	121.3(3.8)	121.7	*R*(22,0)	751.0870	99.1 (2.9)	99.8
P(9,4)	724.5118	122.5(4.4)	121.7	*R*(22,1)	751.0819	104.2(2.3)	99.8
P(9,5)	724.4703	119.8(4.4)	117.9	*R*(23,2)	751.0667	102.0(2.9)	99.7
P(11,0)	722.6676	117.9(4.3)	124.2	*R*(22,3)	751.0415	102.1(2.7)	99.6
P(11,1)	722.6630	118.1(3.4)	124.0	*R*(22,4)	751.0061	101.75(2.4)	99.4

4.3.3.2. Methyl fluoride CH_3F

N_2- and O_2-broadening coefficients of CH_3F in the ν_2 and ν_5 bands (1416–1553 cm^{-1}) were studied in Refs [111, 112] using Voigt and Rautian–Sobel'man profile models; for the latter model these coefficients are given in Table 4.28 (with the quantum numbers J and K referring, as usual, to the ground state). According to Refs [111, 112], the γ-values extracted with Voigt profile are systematically smaller than those obtained with Rautian–Sobel'man profile.

Table 4.28. Experimental and theoretical N_2- and O_2-broadening coefficients (in cm^{-1} atm^{-1}) for the ν_2 and ν_5 bands of $^{12}CH^3F$ obtained with Rautian–Sobel'man model profile at 296 K.

CH_3F–N_2 [111]				CH_3F–O_2 [112]				
Line	ν, cm^{-1}	expt	calc.	Line	ν, cm^{-1}	Band	expt	calc.
^{P}P (7,4)	1416.9913	0.1031(36)	0.1149	^{P}Q(10,5)	1416.9304	ν_5	0.0923(99)	0.0860
^{P}P (5,4)	1420.7170	0.1054(39)	0.1099	^{P}P (7,4)	1416.9913	ν_5	0.0933(47)	0.0932
^{P}Q (25,4)	1420.9824	0.0964(33)	0.0936	^{Q}P (16,1)	1420.4468	ν_2	0.0766(43)	0.0771
^{Q}P (14,0)	1427.0376	0.0997(61)	0.0997	^{P}P (5,4)	1420.7170	ν_5	0.0885(66)	0.0926
^{P}Q (12,4)	1427.3082	0.1024(34)	0.1017	^{P}P(14,2)	1421.0524	ν_5	0.0815(30)	0.0804
^{P}P (27,1)	1427.5181	0.0997(48)	0.0945	^{P}Q (12,4)	1427.3082	ν_5	0.0777(45)	0.0828
^{P}Q (11,4)	1427.6355	0.1000(30)	0.1043	^{P}Q (11,4)	1427.6355	ν_5	0.0815(34)	0.0848
^{P}P (7,2)	1438.0165	0.1073(55)	0.1211	^{P}Q (10,4)	1427.9409	ν_5	0.0860(36)	0.0868
^{P}Q (18,2)	1440.7374	0.1066(63)	0.0940	^{P}Q (7,4)	1428.7079	ν_5	0.0909(28)	0.0926
^{P}P (4,2)	1444.7243	0.1255(65)	0.1247	^{P}Q (6,4)	1428.9103	ν_5	0.0897(40)	0.0938
^{Q}P (4,0)	1451.7479	0.1404(60)	0.1293	^{P}Q (5,4)	1429.0851	ν_5	0.0925(49)	0.0939
^{Q}Q (19,3)	1463.1755	0.0959(57)	0.0936	^{P}Q (4,4)	1429.2316	ν_5	0.0852(39)	0.0914
^{P}Q (17,1)	1471.0718	0.1013(42)	0.0945	^{P}Q (18,2)	1440.7374	ν_5	0.0766(23)	0.0741
^{Q}R (13,3)	1485.6978	0.0983(56)	0.0988	^{P}R (22,6)	1441.7404	ν_5	0.0729(40)	0.0708
^{Q}R(14,6)	1485.7910	0.0962(55)	0.0948	^{P}R (7,4)	1442.0990	ν_5	0.0881(82)	0.0918
^{R}R (3,1)	1493.0238	0.1200(70)	0.1247	^{P}Q (12,2)	1445.6671	ν_5	0.0817(60)	0.0834
^{R}P (9,3)	1493.2230	0.1067(53)	0.1111	^{P}R (12,4)	1449.0870	ν_5	0.0838(90)	0.0819
^{P}R (13,1)	1493.8705	0.1051(47)	0.0995	^{Q}P(4,1)	1451.2116	ν_2	0.0927(48)	0.1020
^{R}Q (4,2)	1497.0582	0.1159(55)	0.1186	^{R}R (1,0)	1478.5198	ν_5	0.1065(103)	0.1055
^{R}Q (5,2)	1497.2228	0.1109(88)	0.1205	^{Q}R (10,5)	1478.9453	ν_2	0.0797(46)	0.0845
^{R}P (21,5)	1497.3345	0.0989(102)	0.0927	^{Q}R (10,4)	1479.1314	ν_2	0.0864(63)	0.0854
^{R}R(9,0)	1497.5023	0.1072(39)	0.1112	^{R}P (19,3)	1479.6244	ν_5	0.0754(45)	0.0735
^{R}Q (22,3)	1513.8218	0.1013(52)	0.0934	^{Q}R (13,6)	1484.0959	ν_2	0.0727(52)	0.0787
^{R}Q (23,3)	1514.3210	0.0922(50)	0.0934	^{Q}R (13,5)	1484.1607	ν_2	0.0709(50)	0.0793
^{R}R(12,3)	1532.0903	0.0977(47)	0.1005	^{R}R(4,0)	1485.1360	ν_5	0.0847(53)	0.1013
^{R}Q (18,5)	1532.3054	0.0933(51)	0.0926	^{P}R (9,1)	1485.3929	ν_5	0.0855(33)	0.0893
^{R}R(13,3)	1546.8515	0.0911(42)	0.0934	^{R}P (7,2)	1485.4984	ν_5	0.0883(44)	0.0951
^{R}R(12,5)	1552.9157	0.0969(35)	0.0984	^{R}R(11,0)	1502.6761	ν_5	0.0914(112)	0.0846
				^{R}Q (20,2)	1503.0300	ν_5	0.0742(52)	0.0723
				^{R}R(8,1)	1503.0973	ν_5	0.0875(62)	0.0914
				^{R}R(8,3)	1524.1069	ν_5	0.0843(49)	0.0899
				^{R}Q (23,4)	1524.1895	ν_5	0.0694(18)	0.0707
				^{R}Q (6,5)	1529.4264	ν_5	0.0880(67)	0.0848
				^{R}Q (7,5)	1529.5654	ν_5	0.0938(89)	0.0872
				^{R}Q (9,5)	1529.9013	ν_5	0.0790(46)	0.0863
				^{R}Q (11,5)	1530.3123	ν_5	0.0840(76)	0.0833
				^{R}R (16,2)	1530.4278	ν_5	0.0817(82)	0.0757
				^{R}Q (12,5)	1530.5453	ν_5	0.0843(105)	0.0816
				^{R}R (22,0)	1531.9189	ν_5	0.0715(50)	0.0711
				^{R}Q (17,5)	1531.9710	ν_5	0.0709(35)	0.0745
				^{R}Q (18,5)	1532.3054	ν_5	0.0742(75)	0.0735
				^{R}Q (15,8)	1562.1280	ν_5	0.0808(64)	0.0753
				^{R}Q (16,8)	1562.3644	ν_5	0.0780(63)	0.0742
				^{R}Q (16,9)	1572.4556	ν_5	0.0808(127)	0.0736
				^{R}Q (17,9)	1572.6916	ν_5	0.0810(114)	0.0728

Theoretical estimations of these coefficients were realised in the framework of the RB formalism for long-range interactions containing, besides the usual electrostatic contributions, the induction and dispersion terms obtained by Leavitt [100] (13 terms in the CH_3F–N_2 interaction potential and 14 terms in the CH_3F–O_2 interaction potential). Two forms were considered for the dispersion contribution: $(3/2)\bar{u}\alpha_1\alpha_2$ with $\bar{u} = u_1u_2/(u_1+u_2)$ defined by the first ionization energies u_1 and u_2, and $(3/2)\bar{u}\alpha_1\alpha_2$ replaced by $4\varepsilon_{12}\sigma_{12}^6$ with the Lennard–Jones parameters ε_{12} and σ_{12} for the colliding pair. For the case of nitrogen-broadening [111] it was found that the agreement between the experimental data and calculations with the first form for the dispersion contribution is generally satisfactory, except for a few lines (these calculations are given in Table 4.28); the second form for the dispersion contribution generally led to overestimated broadening coefficients. In the case of CH_3F–O_2 [112], the theoretical values obtained with the first form of the dispersion contribution were too low in comparison with all measurements whereas the second form yielded the line-broadening coefficients (given in Table 4.28) close to the experimental data obtained with the Rautian–Sobel'man model (slight overestimation was stated for some lines only). In Ref. [112], the averaged value of the narrowing coefficient $\beta_0 = (35 \pm 12)\cdot 10^{-3}$ cm^{-1} atm^{-1} was also calculated and found to be close to the diffusion coefficient $\beta_{diff} = 21\cdot 10^{-3}$ cm^{-1} atm^{-1}.

The self-broadening case was considered in Refs [113–116]. The self-broadening coefficients for 33 lines in the ν_2 and ν_5 bands with $1 \le J \le 19$, $0 \le K \le 6$ at 183 K were measured in Ref. [115] using Voigt, Rautian–Sobel'man and Galatry profiles. In the same reference, the self-broadening coefficients at 296 K and the temperature exponents were obtained for 18 lines (Table 4.29). The γ-values calculated by the RB method (with dipole–dipole, dipole–quadrupole and quadrupole-quadrupole interactions) were greater than the corresponding experimental values, but the J dependence was correctly reproduced.

Line-broadening coefficients of CH_3F perturbed by polar molecules CH_3Br, NH_3 and air were obtained in Refs [114, 116]. In Ref. [114], the line shifts were also observed.

He- and Ar-broadening coefficients of CH_3F lines in the ν_6 band were determined in Refs [117, 118], respectively. H_2-broadening coefficients for absorption lines in the ν_5 band (6.6 μm) at 297, 223 and 183 K were measured in Ref. [97]. The H_2- and He-broadening coefficients are given in Table 4.30. The theoretical values given in this table were obtained by the RB approach with the mean thermal velocity. The CH_3F–He interaction potential was chosen as the sum of electrostatic and atom–atom interactions; the model potential of Eq. (2.8.14) was additionally tested. The parameters A_1, R_1, A_2 and R_2 were adjusted to get the best agreement of the computed line-broadening coefficients with the corresponding experimental values. The obtained values $A_1 = 0.030$, $R_1 = 0.95$, $A_2 = 0.006$, $R_2 = 0.0222$ yielded reasonable predictions for the broadening coefficients except the low J-transitions of the ^QQ branch. It was noted that the repulsive forces play the major role in the broadening coefficients through the parameter R_1. Moreover, the broadening coefficients at high J were found to be very sensitive to the potential parameters values. (*Ab initio* calculated potential energy surfaces for the CH_3F–He system can be found in Ref. [119].)

Table 4.29. Experimental and calculated self-broadening coefficients (in cm^{-1} atm^{-1}) at 296 K and temperature exponents for the ν_2 and ν_5 bands of $^{12}CH^3F$ [115]. Reproduced with permission.

Line	ν, cm^{-1}	γ				n	
		expt (Voigt)	expt (Rautian–Sobel'man)	expt (Galatry)	calc.	expt (Rautian–Sobel'man)	calc.
^PP (5,4)	1420.71697		0.346(12)	0.346(11)	0.3890	1.047(143)	0.971
^PQ (12,4)	1427.30822		0.507(21)	0.507(21)	0.6356	0.822(162)	0.767
^PQ (11,4)	1427.63551		0.510(13)	0.510(13)	0.6249	0.844(126)	0.864
^QP (12,0)	1432.26562	0.5383(181)	0.5494(245)	0.5497(249)	0.6475	1.006(186)	0.824
^PP (3,2)	1446.80935	0.4164(163)	0.4347(167)	0.4349(167)	0.4676	0.730(134)	0.754
^PQ (5,2)	1450.53199	0.3906(174)	0.4063(140)	0.4067(140)	0.4701	1.065(143)	1.038
^PQ (4,2)	1451.07041	0.3862(91)	0.3967(136)	0.3969(138)	0.4567	1.007(123)	0.930
^PQ (12,1)	1469.25077	0.5032(201)	0.5281(195)	0.5287(190)	0.6496	0.805(174)	0.767
^PQ (15,1)	1470.40558	0.4874(161)	0.5135(218)	0.5137(218)	0.6262	0.424(164)	0.437
^RQ (4,2)	1497.05774		0.404(27)	0.404(27)	0.4317	1.086(215)	0.935
^RQ (5,2)	1497.22278		0.417(34)	0.418(36)	0.4538	0.950(238)	1.035
^RQ (7,2)	1497.64887	0.4295(151)	0.4493(231)	0.4495(231)	0.5186	0.989(183)	1.107
^RQ (9,2)	1498.19940	0.4668(112)	0.4782(105)	0.4786(107)	0.5875	1.024(105)	1.026
^RQ (17,2)	1501.50829	0.4491(134)	0.4587(117)	0.4590(114)	0.5759	0.226(132)	0.201
^RR(8,2)	1513.52478	0.4948(121)	0.5108(141)	0.5109(142)	0.5724	1.065(118)	1.062
^RR(5,4)	1529.04570		0.357(14)	0.358(14)	0.3827	1.127(174)	1.047
^RR(9,4)	1536.61289		0.496(16)	0.496(16)	0.5726	1.102(193)	0.995
^RR(19,3)	1546.85149		0.453(16)	0.453(15)	0.5000	−0.051(167)	−0.091

Table 4.30. Room-temperature experimental (Voigt profile) and theoretical H_2- and He-broadening coefficients of CH_3F (in 10^{-3} cm^{-1} atm^{-1}) [97, 117]. Reproduced with permission.

CH_3F–H_2, ν_5 band [97]				CH_3F–He, ν_6 band [117]			
Line	ν, cm^{-1}	expt	calc.	Line	ν, cm^{-1}	expt	calc.
^RR (6,4)	1430.9057	124.3(3.3)	126.9	^RQ(1,0)	1183.9319	69.7(2.3)	67.92
^RR(1,0)	1478.5198	144.9(5.7)	141.3	^RQ(2,0)	1183.9158	71.1(2.6)	67.92
^RR(2,0)	1480.6253	143.6(3.8)	139.5	^RQ(3,0)	1183.8915	70.0(2.6)	67.91
^RR(3,0)	1482.8265	143.5(3.3)	138.3	^RQ(4,0)	1183.8592	70.4(2.9)	67.89
^RR(4,0)	1485.1360	140.9(3.4)	138.0	^RQ(5,0)	1183.8189	70.7(3.4)	67.87
^RR(3,1)	1493.0238	136.3(4.8)	137.2	^RQ(6,0)	1183.7705	70.0(2.3)	67.84
^RQ(4,2)	1497.0577	133.5(5.1)	132.4	^RQ(7,0)	1183.7140	69.2(2.2)	67.79
^RQ(5,2)	1497.2228	135.1(3.6)	135.4	^RQ(8,0)	1183.6494	68.6(1.7)	67.73
^RR(9,0)	1497.5023	136.6(5.1)	136.7	^RQ(10,0)	1183.4963	68.0(1.6)	67.57
^RQ(7,2)	1497.6489	136.4(5.6)	136.8	^RQ(12,0)	1183.3109	67.1(1.6)	67.35
^RQ(13,2)	1499.6505	140.9(5.2)	135.7	^RQ(14,0)	1183.0933	67.4(1.5)	67.08
^RR(11,0)	1502.6761	137.2(3.0)	136.3	^RQ(15,0)	1182.9725	66.2(1.9)	66.93
^RR(3,2)	1501.8711	127.7(3.1)	128.7	^RQ(21,0)	1182.0800	69.4(3.0)	65.9
^RQ(10,3)	1509.0429	132.8(6.0)	136.0	^RQ(29,0)	1180.4455	66.7(1.9)	64.54
^RQ(11,3)	1509.3270	132.5(5.0)	136.0	^RQ(7,6)	1218.0801	62.1(2.6)	62.71
^RR(15,0)	1513.2112	135.0(3.0)	134.4	^RQ(8,6)	1218.0182	60.1(3.5)	64.04
^RR(8,2)	1513.5248	136.7(3.1)	136.7	^RQ(9,6)	1217.9483	64.3(1.6)	64.84
^RR(13,1)	1514.1159	135.2(2.1)	135.3	^RQ(10,6)	1217.8708	64.6(1.7)	65.33
^RR(3,3)	1514.6713	105.5(2.9)	90.2	^RQ(11,6)	1217.7855	64.3(2.7)	65.66
^RR(9,2)	1515.5464	135.9(3.1)	136.6	^RQ(12,6)	1217.6924	64.6(1.8)	65.84
^RR(16,0)	1515.8694	133.5(5.9)	133.8	^RQ(14,6)	1217.4829	64.8(1.9)	66.01
^RR(5,3)	1518.3696	127.7(4.3)	130.3	^RQ(15,6)	1217.3666	64.4(1.7)	66.00
^RR(17,0)	1518.5341	136.0(3.2)	133.2	^RQ(16,6)	1217.2424	64.9(1.5)	65.96
^RQ(9,4)	1519.3674	131.4(4.2)	133.7	^RQ(17,6)	1217.1106	65.1(1.9)	65.89
^RR (7,4)	1532.7872	126.8(2.8)	130.9	^RQ(18,6)	1216.9710	63.9(1.7)	65.80
^RR (14,3)	1536.2138	133.9(3.5)	134.9	^RQ(19,6)	1216.8235	65.1(1.9)	65.67
^RR (9,4)	1536.6129	131.8(4.4)	134.1	^RQ(21,6)	1216.5056	65.0(1.9)	65.43
^RR (19,2)	1537.0352	131.3(3.6)	132.1	^RQ(22,6)	1216.3349	65.2(1.9)	65.30

4.3.3.3. Monodeuterated methane CH_3D

The collisional broadening of vibrotational CH_3D lines by various perturbers and at various temperatures was addressed in Refs [99, 120–125].

In Ref. [99], N_2-broadening and shifting coefficients for 368 transitions ($J \le 20$, $0 \le K \le 6$, $\Delta K = 0$) in the ν_2 band at 296 K were measured ($0.0248 < \gamma < 0.0742$ cm^{-1} atm^{-1}, $-0.0003 < \delta < -0.0094$ cm^{-1} atm^{-1}) and compared with theoretical values obtained by the RB approach. The isotropic potential was chosen as

$$V_{iso}(r) = -C_6/r^6 + C_{14}/r^{14},$$

where $C_6 = 4\varepsilon\sigma^6$ and $C_{14} = 4\varepsilon\sigma^{14}$. The ratios $\Delta C_6/C_6 = 0.008$ and $y = (\Delta C_{14}/C_{14})/(\Delta C_6/C_6) = 1.0$ were estimated from the experimental data for the self-induced line shifts in the ^QQ branch. The theoretical line-broadening and line-shifting coefficients showed good overall agreement with the experimental data (within 8.7%). Moreover, the experimental N_2-broadening coefficients at 298 K were fitted to empirical expressions. In the ^QQ branch, for the transitions with $J = K$ and $J = K + 1$ this empirical expression (in cm^{-1} atm^{-1}) had the form

$$\gamma = c_0 + c_1|m|\,(|m| + 1) + c_2|m|^2\,(|m| + 1)^2, \tag{4.3.8}$$

whereas for the other lines it was

$$\gamma = c_0 + c_1\,|m|\,(|m| + 1) + c_2K^2 \tag{4.3.9}$$

(constants c_0, c_1 and c_2 are given in Table 4.31).

Table 4.31. Parameters of the empirical Eqs (4.3.8) and (4.3.9) for CH_3D–N_2 broadening coefficients (in cm^{-1} atm^{-1}) at 298 K [99]*. Reproduced with permission.

Band	ν_2	ν_3	ν_2	ν_3	ν_2
Line sets	(1)	(1)	(2)	(2)	(3)
c_0	0.0674(4)	0.0679(4)	0.0684(4)	0.0700(9)	0.0680(6)
c_1	$-5.51(21)\cdot10^{-5}$	$-4.98(30)\cdot10^{-5}$	$-4.03(13)\cdot10^{-4}$	$-4.26(19)\cdot10^{-4}$	$-1.80(12)\cdot10^{-4}$
c_2	$6.25(48)\cdot10^{-5}$	$7.96(65)\cdot10^{-5}$	$8.94(48)\cdot10^{-7}$	$9.93(78)\cdot10^{-7}$	$2.14(47)\cdot10^{-7}$
N	346	297	11	13	10

* Line sets: (1) — all lines except $J = K$ in the QQ-branch, (2) — the lines of the QQ-branch with $J = K$, (3) — the lines of the QQ-branch with $J = K + 1$; N is the total number of lines.

O_2-broadening and shifting coefficients in the same ν_2 band were measured and calculated in Ref. [120]. For the ν_3 band CH_3D–O_2-broadening coefficients were obtained in Ref. [121].

The case of self-broadening was considered by Lance *et al.* [122]. H_2-broadening coefficients at room and low temperatures were examined in Refs [123–124].

RB calculations for the line-broadening coefficients in the ν_2 band of CH_3D perturbed by helium were made by Féjard *et al.* [125] with two model potentials. For the lines with $0 \le J \le 14$ (the most populated states at room temperature) the atom–atom potential was used. Since the Lennard–Jones parameters available in the literature for the atom–atom interactions [126] did not give a good agreement between calculated and experimental line widths, the authors kept these parameters free. Although the order of magnitude of the experimental line widths was correctly reproduced, the differences between theoretical and experimental values were quite pronounced. The second potential of *ab initio* type was employed for a smaller set of lines (with lower J-values) and resulted in a similar global agreement between the calculations and measurements. Failures of calculations for high values of the rotational quantum numbers were evidenced and a need for improvements in the theoretical model was evoked.

4.4. Broadening coefficients of vibrotational lines of methane

Methane CH_4 is the principle component of natural gas, not toxic but highly flammable: a concentration of 5–14% in air gives an explosive mixture. Like CO_2, it is responsible for the greenhouse effect and global warming. The control of methane concentration in the Earth's atmosphere therefore plays a major role in the thermal balance prediction. The precise knowledge of its line-broadening coefficients is also necessary for remote sensing of outer planets (Saturn, Jupiter) and their moons (Titan). In particular, for the atmosphere of Titan (98% of N_2 and 1.6% of CH_4) the case of CH_4 line broadening by nitrogen is of major importance.

Methane molecule is a spherical top and belongs to the tetrahedral point group T_d. It has four normal vibrational modes: one fully symmetric (A_1) and non-degenerate mode ν_1, one doubly degenerate (E) mode ν_2 and two triply degenerate (F) modes ν_3 and ν_4. Vibrational wave functions are defined therefore by a set of four vibrational quantum numbers ($v_1v_2v_3v_4$). The normal vibration frequencies are given in Table 4.32.

Table 4.32. Normal frequencies and other spectroscopic parameters of methane [1–3, 127–129]*.

Parameter	Value	Parameter	Value	Parameter	Value
ν_1, cm^{-1}	2916.7	ξ_{23}	–0.7974	μ, D	0.0
ν_2, cm^{-1}	1533.6	ξ_{24}	0.6034	Q, DÅ	0.0
ν_3, cm^{-1}	3018.9	ξ_{34}	0.72	Ω, DÅ^2	2.6
ν_4, cm^{-1}	1306.2	ξ_3	0.0462	Φ, DÅ^3	
B, cm^{-1}	5.321	ξ_4	0.4538	α, Å^3	2.564

* Additionally, $A_{\rm II} = A_{\perp} = 0.82$ Å4.

The energy levels $E_0(J)$ of a rigid spherical top with the rotational Hamiltonian $H_0 = BJ^2$ are defined through the relation $E_0(J) = BJ(J+1)$, where B is the rotational constant. These levels are $(2J+1)^2$-times degenerate with the respect to the projections m and K of the total angular momentum $\vec{J}$ on the laboratory- and molecule-fixed z-axes. The wave function corresponding to the rotational state with the energy $E_0(J)$ is a wave function of symmetric top. Taking into account the vibrotational interaction, the centre of the rotational manifold is given by

$$E(J) = BJ(J+1) - D[J(J+1)]^2 + H[J(J+1)]^3 + \cdots, \tag{4.4.1}$$

where D, H, ... are constants. For the vibrational states (0010) and (0001), additional terms $\Delta E_t = B\xi_t[R(R+1) - J(J+1) - l_t(l_t+1)]$ due to the Coriolis interaction of the first order should be added to this formula:

$$\begin{aligned} &2B\xi_t J \text{ (for the states with } R = J+1), \\ &-2B\xi_t J \text{ (for the states with } R = J), \\ &-2B\xi_t(J+1) \text{ (for the states with } R = J-1), \end{aligned} \tag{4.4.2}$$

where $t = 3, 4$ and R is the rotational quantum number defined, for $J = 0, 1, 2, \ldots, l = 0, 1, 2, \ldots$, by the angular momenta addition rules: $R = |J - l|, |J - l + 1|, \ldots, J + l$. The vibrotational energy levels are obtained by a numerical diagonalization of an effective Hamiltonian which is built for the polyad of interacting states for a given J-value and a given symmetry type, so that each rotational level ($P\ J\ C\ N$) is characterised by the polyad number P, rotational quantum number J, symmetry type $C = A_1, A_2, E, F_1, F_2$ and current number N (energy levels are numbered in energy increasing order); more details can be found in Ref. [129]. The dipole moment appearing during molecular vibrations is transformed according to the representation A_2 of the group T_d, so that the general selection rules for the dipolar transitions in the absorption spectra are $A_1 \leftrightarrow A_2$, $E \leftrightarrow E$, $F_1 \leftrightarrow F_2$, $\Delta J = 0$ (Q-branch), $\Delta J = -1$ (P-branch) and $\Delta J = 1$ (R-branch). More detailed selection rules can be found in Refs [128, 129]. In particular, for the transitions from the rotational levels of the ground vibrational state ($v_1 = v_2 = v_3 = v_4 = 0$) to the rotational levels of the triply degenerate states ($v_t = l_t = 1$) one has $\Delta R = 0$.

The spectrum of methane is investigated up to 9000 cm^{-1}. The HITRAN database for methane polyads includes the region of rotational transitions (0.01–815 cm^{-1}), the dyad ν_2/ν_4 (855–2080 cm^{-1}), the pentad $\nu_1/\nu_3/2\nu_4/2\nu_2/\nu_2+\nu_4$ (1980–3450 cm^{-1}), the octad $3\nu_4/\nu_2+2\nu_4/\nu_1+\nu_4/\nu_3+\nu_4/2\nu_2+\nu_4/\nu_1+\nu_2/\nu_2+\nu_3/3\nu_2$ (3370–4819 cm^{-1}) and the tetradecad (fourteen vibrational bands) in the region 4800–6250 cm^{-1}. Many hundreds of articles are devoted to the experimental and theoretical studies of absorption line shapes in various bands and branches of CH_4 mixed with N_2, O_2, Ar, H_2, He, air and itself [129–150] (see also references cited therein). The temperature dependence of line widths for more than 600 CH_4 lines is studied for the fundamental ν_3, ν_4, for the overtone $6\nu_1$, $5\nu_3$ and for the combination $\nu_1+\nu_3$, $\nu_2+\nu_3$, $\nu_1+\nu_4$, $\nu_2+\nu_4$ bands. A summary of experimental results for the temperature-dependent line widths and shifts for various spectral ranges is given in Ref. [150]. All these studies demonstrated the dependence of the line shape parameters on the symmetry types C of the connected states: for a given J-value the lines of E-type (symmetry type and current number are put for the lower rotational state) are narrower than the lines of A and F-types. The behaviour of γ coefficients as a function of J, C and N in the Q-branch of the ν_3 band was found to be identical for N_2, O_2 and Ar buffer gases [130]: they differ only by a constant factor.

An analysis of the ν_3 and ν_4 bands of CH_4 in mixture with various perturbers also showed that the spectral lines can not be considered as isolated. The interference of P- and R-branch lines in the ν_3 band ($J \leq 10$) of CH_4–N_2 and CH_4–Ar mixtures at buffer pressures of 50, 100, 200, 500 Torr and 295 K was investigated in Ref. [131]. This interference was characterised by the parameters ζ_k in the absorption coefficient expression:

$$K(\nu) = \frac{1}{\sqrt{\pi}\Delta\nu'_D} \sum_k I_k [\mathrm{Re}\, F_R(x, y+z) + \zeta_k \, \mathrm{Im}\, F_R(x, y+z)], \tag{4.4.3}$$

where $\Delta\nu'_D$ is the Doppler half-width at the level $1/e$ (see Eq. (1.2.7)), I_k are the intensities of the lines, and the real and imaginary parts are taken for the Rautian–Sobel'man profile. For weakly overlapping lines these parameters ζ_k can be expressed through the off-diagonal relaxation matrix elements Γ_{lk}, as given by Eq. (1.4.3); this equation is valid if the elements Γ_{lk} are much smaller than the differences $\omega_l - \omega_k$. For binary collisions γ, δ, β_0 and ζ_k are proportional to the pressure P. If all ζ_k are identical, then $\Sigma_k \zeta_k = 0$. Fitting of Eq. (4.4.3) to the experimental values yielded the parameters I_k, γ, δ, β_0 and ζ_k for each CH_4 line perturbed by N_2 and Ar [131]. A non-linear dependence of these parameters on pressure observed for $J \geq 6$ in the R-branch and for $J \geq 9$ in the P-branch at $P > 30$ kPa means that the Rozenkranz's formula (1.4.2), derived for the mixing coefficients in the first order, is invalid. The obtained values of γ, δ, β_0 and ζ_k for the ν_3 band of CH_4–N_2 and CH_4–Ar at 295 K [131] are given in Tables 4.33, 4.34. An analysis of 164 γ-values for the lines with the same J, C, N sets broadened by N_2 and Ar showed that the mean ratio $\langle \gamma(\mathrm{Ar})/\gamma(N_2) \rangle = 0.879 \pm 0.010$ (this empirical relation is valid also for the ν_4 band). This fact is very interesting since N_2-broadening and Ar-broadening are due to different parts of the intermolecular interaction potential, and it means that there are no rotational resonances in CH_4 colliding with N_2. An analogous empirical relation $\langle \gamma(O_2)/\gamma(N_2) \rangle = 0.937 \pm 0.015$ is also found for the Q-branch of the ν_3 band. It follows from this last relation that 99% of the air-broadening of CH_4 lines is due to N_2.

Table 4.33. Experimental line shape parameters (in cm^{-1} MPa^{-1}, except ζ which are in MPa^{-1}) for the *R*-branch of the ν_3 band of CH_4 at 295 K [131]. Reproduced with permission.

J	*C*	*N*	γ(Ar)	$\gamma(N_2)$	δ(Ar)	$\delta(N_2)$	β_0(Ar)	$\beta_0(N_2)$	ζ(Ar)	$\zeta(N_2)$
*R*0	A_1	1	0.4989	0.5797	−0.0538	−0.0593	0.2932	0.2604		
*R*1	F_1	1	0.5563	0.6383	−0.0496	−0.0457	0.1832	0.1490		
*R*2	*E*	1	0.5072	0.5829	−0.0468	−0.0491	0.2603	0.1961		
*R*2	F_2	1	0.5783	0.6618	−0.0487	−0.0480	0.2156	0.1945		
*R*3	F_1	1	0.5960	0.6689	−0.0596	−0.0505	0.2252	0.1642	−2.892	−3.282
*R*3	F_2	1	0.5855	0.6561	−0.0431	−0.0408	0.2263	0.1804	2.892	3.282
*R*3	A_2	1	0.5137	0.5891	−0.0537	−0.0464	0.2845	0.2300		
*R*4	A_1	1	0.5144	0.5918	−0.0588	−0.0536	0.2460	0.2501		
*R*4	F_1	1	0.5922	0.6683	−0.0588	−0.0481	0.2218	0.2122	−1.299	−1.535
*R*4	*E*	1	0.4757	0.5420	−0.0616	−0.0520	0.2622	0.2382		
*R*4	F_2	1	0.5388	0.6192	−0.0468	−0.0410	0.2190	0.2358	1.292	1.528
*R*5	F_1	2	0.5301	0.6008	−0.0630	−0.0549	0.2441	0.2341	−0.480	−0.473
*R*5	F_2	1	0.5668	0.6369	−0.0622	−0.0536	0.1885	0.2312	−1.190	−1.378
*R*5	*E*	1	0.4664	0.5361	−0.0521	−0.0425	0.2872	0.2897		
*R*5	F_1	1	0.5376	0.6154	−0.0482	−0.0427	0.2417	0.2322	1.659	1.844
*R*6	*E*	1	0.4905	0.5518	−0.0665	−0.0602	0.3595	0.2989		
*R*6	F_2	2	0.5469	0.6160	−0.0659	−0.0558	0.2450	0.2428	−0.394	−0.638
*R*6	A_2	1	0.5008	0.5683	−0.0694	−0.0563	0.2175	0.2022	− 1.614	−1.780
*R*6	F_2	1	0.5616	0.6300	−0.0735	−0.0665	0.1947	0.1832	−5.234	−5.227
*R*6	F_1	1	0.5414	0.6124	−0.0379	−0.0241	0.2207	0.2218	5.636	5.858
*R*6	A_1	1	0.5135	0.5833	−0.0444	−0.0369	0.2643	0.2299	1.569	1.735
*R*7	F_1	2	0.4808	0.5472	−0.0700	−0.0594	0.2743	0.2237	−1.123	−1.531
*R*7	F_2	1	0.5191	0.5851	−0.0616	−0.0550	0.2311	0.2068	0.026	0.308
*R*7	A_2	1	0.4673	0.5319	−0.0556	−0.0469	0.2969	0.2169		
*R*7	F_2	1	0.5536	0.6216	−0.0196	−0.0527	0.2322	0.2354	−2.193	−2.496
*R*7	*E*	1	0.3730	0.4284	−0.0542	−0.0417	0.3397	0.2776		
*R*7	F_1	1	0.5249	0.5940	−0.0399	−0.0360	0.2447	0.2580	3.251	3.679
*R*8	A_1	1	0.4205	0.4827	−0.0730	−0.0620	0.3186	0.2564		
*R*8	F_1	2	0.4903	0.5735	−0.0810	−0.0593	0.2772	0.2403	−0.424	−0.826
*R*8	*E*	2	0.4736	0.5296	−0.0695	−0.0406	0.2772	0.2403	0.0(fix)	0.0(fix)
*R*8	F_2	2	0.4790	0.5463	−0.0603	−0.0519	0.2880	0.2282	−0.015	−0.161
*R*8	F_1	1	0.5165	0.5812	−0.0527	−0.0488	0.1999	0.1999	−4.096	−3.980
*R*8	*E*	1	0.4103	0.4731	−0.0497	−0.0394	0.3213	0.2650	0.0(fix)	0.0(fix)
*R*8	F_2	1	0.5173	0.5884	−0.0457	−0.0381	0.2046	0.2219	4.475	4.904
*R*9	F_1	3	0.4266	0.4796	−0.1001	−0.0783	0.2817	0.2474	−0.582	−0.578
*R*9	F_2	2	0.4708	0.5330	−0.0576	−0.0646	0.2817	0.2474	0.0(fix)	0.0(fix)
*R*9	*E*	1	0.4221	0.4747	−0.0659	−0.0612	0.3495	0.2857		
*R*9	F_1	2	0.5053	0.5742	−0.0725	−0.0625	0.2733	0.2205	−0.034	−0.619
*R*9	A_1	1	0.4734	0.5315	−0.0537	−0.0471	0.2712	0.2373	−4.141	−4.400
*R*9	F_1	1	0.5165	0.5724	−0.0794	−0.0770	0.1772	0.1313	−8.384	−8.512
*R*9	F_2	1	0.5025	0.5663	0.0053	0.0097	0.2103	0.1985	8.665	9.349
*R*9	A_2	1	0.5077	0.5716	−0.0430	−0.0396	0.2605	0.2805	4.051	4.299
*R*10	EF^*		0.3202	0.3879	−0.0450	−0.0362	0.2097	0.2023	0.0(fix)	0.0(fix)
*R*10	A_2	1	0.5402	0.5956	−0.0496	−0.0465	0.2097	0.2023	−0.912	−1.137
*R*10	F_2	2	0.3999	0.4854	−0.1147	−0.0921	0.2097	0.2023	−0.882	−2.038
*R*10	F_1	2	0.5732	0.6058	−0.0364	−0.0404	0.2097	0.2023	0.0(fix)	0.0(fix)
*R*10	*F* +		0.6345	0.7361	−0.0555	−0.0414	0.2097	0.2023	0.0(fix)	0.0(fix)
*R*10	*F* +		0.6345	0.7361	−0.0570	−0.0411	0.2097	0.2023	0.0(fix)	0.0(fix)
*R*10	A_1	1	0.4575	0.5280	−0.0623	−0.0556	0.2997	0.2464	0.871	1.089
*R*10	F_1	1	0.5369	0.5967	−0.0048	−0.0088	0.2097	0.2023	−5.598	−4.756
*R*10	*E*	1	0.2867	0.3404	−0.0409	−0.0327	0.2097	0.2023	0.0(fix)	0.0(fix)
*R*10	F_2	1	0.4987	0.5588	−0.0271	−0.0162	0.2097	0.2023	5.466	5.818

Table 4.34. Experimental line shape parameters (in cm^{-1} MPa^{-1}, except ζ which are in MPa^{-1}) for the *P*-branch of the ν_3 band of CH_4 at 295 K [131]. Reproduced with permission.

J	C	N	γ(Ar)	$\gamma(N_2)$	δ(Ar)	$\delta(N_2)$	β_0(Ar)	$\beta_0(N_2)$	ζ(Ar)	$\zeta(N_2)$
*P*1	F_1	1	0.4999	0.5892	−0.0875	−0.0712	0.2647	0.2172		
*P*2	E	1	0.5666	0.6546	−0.0869	−0.0810	0.1395	0.1397		
*P*2	F_2	1	0.5608	0.6499	−0.0885	−0.0752	0.1668	0.1865		
*P*3	F_1	1	0.5930	0.6762	−0.0839	−0.0734	0.1814	0.1814	−0.568	−0.728
*P*3	F_2	1	0.5829	0.6685	−0.0719	−0.0666	0.2058	0.1937	0.573	0.732
*P*3	A_2	1	0.5355	0.6212	−0.0763	−0.0709	0.2411	0.2331		
*P*4	A_1	1	0.5646	0.6478	−0.0811	−0.0743	0.2130	0.1984		
*P*4	F_1	1	0.6004	0.6805	−0.0788	−0.0725	0.1874	0.1890	−0.286	−0.326
*P*4	E	1	0.5228	0.5943	−0.0869	−0.0789	0.2294	0.1953		
*P*4	F_2	1	0.5449	0.6252	−0.0719	−0.0678	0.2610	0.2339	0.284	0.328
*P*5	F_1	2	0.5685	0.6415	−0.0812	−0.0741	0.2101	0.1741	−0.032	−0.053
*P*5	F_2	1	0.5904	0.6629	−0.0721	−0.0675	0.1991	0.2295	−0.246	−0.436
*P*5	E	1	0.4948	0.5672	−0.0815	−0.0800	0.2346	0.2073		
*P*5	F_1	1	0.5391	0.6165	−0.0653	−0.0614	0.2513	0.2199	0.283	0.494
*P*6	E	1	0.5285	0.6032	−0.0823	−0.0724	0.2551	0.2459		
*P*6	F_2	2	0.5649	0.6371	−0.0756	−0.0676	0.2246	0.2013	−0.158	−0.183
*P*6	A_2	1	0.5368	0.6067	−0.0660	−0.0597	0.2389	0.1982	−0.461	−0.521
*P*6	F_2	1	0.5769	0.6527	−0.0794	−0.0750	0.2224	0.2398	−1.444	−1.576
*P*6	F_1	1	0.5499	0.6267	−0.0653	−0.0643	0.2379	0.2339	1.593	1.751
*P*6	A_1	1	0.5124	0.5896	−0.0602	−0.0577	0.2671	0.2484	0.461	0.521
*P*7	F_1	2	0.5178	0.5863	−0.0738	−0.0690	0.2441	0.2237	−0.334	−0.304
*P*7	F_2	2	0.5377	0.6096	−0.0600	−0.0596	0.2165	0.2032	−0.124	−0.071
*P*7	A_2	1	0.5164	0.5872	−0.0827	−0.0761	0.2519	0.2423		
*P*7	F_2	1	0.5621	0.6302	−0.0748	−0.0681	0.1861	0.1666	−0.616	−0.827
*P*7	E	1	0.4119	0.4715	−0.0802	−0.0737	0.3235	0.2783		
*P*7	F_1	1	0.5269	0.5979	−0.0697	−0.0663	0.2385	0.2158	1.078	1.199
*P*8	A_1	1	0.4576	0.5271	−0.0771	−0.0681	0.2679	0.2629		
*P*8	F_1	2	0.4974	0.5680	−0.0632	−0.0574	0.2479	0.2236	−0.233	−0.173
*P*8	E	2	0.4897	0.5551	−0.0666	−0.0621	0.2740	0.2302	−0.150	−0.301
*P*8	F_2	2	0.5087	0.5817	−0.0710	−0.0718	0.2230	0.2209	−0.225	0.086
*P*8	F_1	1	0.5423	0.6105	−0.0708	−0.0641	0.2611	0.2663	−0.778	−1.229
*P*8	E	1	0.4147	0.4784	−0.0715	−0.0674	0.2868	0.2456	0.154	0.304
*P*8	F_2	1	0.5180	0.5883	−0.0608	−0.0591	0.1870	0.2005	1.240	1.323
*P*9	F_1	3	0.4467	0.5093	−0.0698	−0.0588	0.2882	0.2232	−0.199	−0.473
*P*9	F_2	2	0.4725	0.5378	−0.0613	−0.0555	0.2393	0.2147	−0.173	−0.402
*P*9	E	1	0.4631	0.5270	−0.0759	−0.0713	0.3138	0.3132		
*P*9	F_1	2	0.5127	0.5816	−0.0699	−0.0650	0.2367	0.2299	−0.364	−0.267
*P*9	A_1	1	0.4994	0.5615	−0.0675	−0.0606	0.2446	0.2421	−0.898	−0.999
*P*9	F_1	1	0.5208	0.5833	−0.0772	−0.0734	0.1906	0.1969	−2.502	−2.449
*P*9	F_2	1	0.5135	0.5778	−0.0627	−0.0579	0.2006	0.2160	3.216	3.577
*P*9	A_2	1	0.5050	0.5676	−0.0637	−0.0533	0.2066	0.2021	0.898	0.999
*P*10	E	2	0.4020	0.4594	−0.0714	−0.0705	0.2097	0.2023	0.0(fix)	0.0(fix)
*P*10	F_2	3	0.4338	0.4983	−0.0632	−0.0572	0.2097	0.2023	0.0(fix)	0.0(fix)
*P*10	A_2	1	0.4378	0.5058	−0.0545	−0.0492	0.2869	0.2349	−0.733	−0.612
*P*10	F_2	2	0.4771	0.5420	−0.0654	−0.0549	0.3078	0.2838	−1.075	−1.405
*P*10	F_1	1	0.5040	0.5692	−0.0625	−0.0562	0.2777	0.1642	0.068	−0.143
*P*10	A_1	1	0.4915	0.5554	−0.0762	−0.0678	0.2715	0.2026	0.759	0.631
*P*10	F_1	1	0.5039	0.5642	−0.0808	−0.0748	0.2481	0.1377	−0.135	−0.590
*P*10	E	1	0.3062	0.3577	−0.0692	−0.0602	0.3591	0.2809	0.0(fix)	0.0(fix)
*P*10	F_2	1	0.5066	0.5644	−0.0538	−0.0563	0.2406	0.2821	1.120	2.097

The temperature exponents for the nitrogen-broadening (n_γ), narrowing (n_β), and line-mixing (n_ζ), coefficients of four lines in the *P*9 manifold appearing in Table 4.34 were experimentally obtained in Ref. [149]: n_γ = 0.836(35), 0.841(14), 0.858(12), 0.839(35) for (*P*9 A_2 1), (*P*9 F_2 1), (*P*9 F_1 1), and (*P*9 A_1 1) lines, respectively; n_β = 1.12(2) for all lines, and n_ζ = 1.22(9), 1.14(9), 1.47(13) for (*P*9 A_2 1), (*P*9 F_2 1), and (*P*9 F_1 1) lines, respectively. These results were obtained by considering the measurements at 90, 140 and 250 K completed by the data of Pine [130–131] at 295 K. It was noted that the measured n_γ-values (between 0.836 and 0.858) are very different from the value of 0.65 given by the HITRAN2004 database.

Systematic studies of CH_4 line shapes perturbed by N_2 and O_2 in the region 5550–6236 cm^{-1} were undertaken in Ref. [146] using the Voigt profile with theoretically calculated Doppler width. Broadening and shifting coefficients as well as the exponents of their temperature dependence were determined for 452 lines. Although the obtained γ-values varied, from line to line, in a rather large range, the line widths generally decreased with increasing |*m*|. To generate a complete line list for databases, an empirical polynomial for the rotational dependence of the broadening coefficients (valid for $|m| \le 14$) was proposed:

$$\gamma = \gamma_0 + a|m| + b|m|^2 \tag{4.4.4}$$

(γ_0-, *a*- and *b*-values are given in Table 4.35; for the air-broadening the usual relation $\gamma_{air} = 0.79\gamma_{N_2} + 0.21\gamma_{O_2}$ is used). For the room-temperature line-shift coefficients the mean values $\bar{\delta}(N_2)$ = −0.0110(34) cm^{-1} atm^{-1}, $\bar{\delta}(O_2)$ = −0.0105(29) cm^{-1} atm^{-1} and $\bar{\delta}$(air) = −0.0109(33) cm^{-1} atm^{-1} (within one standard deviation given in parentheses) were recommended.

Table 4.35. Parameters of the polynomial (4.4.4) (in cm^{-1} atm^{-1}, within one standard deviation) at 296 K [146]. Reproduced with permission.

Gas	γ_0	$a\cdot10^3$	$b\cdot10^4$
N_2	0.0641(10)	1.51(33)	−2.22(25)
O_2	0.0595(8)	1.56(27)	−2.01(21)
air	0.0631(10)	1.52(32)	−2.18(24)

Self-broadening and self-shifting room-temperature coefficients for 1423 methane lines ($J \le 16$) in the octad bands $\nu_2 + 2\nu_4$, $\nu_1 + \nu_4$, $\nu_3 + \nu_4$, $2\nu_2 + \nu_4$ and $\nu_2 + \nu_3$ (located between 4100 and 4635 cm^{-1}) were measured by Predoi-Cross *et al.* [142] (Table 4.36). The line widths extracted with a Lorentzian profile varied from 0.045 to 0.090 cm^{-1} atm^{-1} at 296 K and the line-shift coefficients ranged between −0.005 and −0.020 cm^{-1} atm^{-1} at 298.3 ± 1.2 K. It was noted that the line widths change, in average, within 5% when passing from one vibrational band to another: at low *J*-values the line widths are almost identical (to 1–2%) but at high *J*-values, for a complex band as $\nu_3 + \nu_4$, the differences between them can reach 10–15%. The measurements obtained for a simple band can therefore not be used in a complex band with the same accuracy.

Table 4.36. Experimental room-temperature CH_4 self-broadening coefficients (in 10^{-3} cm^{-1} atm^{-1}) [142]. Reproduced with permission.

|*m*|	Branch	Band	*J*'	*C*'	*N*'	*J*"	*C*"	*N*"	ν, cm^{-1}	γ	δ
1	*P*	$\nu_3+\nu_4$	0	*F*2	5	1	*F*1	1	4308.72886	80.6 (0.5%)	−14.2(3)
1	*Q*	$\nu_2+2\nu_4$	1	*F*2	6	1	*F*1	1	4138.38588	82.6 (5.0%)	
1	*Q*	$\nu_1+\nu_4$	1	*F*2	9	1	*F*1	1	4218.41491	81.0 (0.6%)	−12.7(4)
1	*Q*	$\nu_3+\nu_4$	1	*F*2	10	1	*F*1	1	4316.35433	80.3 (0.5%)	−13.4(4)
1	*Q*	$\nu_3+\nu_4$	1	*F*2	11	1	*F*1	1	4324.93383	80.4 (0.5%)	−12.4(2)
2	*R*	$\nu_1+\nu_4$	2	*F*2	15	1	*F*1	1	4229.19393	80.0 (0.5%)	−12.4(2)
2	*R*	$\nu_3+\nu_4$	2	*F*2	19	1	*F*1	1	4337.54960	80.5 (0.5%)	−11.0(2)
2	*R*	$\nu_3+\nu_4$	2	*F*2	20	1	*F*1	1	4348.93814	81.1 (0.5%)	−10.1(1)
2	*R*	$2\nu_2+\nu_4$	2	*F*2	23	1	*F*1	1	4360.17114	79.4 (1.0%)	−10.8(8)
2	*R*	$2\nu_2+\nu_4$	2	*F*2	23	1	*F*1	1	4360.17116	79.0 (1.6%)	−11.3(12)
2	*R*	$\nu_2+\nu_3$	2	*F*2	33	1	*F*1	1	4566.99632	80.4 (0.5%)	−11.4(4)
5	*P*	$\nu_1+\nu_4$	4	*F*2	29	5	*F*1	1	4190.45571	80.1 (0.5%)	−13.2(2)
5	*P*	$\nu_3+\nu_4$	4	*F*2	31	5	*F*1	1	4242.20557	84.5 (1.1%)	−14.0(7)
5	*P*	$\nu_3+\nu_4$	4	*F*2	32	5	*F*1	1	4245.85513	86.0 (0.6%)	−15.2(5)
5	*P*	$\nu_3+\nu_4$	4	*F*2	33	5	*F*1	1	4259.79647	83.2 (0.6%)	−14.1(4)
5	*P*	$\nu_3+\nu_4$	4	*F*2	34	5	*F*1	1	4262.64614	86.2 (0.5%)	−15.7(4)
5	*P*	$\nu_3+\nu_4$	4	*F*2	41	5	*F*1	1	4286.94940	80.9 (1.3%)	−13.6(10)
5	*P*	$\nu_2+\nu_3$	4	*F*2	54	5	*F*1	1	4485.79031	86.0 (2.6%)	−16.7(21)
5	*P*	$\nu_2+\nu_3$	4	*F*2	57	5	*F*1	1	4491.13048	79.5 (0.5%)	−14.7(4)
5	*Q*	$\nu_1+\nu_4$	5	*F*2	32	5	*F*1	1	4215.62750	81.5 (0.5%)	−12.3(3)
5	*Q*	$\nu_3+\nu_4$	5	*F*2	35	5	*F*1	1	4286.42876	83.3 (1.4%)	−16.7(12)
5	*Q*	$\nu_3+\nu_4$	5	*F*2	36	5	*F*1	1	4288.80000	85.3 (0.5%)	−13.6(3)
5	*Q*	$\nu_3+\nu_4$	5	*F*2	38	5	*F*1	1	4311.84141	83.3 (1.0%)	−13.7(8)
5	*Q*	$\nu_3+\nu_4$	5	*F*2	47	5	*F*1	1	4342.54067	81.9 (0.7%)	−13.7(5)
5	*Q*	$\nu_3+\nu_4$	5	*F*2	50	5	*F*1	1	4346.39721	80.0 (0.5%)	−11.5(2)
5	*Q*	$\nu_2+\nu_3$	5	*F*2	65	5	*F*1	1	4540.50717	75.9 (5.0%)	−13.1(17)
5	*R*	$\nu_1+\nu_4$	5	*F*2	28	4	*F*1	1	4244.56693		
5	*R*	$\nu_1+\nu_4$	5	*F*2	28	4	*F*1	1	4244.56693	81.7 (0.5%)	−11.0(2)
5	*R*	$\nu_3+\nu_4$	5	*F*2	36	4	*F*1	1	4341.15016	83.4 (2.8%)	−10.4(23)
5	*R*	$\nu_3+\nu_4$	5	*F*2	38	4	*F*1	1	4364.19156	84.3 (9.2%)	
5	*R*	$\nu_3+\nu_4$	5	*F*2	39	4	*F*1	1	4364.94279	82.4 (1.3%)	−12.3(11)
5	*R*	$\nu_3+\nu_4$	5	*F*2	39	4	*F*1	1	4364.94329	82.8 (1.1%)	−9.8(9)
5	*R*	$\nu_3+\nu_4$	5	*F*2	40	4	*F*1	1	4366.82692	80.3 (0.5%)	−10.8(2)
5	*R*	$\nu_3+\nu_4$	5	*F*2	45	4	*F*1	1	4390.87627	80.4 (2.0%)	−10.6(15)
5	*R*	$\nu_3+\nu_4$	5	*F*2	46	4	*F*1	1	4394.30957	82.1 (0.5%)	−11.5(1)
5	*R*	$\nu_3+\nu_4$	5	*F*2	50	4	*F*1	1	4398.74640	80.6 (7.7%)	
5	*R*	$\nu_2+\nu_3$	5	*F*2	62	4	*F*1	1	4587.72014	80.9 (4.3%)	
5	*R*	$\nu_2+\nu_3$	5	*F*2	66	4	*F*1	1	4594.59972	69.8 (7.0%)	
5	*R*	$\nu_2+\nu_3$	5	*F*2	69	4	*F*1	1	4602.72196	81.7 (0.6%)	−12.4(5)

An analogous study by the same authors was also realised for the air-broadening and shifting coefficients of 1011 lines in the same spectral region [143]. Like the self-broadening results, it was found that for low *J*-values the air-broadened widths in different bands are very similar (within 0.5–2%) but diverge (up to 5–15%) for higher values of *J*. Both air-broadened widths and air-induced shifts depend on the considered transitions, but no systematic differences for the *A*, *E*, and *F* symmetry lines are observed. For the *Q*-branch lines in the $\nu_1+\nu_4$ band the average shift coefficient is $\bar{\delta} = -8.4\ (1.1)\cdot 10^{-3}\ \mathrm{cm^{-1}\ atm^{-1}}$, for the *R*- and *P*-branches it is $-8.8\ (2.0)\cdot 10^{-3}$ and $-9.2\ (1.2)\cdot 10^{-3}\ \mathrm{cm^{-1}\ atm^{-1}}$, respectively. For the other bands these coefficients are: $\bar{\delta}\,(\nu_3) = -0.0065$, $\bar{\delta}\,(\nu_4) = -0.0028$, $\bar{\delta}\,(\nu_3+\nu_4) = -0.0086$, $\bar{\delta}\,(\nu_2+\nu_3) = -0.0098$, $\bar{\delta}\,(\nu_1+\nu_4) = -0.0087$, $\bar{\delta}\,(\nu_2+2\nu_4) = -0.0074$, $\bar{\delta}\,(\nu_2+\nu_4) = -0.0051$, $\bar{\delta}\,(2\nu_4) = -0.0033$.

Many calculations of CH_4 line-broadening coefficients for mixtures with various perturbers were performed by semi-classical methods. The ATC computations [144, 145] of N_2- and O_2-broadening coefficients accounting for the octupole and hexadecapole moments compared favourably with experimental data.

The RB formalism was applied in Refs [136, 138, 150] for calculation of N_2-, O_2- and Ar-broadened line shape parameters for the strongest ν_3 band of methane. N_2- and O_2-broadening line widths and shifts were calculated [150] for more than 4,000 transitions in the spectral range 2726–3200 $\mathrm{cm^{-1}}$ (with a particular attention paid to the transitions of *A* and *F* symmetries). The calculations were made at 225 and 296 K, in order to determine the temperature dependence of the line widths. The vibrational dependence of the intermolecular potential was described by the vibrational dependence of the mean polarizability of CH_4 molecule in the isotropic dispersion interaction (Eq. (2.5.18)). The complex second-order contribution S_2 was calculated for the anisotropic interaction potential composed of the leading electrostatic components and atom–atom terms in the form of Eq. (2.5.20). The wave functions were obtained by a diagonalization of an effective Hamiltonian built for the ground state and for the pentad of interacting states, accounting for all symmetry-allowed vibrational couplings. It was found that the atom–atom part of the intermolecular potential converges in the 12th order of expansion for the *A*- and *F*-symmetry transitions, and in the 14th order for the *E*-symmetry transitions (in the latter case no results were reported). Some parameters of the atom–atom potential were adjusted in order to obtain a better correlation with the experimental data [130]. As a result, for 52 *A*- and *F*-symmetry transitions the average difference between the calculated and the experimental values [130] was 0.15% in the case of CH_4-N_2 and −2.1% in the case of CH_4-O_2. The calculated values were also used to obtain the temperature exponents of line widths (Eq. (3.2.1)) for each of 4120 *A*- and *F*-symmetry transitions in the ν_3 band. Since many obtained *n*-values were small and sometimes even negative, the power model (3.2.1) seems to be very questionable for methane.

4.5. Linear molecules

Carbon dioxide CO_2, nitrous oxide N_2O, hydrogen cyanide HCN, and acetylene H_2C_2 are the most important polyatomic linear molecules for which accurate broadening and shifting coefficients are required for atmospheric applications in a wide temperature range and for collisions with various partners.

The vibrotational states of a three-atom linear molecule are denoted as $(v_1\, v_2^{l_2}\, v_3) \equiv (V, l)$ where the quantum number $l_2 = \pm v_2, \pm (v_2 - 2), \ldots, \pm 1$ or 0 characterises the degenerate bending vibration. The rotational states of an isolated vibrational state are defined by the single rotational quantum number J. The corresponding wave functions are simply the wave functions $|JK = 0\rangle$ of symmetric top with $K = 0$ (spherical harmonics). The vibrotational energy levels are characterised by both vibrational and rotational quantum numbers as well as by the symmetry type *e* ($\varepsilon = 1$) or *f* ($\varepsilon = -1$) issued from the following combinations of wave functions:

$$|VlJ\varepsilon\rangle = \frac{1}{\sqrt{2}}\left(|Vl\rangle|JK = l\rangle + \varepsilon|V - l\rangle|JK = -l\rangle\right),$$
$$|V\,l = 0\,J\,\varepsilon = 1\rangle = |V\,l = 0\rangle|JK = 0\rangle. \qquad (4.5.1)$$

The general selection rules for vibrotational transitions are given by $\Delta J = 0$ (*Q*-branch, $e \leftrightarrow f$), $\Delta J = \pm 1$ (*P*-, *R*-branches, $e \leftrightarrow e, f \leftrightarrow f$). Sometimes, to put together on the same graph the *P*- and *R*-branch spectral parameters, the specific quantum number *m* ($m = -J$ for *P*(*J*)-lines and $m = J+1$ for *R*(*J*)-lines) is introduced. Table 4.37 gives spectroscopic parameters of some linear molecules.

Table 4.37. Spectroscopic parameters of some three-atom linear molecules.

Molecule	ν_1, cm^{-1}	ν_2, cm^{-1}	ν_3, cm^{-1}	*B*, cm^{-1}	μ, D	*Q*, DÅ	α, Å^3
CO_2	1388	667	2349	0.39	0	−4.3	2.59
N_2O	2224	1285	589	0.419	0.1608	−3.0	2.92
HCN	3311	2097	712	1.48	2.98	2.38	2.59

4.5.1. CO_2 molecule

The CO_2 molecule plays an important role in the thermal balance of the Earth's atmosphere because of the greenhouse effect. Its line-broadening coefficients in mixtures with various gases were extensively studied [151–159] both experimentally and theoretically (see Refs [160–168] for recent measurements).

Systematic computations of these coefficients for infrared absorption and Raman scattering spectra of $^{12}C^{16}O_2$-CO_2(H_2O, N_2, O_2) in the 300–2400 K temperature range were made by Rosenmann *et al.* [151] in the framework of the semi-classical RB formalism (Table 4.38). The temperature dependence of the broadening coefficients was represented by Eq. (3.2.1) with T_0 = 300 K, and the quantities γ(300 K) and n were deduced from least-squares fitting of the calculated values to the experimental data at 300, 600, 1200 and 2400 K. For model calculations, analytical expressions for the m-dependence were proposed:

$$\gamma_{|m|}(300\text{ K}) = a_0 + a_1|m| + a_2|m|^2, \quad 1 \le |m| \le 101, \qquad (4.5.2)$$

$$n_{|m|} = b_0 + b_1|m| + b_2|m|^2 \qquad (4.5.3)$$

(values of the coefficients a_i and b_i [151] are given Table 4.39).

Table 4.38. CO_2 γ coefficients and temperature exponents [151]. Reproduced with permission.

$\lvert m\rvert$	CO_2–CO_2		CO_2–H_2O		CO_2–N_2		CO_2–O_2	
	$\gamma_{\lvert m\rvert}$ (300 K)	$n_{\lvert m\rvert}$	$\gamma_{\lvert m\rvert}$ (300 K)	$n_{\lvert m\rvert}$	$\gamma_{\lvert m\rvert}$ (300 K)	$n_{\lvert m\rvert}$	$\gamma_{\lvert m\rvert}$ (300 K)	$n_{\lvert m\rvert}$
1	119.9	0.723	117.7	0.726	95.0	0.737	85.9	0.721
3	115.8	0.729	113.9	0.754	91.4	0.745	83.0	0.730
5	116.9	0.735	116.7	0.764	91.6	0.747	82.7	0.729
7	117.0	0.731	119.6	0.771	91.4	0.744	82.0	0.723
9	116.1	0.722	123.5	0.782	90.7	0.735	80.8	0.712
11	113.3	0.701	126.5	0.789	89.2	0.722	79.0	0.700
13	111.3	0.689	129.6	0.797	87.2	0.705	76.7	0.682
15	109.2	0.679	132.6	0.802	84.7	0.687	74.1	0.661
17	107.1	0.671	135.6	0.811	82.2	0.669	71.1	0.637
19	106.0	0.670	136.6	0.808	79.6	0.653	68.3	0.617
21	104.9	0.669	136.7	0.800	77.4	0.641	65.9	0.603
23	103.7	0.669	136.7	0.793	75.8	0.635	64.2	0.595
25	102.7	0.669	137.7	0.791	74.6	0.633	63.0	0.595
27	100.8	0.663	140.6	0.802	73.9	0.636	62.5	0.601
29	98.9	0.655	143.5	0.812	73.5	0.642	62.1	0.609
31	97.0	0.646	144.5	0.811	73.2	0.648	61.8	0.615
33	94.8	0.635	144.5	0.804	73.1	0.656	61.6	0.624
35	92.5	0.623	144.5	0.800	73.0	0.661	61.4	0.628
37	90.2	0.609	144.5	0.795	72.9	0.666	61.3	0.633
39	87.8	0.596	145.6	0.795	72.9	0.671	61.2	0.638
41	85.5	0.583	147.6	0.800	72.7	0.674	61.0	0.640
43	83.3	0.570	147.7	0.795	72.5	0.677	60.8	0.642

45	81.2	0.557	146.8	0.785	72.3	0.676	60.6	0.642
47	79.2	0.547	146.9	0.782	72.1	0.678	60.3	0.642
49	77.5	0.537	147.8	0.783	71.9	0.678	60.1	0.643
51	75.7	0.529	148.8	0.784	71.6	0.676	59.9	0.641
53	74.2	0.521	150.8	0.791	71.3	0.677	59.6	0.640
55	72.8	0.515	150.8	0.787	70.9	0.672	59.4	0.640
57	71.4	0.508	150.8	0.784	70.4	0.670	59.1	0.637
59	70.2	0.504	149.9	0.776	70.0	0.666	59.0	0.638
61	69.1	0.500	149.9	0.773	69.5	0.663	58.6	0.635
63	68.0	0.497	148.0	0.761	69.0	0.659	58.3	0.632
65	67.0	0.493	147.0	0.752	68.5	0.655	58.0	0.632
67	66.0	0.492	146.2	0.744	67.9	0.650	57.7	0.630
69	65.1	0.488	145.3	0.736	67.3	0.644	57.3	0.626
71	64.3	0.488	144.4	0.728	66.8	0.639	56.9	0.621
73	63.6	0.487	143.5	0.720	66.1	0.634	56.6	0.619
75	62.9	0.486	141.6	0.709	65.5	0.628	56.2	0.615
77	62.1	0.485	139.7	0.699	64.9	0.622	55.9	0.615
79	61.5	0.486	137.7	0.688	64.2	0.616	55.6	0.610
81	60.9	0.485	136.7	0.683	63.6	0.612	55.2	0.607
83	60.3	0.485	135.7	0.678	62.9	0.606	54.9	0.605
85	59.8	0.485	135.7	0.677	62.3	0.599	54.5	0.600
87	59.3	0.487	135.7	0.677	61.6	0.593	54.1	0.598
89	58.8	0.487	134.7	0.671	60.9	0.588	53.7	0.594
91	58.3	0.487	134.7	0.671	60.3	0.581	53.3	0.591
93	57.7	0.487	133.7	0.666	59.6	0.575	52.9	0.588
95	57.2	0.486	131.7	0.658	58.9	0.568	52.6	0.586
97	56.6	0.486	128.8	0.642	58.3	0.563	52.2	0.583
99	56.0	0.484	127.8	0.639	57.6	0.556	51.9	0.578
101	55.4	0.482	126.7	0.634	56.9	0.550	51.4	0.573

Table 4.39. Parameters a_i (in 10^{-3} cm^{-1} atm^{-1}) and b_i for Eqs (4.5.2) and (4.5.3) [151]. Reproduced with permission.

System	a_0	a_1	a_2	b_0	b_1	b_2
CO_2–H_2O	$1.142 \cdot 10^2$	$1.266 \cdot 10^0$	$-1.163 \cdot 10^{-2}$	$7.618 \cdot 10^{-1}$	$2.354 \cdot 10^{-3}$	$-3.785 \cdot 10^{-5}$
CO_2–CO_2	$1.253 \cdot 10^2$	$-1.172 \cdot 10^0$	$4.660 \cdot 10^{-3}$	$7.706 \cdot 10^{-1}$	$-6.017 \cdot 10^{-3}$	$3.078 \cdot 10^{-5}$
CO_2–N_2	$9.169 \cdot 10^1$	$-5.323 \cdot 10^{-1}$	$2.149 \cdot 10^{-3}$	$7.146 \cdot 10^{-1}$	$-6.306 \cdot 10^{-4}$	$-7.544 \cdot 10^{-6}$
CO_2–O_2	$8.293 \cdot 10^1$	$-6.797 \cdot 10^{-1}$	$3.982 \cdot 10^{-3}$	$6.947 \cdot 10^{-1}$	$-1.798 \cdot 10^{-3}$	$8.167 \cdot 10^{-6}$

It should be kept in mind that Eq. (4.5.2) has no physical meaning and can not be used for $|m| > 101$. The maximal disagreements between the directly calculated coefficients and the model values of Eqs (4.5.2) and (4.5.3) are 12.6, 5.9, 6.7 and 8.6% respectively for H_2O, CO_2, N_2 and O_2 partners. For high temperatures, γ-values are practically independent of $|m|$, and a very simple and accurate empirical relation can be used for their mean values at a given temperature T:

$$\bar{\gamma}(T) = \bar{\gamma}(300\,\mathrm{K})[300/T]^{\bar{n}(T)}, \quad \bar{n}(T) = \bar{n}(300\,\mathrm{K})[300/T]^{\alpha} \tag{4.5.4}$$

(necessary parameters from Ref. [151] are gathered in Table 4.40).

Table 4.40. Parameters of Eq. (4.5.4) for CO_2 [151]. Reproduced with permission.

System	$\bar{\gamma}(300\,\mathrm{K})$, 10^{-3} cm^{-1} atm^{-1}	$\bar{n}(300\,\mathrm{K})$	α
CO_2–H_2O	135.0	0.750	0.013
CO_2–CO_2	105.0	0.773	0.050
CO_2– O_2	69.7	0.722	0.017
CO_2– N_2	80.6	0.736	0.017

For some high |*m*|-values (|*m*| = 121, 161, 201) the broadening coefficients were also computed for T = 300, 600, 1200 and 2400 K (Table 4.41); they can be used for the calculation of long-path or high-temperature spectra. More accurate fitting laws and extrapolations for the room-temperature broadening of lines corresponding to high J-values can be found in Ref. [151]

Table 4.41. Calculated CO_2 line-broadening coefficients for high |*m*| = 121, 161, 201 [151]. Reproduced with permission.

T, K	CO_2–CO_2	CO_2–H_2O	CO_2–N_2	CO_2–O_2
	48.9	110	49.3	46.3
300	–	–	–	–
	–	–	–	–
	35.6	77.9	36.8	32.8
600	29.4	66.2	30.8	28.5
	–	–	–	–
	26.2	51.2	25.7	22.5
1200	23.3	46.4	23.8	21.2
	20.5	41.4	21.4	19.5
	19.9	32.6	17.4	15.5
2400	17.6	30.9	16.9	15.0
	16.2	28.4	16.2	14.5

Line-broadening and shifting coefficients of CO_2 perturbed by N_2 and air as well as the temperature exponents were also calculated using the semi-empirical approach of Sec. 3.4 [159, 160]. The line shape parameters obtained for 1000 K are included in a freely-available carbon dioxide spectroscopic data bank [169].

Since no significant dependence of CO_2-broadening coefficients on vibrational quantum numbers has been observed, the theoretical values of Refs [151, 159, 160] can be used for any vibrational band.

4.5.2. N_2O molecule

Nitrous oxide is principally known as a pollutant involved in important chemical reactions in the terrestrial atmosphere. Together with carbon dioxide it is used for the validation of spectroscopic data obtained by remote sensing techniques and is included in numerous spectroscopic molecular databases.

Experimental values of its N_2- and O_2-broadening coefficients in the $\nu_1 + \nu_3$ band (3390–3510 cm^{-1}) at 296 K were obtained in Ref. [170]. They were further used to determine parameters a_i and b_i of the 4, 4 Pade-approximant

$$\gamma = \sum_{i=0}^{4} a_i |m|^i \Big/ \left(1 + \sum_{i=0}^{4} b_i |m|^i\right) \tag{4.5.5}$$

(in $cm^{-1}\,bar^{-1}$). These parameters [170] (presented in Table 4.42) are also valid for other vibrational bands.

Table 4.42. Parameters of 4,4 Pade-approximant (4.5.5) for N_2O–N_2 and N_2O–O_2 line-broadening coefficients in the $\nu_1 + \nu_3$ band at 296 K ($a_3 = b_3 = 0$) [170]. Reproduced with permission.

Parameters	N_2O–O_2	N_2O–N_2
a_0 ($cm^{-1}\,bar^{-1}$)	0.10586302	0.10759947
a_1 ($cm^{-1}\,bar^{-1}$)	$-0.28739300\cdot10^{-2}$	$0.32958014\cdot10^{-2}$
a_2 ($cm^{-1}\,bar^{-1}$)	$-0.64065470\cdot10^{-3}$	$0.12348761\cdot10^{-3}$
a_4 ($cm^{-1}\,bar^{-1}$)	$0.18503999\cdot10^{-5}$	$0.10145845\cdot10^{-5}$
b_1	$0.61625862\cdot10^{-1}$	$0.85164796\cdot10^{-1}$
b_2	$-0.14935469\cdot10^{-1}$	$-0.61306262\cdot10^{-3}$
b_4	$0.31648963\cdot10^{-4}$	$0.15287607\cdot10^{-4}$

Rotational N_2O lines $J = 8 \leftarrow 7$, $J = 22 \leftarrow 21$, and $J = 23 \leftarrow 22$ (near 201, 552, and 577 GHz, respectively) broadened by N_2 and O_2 pressure at 235–350 K were studied both experimentally and theoretically in Ref. [171]. Significant deviations of the experimental line shapes from the Voigt profile model (typical for the line-narrowing processes) were clearly observed; at the same time the SDV profile almost perfectly matched the recorded lines. Line-broadening coefficients were also calculated by the RB approach with the interaction potential containing atom–atom and electrostatic interactions (up to hexadecapoles) and showed a good agreement with measurements.

Two other millimetre-absorption lines $J = 20 \leftarrow 19$, and $J = 24 \leftarrow 23$ of room temperature N_2O–N_2 and N_2O–N_2 were analysed from both experimental and theoretical viewpoints in Ref. [172]. Voigt, Galatry and SDV profiles were employed for the determination of collisional broadening coefficients. Clear departures of the experimental intensities from the Voigt profile and a non-linear behaviour of the extracted line widths (especially pronounced at low densities) argued that no reliable line-broadening coefficients could be extracted with this model. When indistinguishable on the residuals level, the Galatry and SDV profiles gave very close γ-values. A slight preference for the speed dependence of the relaxation rates was, however, evoked to explain the observed line narrowing, since for high pressures the Galatry parameter β started to increase in a non-linear manner. Calculations realised for the line-broadening coefficients by the RBE approach with the interaction potential composed of atom–atom, dipole–quadrupole and quadrupole-quadrupole terms showed good agreement with the measured values as well as with experimental data for infrared absorption.

4.5.3. HCN molecule

Hydrogen cyanide is a key molecule for organic synthesis. It is present in the atmospheres of Earth and Titan and is expected to be found in the atmospheres of Jupiter and Saturn. Its line broadening and shifting by various buffer gases have been extensively studied.

In Ref. [173], HCN–N_2-broadening coefficients were measured for 28 lines in the ν_2 band at 20, 30, 40 and 50 mbar and 203 K using a tuneable diode laser spectrometer. The collisional half-widths were extracted through the relation

$$P(\mathrm{HCN}) = \pi\gamma K(\nu_0)/(lI_0) \qquad (4.5.6)$$

where P(HCN) is the partial pressure of HCN (varying from 0.04 to 0.40 mbar), $K(\nu_0)$ is the absorption coefficient at the line centre ν_0, I_0 is the calculated line intensity and l is the absorption path length. For each line the self-broadening contribution (inferior to $0.8\cdot10^{-3}$ cm^{-1}) to the total width was also taken into account. The N_2-broadening coefficients were obtained from the slopes of the linear dependences of line widths on $P(N_2)$ pressure (Table 4.43). The temperature exponents n were also determined from the experimental and calculated line-broadening coefficients for 297 and 203 K. It was found that the temperature exponent depends strongly on $|m|$: $n = 0.84$ for $|m| = 1$ but for $|m| = 6$ it decreases to $n = 0.59$, then for $|m| = 13$ it increases again to $n = 0.86$ and decreases slightly for higher $|m|$-values. The absolute error limit for n was estimated at 20%.

The same reference also contains theoretical RB estimations of HCN–N_2-broadening coefficients (for $|m| \le 30$) for two models of intermolecular potential. In the first model, the electrostatic interactions were completed by a dispersion term dependent on the HCN polarizability anisotropy ($\gamma_1 = 0.257$):

$$V_{aniso} = V^{(elec)} + V^{(disp)} = V_{\mu_1 Q_2} + V_{Q_1 Q_2} + 4\varepsilon\gamma_1 \left[\left(\frac{\sigma}{r}\right)^{12} - \left(\frac{\sigma}{r}\right)^{6}\right] P_2(\cos\theta) \qquad (4.5.7)$$

The classical trajectory was governed by the Lennard–Jones potential with $\varepsilon_{12} = 86.92$ K and $\sigma_{12} = 3.867$ Å. The calculated line widths were overestimated for $|m| > 10$. In the second model, an extended electrostatic potential was used:

$$V^{(elec)} = V_{\mu_1 Q_2} + V_{Q_1 Q_2} + V_{\mu_1 \alpha_2 \mu_1} + V_{\mu_1 \gamma_2 \mu_1} + V_{\mu_1 \alpha_2 \theta_1} + V_{\mu_1 \gamma_2 \theta_1} + V_{\alpha_1 \gamma_1 \alpha_2 \gamma_2}, \quad (4.5.8)$$

where $\gamma_2 = 0.136$ is the polarizability anisotropy of the N_2 molecule and the other parameters are given in Table 4.37 and in Appendix B). The extended potential allowed an improvement in the agreement with experimental data (since electrostatic contributions are dominant), but an overestimation of the broadening coefficient for large *J*-values persisted.

Table 4.43. N_2-broadening coefficients (in 10^{-3} cm^{-1} atm^{1}) for the ν_2 band of HCN at 203 K and temperature exponents for $T_0 = 297$ K [173]. Reproduced with permission.

$\|m\|$	*P*-branch	*R*-branch	Smoothed	*J*	*Q*-branch	Smoothed	n_{expt}	n_{calc}
3		200.8	201	3	182.1	182	0.745	0.727
4		183.9	182	4	169.8	171	0.642	0.651
5		168.2	169	5		161	0.574	0.603
6	158.0	160.2	160	6	154.0	154	0.564	0.590
7	154.5		154	7		149	0.575	0.610
8	148.0	153.9	151	8	146.2	145	0.602	0.648
9	150.2	149.5	149	9	142.8	142	0.639	0.700
10		148.8	148	10	140.3	141	0.671	0.761
11	146.2	146.7	147	11	140.6	141	0.696	0.818
12	146.8		147	12		142	0.708	0.851
13	144.5		146	13	142.6	142	0.706	0.863
14			145	14	141.5	141	0.697	0.861
15			144	15	139.3	140	0.682	0.849
16			143				0.669	0.844

Experimental shapes of rotational $J = 2 \leftarrow 1$ (172 GHz) and $J = 7 \leftarrow 6$ (602 GHz) lines of $HC^{15}N$ colliding with H_2, N_2, O_2, CH_3CN, Ne, Kr, Xe and Ar were exhaustively studied in Ref. [174]. Except for light perturbers H_2 and He, non-linear pressure dependences of relaxation rates were obtained with the Galatry model profile, so that the SDV profile was strongly recommended for the extraction of the collisional line widths. For the HCN–N_2 system the RBE computations with the model potential of Ref. [173] gave a good estimation of the line-broadening coefficients, whereas for the HCN–Ar system the values calculated with the model potential composed of atom–atom interactions were slightly larger then the measurements.

The case of O_2-broadening was considered by Yang *et al.* [175] for hydrogen cyanide lines with $5 \leq J \leq 36$ in the 0.5–3 THz (17–100 cm^{-1}) frequency region. RBE computations for this systems resulted in a good agreement with these measurements. Nitrogen- and oxygen-broadening coefficients were further combined to provide air-broadening coefficients as requested by the HITRAN database. For transitions with high J-values a significant difference was observed between the measured and tabulated line-broadening coefficients. A new fourth-order polynomial valid for $|m| > 29$ was suggested for the HITRAN:

$$\gamma_{air} = A_0 + A_1|m| + A_2|m|^2 + A_3|m|^3 + A_4|m|^4 \tag{4.5.9}$$

with $A_0 = 0.1569(14)$, $A_1 = -9.9(5)\cdot 10^{-3}$, $A_2 = 6.29(57)\cdot 10^{-4}$, $A_3 = -1.71(24)\cdot 10^{-5}$, $A_4 = 1.48(35)\cdot 10^{-7}$. For the N_2-broadening the corresponding values of the coefficients were $A_0 = 0.1740(11)$, $A_1 = -9.97(27)\cdot 10^{-3}$, $A_2 = 5.04(18)\cdot 10^{-4}$, $A_3 = -8.58(35)\cdot 10^{-6}$.

The H_2-broadening coefficients of 34 ($2 \le J \le 26$) $H^{12}C^{14}N$ absorption lines in the P-, Q- and R-branches of the ν_2 band (658–780 cm^{-1}) were measured in Ref. [176]. Experimental line shapes were fitted to Voigt and Rautian–Sobel'man profiles. The corresponding theoretical RB values were also computed and showed satisfactory agreement with experimental data. The calculated coefficients γ(in 10^{-3} cm^{-1} atm^{-1}) took values between 113.4 and 137.4 for $2 \le |m| \le 30$ in the P-branch, between 106.5 and 133.3 for $3 \le |m| \le 24$ in the R-branch and between 103.8 and 131.3 for $8 \le J \le 26$ in the Q-branch.

Multispectrum analyses of the $2\nu_2$ and ν_2 bands of $H^{12}C^{14}N$ at room temperature are described in Refs [177, 178] respectively. Four high-purity (99.8%) HCN spectra and three spectra of lean mixtures (about 3%) of HCN in dry air were simultaneously fitted by a non-linear least-squares procedure. The self- and air-broadening coefficients, as well as self- and air-induced shift coefficients for numerous lines in the $2\nu_2^0$ band of $H^{12}C^{14}N$ and the self- and air-broadening coefficients for several lines of the $3\nu_2^1$–ν_2^1 hot band were obtained.

4.5.4. C_2H_2 molecule

Acetylene is a trace constituent of the Earth's atmosphere mainly formed by the photo-dissociation of methane. In the near future, an important increase in its concentration is expected due to the extensive development of car traffic [179].

The acetylene molecule has five normal vibrational modes with frequencies $\nu_1 = 3374$ cm^{-1}, $\nu_2 = 1974$ cm^{-1}, $\nu_3 = 3287$ cm^{-1}, $\nu_4 = 612$ cm^{-1} and $\nu_5 = 729$ cm^{-1} [1]. The vibrations ν_1 and ν_3 are symmetric and anti-symmetric respectively, the vibrations ν_4 and ν_5 are doubly degenerate. The vibrational wave functions are denoted as $(\nu_1\nu_2\nu_3\nu_4^{l_4}\nu_5^{l_5}) \equiv (V, l)$. The vibrotational energy levels are characterised by vibrational and rotational quantum numbers and by the symmetry types e or f:

$$\left|v_1v_2v_3v_4^{l_4}v_5^{l_5}|l|J\right\rangle_{e,f} = \frac{1}{\sqrt{2}}\left[\left|v_1v_2v_3v_4^{l_4}v_5^{l_5}lJ\right\rangle \pm (-1)^l\left|v_1v_2v_3v_4^{-l_4}v_5^{-l_5}lJ\right\rangle\right], \quad (4.5.10)$$

where $l = |l_4 - l_5|, \ldots, l_4 + l_5$ are the quantum numbers for the total vibrational angular momentum. The general selection rules for vibrotational transitions are the same as for three-atom linear molecules: $\Delta J = 0$ (Q-branch, $e \leftrightarrow f$), $\Delta J = \pm 1$ (P-, R-branches, $e \leftrightarrow e, f \leftrightarrow f$).

The measured N_2-, O_2- and self-broadening coefficients for the ν_5, $5\nu_3$, $2\nu_1+\nu_2+\nu_3$, $\nu_2+\nu_3$, and $\nu_1+\nu_3$ bands of C_2H_2 at 298 K are gathered in Ref. [179]. In the same reference, the line shapes of 20 lines ($0 \le J \le 22$) in the $\nu_1 + 3\nu_3$ band (12620–12730 cm^{-1}) of C_2H_2 broadened by N_2, O_2, C_2H_2, He, Ne, Ar, Kr, Xe and air were investigated with Voigt, Galatry and Rautian profile models (see Table 4.44). Two last profiles resulted in almost identical values of γ and δ coefficients but slightly different values of the coefficient β_0. In comparison with the results previously reported by other authors for other vibrational bands, no vibrational dependence of line-broadening coefficients was stated (within the experimental error bars).

Table 4.44. Experimental line-broadening coefficients 2γ(in 10^{-3} cm^{-1} bar^{-1}, within one standard deviation) for the $\nu_1 + 3\nu_3$ band of H_2C_2 perturbed by various gases at 298 K [179].

Line	C_2H_2	N_2	air	O_2	He	Ne	Ar	Kr	Xe
$P(17)$	272(4)	162(3)	153(4)	119(5)	89(5)	71(5)	106(6)	114(5)	142(6)
$P(11)$	302(5)	173(3)	162(3)	132(5)	83(5)	74(5)	117(3)	128(5)	154(5)
$P(10)$	301(7)	173(7)	165(5)	128(5)	85(5)	77(6)	120(11)	134(6)	160(6)
$P(9)$	320(8)	183(7)	174(5)	135(5)	88(5)	77(5)	124(11)	133(5)	158(5)
$P(8)$	306(8)	187(15)	175(6)	144(6)	83(5)	75(5)	122(13)	139(7)	166(6)
$P(7)$	334(9)	189(13)	175(6)	145(6)	86(5)	83(5)	138(13)	141(5)	166(6)
$P(6)$	326(5)	188(6)	180(6)	147(6)	84(4)	74(5)	135(6)	141(6)	165(6)
$P(4)$	354(4)	196(2)	186(6)	150(6)	81(5)	82(6)	145(7)	148(6)	177(6)
$R(1)$	384(5)	204(5)	201(6)	173(6)	74(5)	96(5)	162(8)	167(6)	190(6)
$R(3)$	346(5)	195(4)	184(5)	153(5)	87(4)	86(5)	139(5)	151(5)	169(6)
$R(5)$	323(8)	187(6)	174(5)	142(5)	86(4)	81(4)	136(2)	136(5)	164(5)
$R(6)$	313(5)	178(4)	174(6)	142(5)	82(5)	72(5)	127(7)	130(6)	158(6)
$R(7)$	323(7)	188(9)	166(5)	139(6)	88(5)	77(5)	127(9)	134(5)	161(6)
$R(8)$	318(4)	176(4)	156(5)	125(6)	95(5)	71(5)	121(9)	120(5)	148(6)
$R(9)$	296(8)	171(15)	164(5)	130(5)	86(5)	77(5)	119(12)	129(5)	150(6)
$R(10)$	301(7)	172(13)	166(5)	130(5)	84(6)	72(5)	122(8)	125(4)	147(6)
$R(12)$	286(6)	165(8)	156(5)	125(5)	95(5)	71(5)	113(11)	121(4)	144(4)
$R(13)$	294(4)	169(2)	158(5)	128(5)	85(5)	75(5)	112(11)	122(3)	148(4)
$R(15)$	273(5)	162(6)	151(5)	121(5)	87(5)	72(5)	111(2)	116(4)	143(4)
$R(22)$	244(5)	147(5)	147(8)	106(6)	77(6)	57(5)	102(4)	110(6)	131(6)

Pressure-broadening and shift coefficients for 59 lines in the $\nu_1+\nu_3$ band (6470–6612 cm^{-1}) of C_2H_2 perturbed by N_2, H_2, D_2, air, He, Ne, Ar, Kr and Xe at 295 K were obtained by Arteaga *et al.* [180] using the Voigt profile model. The γ-values were found to be in general agreement with those available in the literature for other vibrational bands whereas the δ-values demonstrated a substantial dependence on the vibrational quantum number.

N_2-broadening coefficients for 12 absorption lines in the $\nu_4+\nu_5$ band at room temperature were recently studied by Fissiaux *et al.* [181] using a tuneable diode-laser spectrometer. Line shapes were fitted separately at 4–16 nitrogen pressures (4.03–179.5 mbar) with Voigt, Galatry and Rautian–Sobel'man profiles. The same authors and J.C. Populaire [182] also reported the temperature dependence of seven N_2-broadening coefficients obtained from measurements at 173.2, 198.2, 223.2, 248.2, 273.2 K. Both N_2-broadened line widths and N_2-induced line shifts from low to room temperatures were reported in Ref. [183].

Self-broadening coefficients for 30 lines in the *P*- and *R*-branches of the $\nu_4+\nu_5$ band of acetylene (located in the spectral range 1275–1390 cm^{-1}) were measured by Lepère *et al.* [184] using Voigt and Rautian–Sobel'man profiles. The authors also provided theoretical values of these coefficients calculated by the semi-classical RB approach with the anisotropic interaction potential

$$V_{aniso} = V_{Q_1Q_2} + V_{Q_1\Phi_2} + V_{\Phi_1Q_2} + V_{\Phi_1\Phi_2} - 4\varepsilon A_2(\sigma/r)^6 P_2(\cos\theta), \qquad (4.5.11)$$

where $A_2 = 0.6$ [185] is an adjustable effective parameter and $Q = 4.5$ D Å; $\Phi = 0$ (calc. I) and $\Phi = 21.8$ D Å^3 (calc. II) values for the hexadecapole moment of C_2H_2 were tested. These results are gathered in Table 4.45.

For the temperature dependence of the self-broadening coefficients in the 5 μm region an empirical formula was proposed in Ref. [186]:

$$\gamma_{self}(T) = [0.2031(19) - 0.00642(45)|m| + 0.000181(29)|m|^2 - 2.93(58)\cdot 10^{-6}|m|^3]\,(T/T_0)^{0.75} \qquad (4.5.12)$$

(with one standard deviation given in parentheses).

Despite the fact that the broadening of C_2H_2 lines by noble gases is of minor importance for atmospheric applications, theoretical studies of this case are valuable for a better understanding of the pressure-broadening mechanism. Semi-classical RB calculations were reported for C_2H_2–Xe [187] and C_2H_2–Ar [188] line-broadening coefficients. Semi-classical RBE and purely classical, Gordon-type computations were realised for the C_2H_2–Ar [189, 190] and C_2H_2–He [191] systems. It was noted that the decoupling of rotational and translational motions (trajectory defined by the isotropic potential) in the framework of the RB(E) formalism becomes questionable for long anisotropic molecules like C_2H_2 and leads to an overestimation of the line broadening when a refined *ab initio* potential is employed. At the same time the pure classical approach accounting for the energy transfer between the rotations and translations (trajectory governed by the full anisotropic potential) yields for the same potential much more realistic results (except the lines with too low-rotational quantum numbers).

Table 4.45. Self-broadening coefficients (in 10^{-3} cm^{-1} atm^{-1}) for the $\nu_4 + \nu_5$ band of C_2H_2 at 297 K [184]. Reproduced with permission.

$\|m\|$	Expt (Voigt)		Expt (Rautian)		Expt mean (Rautian)	calc. I	calc. II
	R-branch	*P*-branch	*R*-branch	*P*-branch			
1	–	204.3(4.7)	–	206.0(7.6)	206.8	198.3	198.7
2	–	185.7(6.0)	–	188.2(9.5)	188.2	185.1	185.6
3	–	174.4(4.7)	–	175.7(4.9)	175.7	177.2	177.8
4	–	–	–	–	–	169.1	169.8
5	160.0(3.8)	162.4(5.1)	161.4(4.7)	170.5(6.9)	166.0	161.1	162.0
6	156.7(4.8)	155.6(3.8)	166.4(6.2)	158.3(6.2)	162.4	154.6	155.8
7	152.4(4.1)	152.8(6.1)	157.2(4.4)	159.3(4.9)	158.3	150.5	151.9
8	–	149.1(3.8)	–	150.7(5.9)	150.7	148.6	150.2
9	–	149.3(4.4)	–	152.8(4.6)	152.8	148.0	149.6
10	–	145.1(4.3)	–	148.2(4.0)	148.2	147.7	149.4
11	–	143.5(4.3)	–	144.0(4.3)	144.0	147.3	148.9
12	138.5(5.8)	142.7(3.3)	140.5(5.8)	145.2(4.5)	142.9	146.4	148.1
13	–	139.8(4.4)	–	141.5(5.5)	141.5	144.9	146.7
14	134.0(4.4)	139.0(5.0)	147.8(4.7)	145.4(6.9)	146.6	143.0	144.9
15	134.3(3.9)	133.6(5.2)	140.3(4.3)	134.6(4.7)	137.5	140.5	142.6
16	–	131.5(4.5)	–	134.2(3.5)	134.2	137.7	140.1
17	127.5(3.0)	–	131.8(4.6)	–	131.8	134.5	137.2
18	–	127.3(4.6)	–	134.9(4.4)	134.9	131.1	134.2
19	–	124.7(3.7)	–	131.0(3.4)	131.0	127.5	131.0
20	115.3(3.5)	117.9(2.9)	117.5(4.5)	121.5(3.2)	119.5	123.8	127.9
21	–	114.5(3.3)	–	116.1(4.3)	116.1	120.0	124.7
22	109.8(3.4)	109.9(3.2)	112.5(3.1)	111.7(3.1)	112.1	116.3	121.5
23	–	–	–	–	–	112.6	118.5
24	103.3(3.2)	–	103.8(4.1)	–	103.8	109.0	115.6
25	–	–	–	–	–	105.5	112.7
26	–	–	–	–	–	102.0	110.0
27	–	–	–	–	–	98.7	107.4
28	–	–	–	–	–	95.5	105.0
29	–	–	–	–	–	92.3	102.6
30	–	–	–	–	–	89.2	100.3

The broadening of C_2H_2 lines by H_2 was considered both experimentally and theoretically in Refs [192, 193]. In Ref. [193], these coefficients were measured for 29 lines in the *P*- and *R*-branches of the ν_5 band at 173.2 K using the Voigt and Rautian–Sobel'man profiles. Semi-classical RB calculations with a model potential composed of electrostatic interactions and atom–atom terms gave a satisfactory agreement with experimental results only for low and middle values of the rotational quantum number. The experimental and theoretical values of the temperature exponents were also determined. Recent measurements and quantum-mechanical calculations for C_2H_2–H_2 line widths can be found in Ref. [194].

4.6. Diatomic molecules

A study of line-broadening coefficients of diatomic molecules is important for practical applications as well as for fundamental research. From the practical point of view, knowledge of CO, NO, OH, HCl,... line-broadening coefficients is necessary for remote sensing of these species in the Earth's and planetary atmospheres. From a fundamental science point of view, these diatomic molecules are the simplest molecules with infrared absorption spectra, and the precise experimental values of their line-broadening and line-shift coefficients constitute an excellent basis for the development of line shape theories and testing of various models of intermolecular potential. However, before starting the analysis of particular molecules some general notions on diatoms are worthy of consideration.

4.6.1. Vibrotational energy levels and wave functions

The vibration-rotation Hamiltonian of a diatomic molecule in the Born–Oppenheimer approximation is written as

$$H = H_0 + H', \tag{4.6.1}$$

where $H_0 = hc\nu(p^2 + q^2)/2$ is the zero-order Hamiltonian of the harmonic vibration of the wave number ν and the momentum $p = -i\hbar\partial/\partial q$ is conjugated to the coordinate $q = (r - r_e)/r_e(\nu/2B_e)^{1/2}$ (r and r_e are the instantaneous and the equilibrium distances between the atoms, respectively, whereas B_e (in cm^{-1}) is the rotational constant in the equilibrium position). The perturbation H'has the form

$$\begin{aligned} H' &= H_V + H_R, \\ H_V &= hc\nu(\alpha_3 q^3 + \alpha_4 q^4 + \alpha_5 q^5 + \cdots), \\ H_R &= hc\nu(\beta_0 + \beta_1 q + \beta_2 q^2 + \cdots), \end{aligned} \tag{4.6.2}$$

where $\alpha_3 = a_1\gamma^{1/2}/2$, $\alpha_4 = a_2\gamma/2,\ldots$; $\beta_0 = J(J+1)\gamma/2$, $\beta_1 = -J(J+1)\gamma^{3/2}$, ... with $\gamma = 2B_e/\text{v}$ and a_1, a_2, ... standing for the Dunham anharmonic coefficients. The vibrotational energies are given by

$$E = \sum_{i,j} Y_{ij}\left(\text{v}_i + \frac{1}{2}\right)^i [J(J+1)]^j, \qquad (4.6.3)$$

where Y_{ij} are spectroscopic constants. The vibrotational wave functions $|\text{v}J\rangle$ represent the products of the unperturbed harmonic-oscillator wave functions $|n\rangle$ and the rigid rotor wave functions $|J\rangle$ (spherical harmonics):

$$|\text{v}J\rangle = \sum_n C_{\text{v}n}|n\rangle|J\rangle, \qquad (4.6.4)$$

and the coefficients $C_{\text{v}n}$ are obtained by the perturbation theory [195–197]. Equations (4.6.3) and (4.6.4) are valid for CO, HF, and HCl molecules.

In NO and OH molecules, the strong coupling of the electronic orbital angular momentum along the intermolecular axis $\Lambda = 1$ with the corresponding spin angular momentum components $\Sigma = \pm 1/2$ (Λ–Σ-interaction) splits the fundamental electronic state into two components $^2\Pi_K$, $K = \Lambda + \Sigma$; that is, into $^2\Pi_{1/2}$ and $^2\Pi_{3/2}$. For NO the $^2\Pi_{1/2}$ levels lie lower than the $^2\Pi_{3/2}$ ones (normal structure) whereas for OH the $^2\Pi_{1/2}$ levels have higher energies than the $^2\Pi_{3/2}$ levels (inverted structure). The energy levels $E(J, K)$ are defined by two rotational quantum numbers: J and $K = 1/2$ or $3/2$. In general, the interaction between the electronic angular momentum and the rotational momentum removes the degeneracy due to two different orientations of the electronic orbital angular momentum along the molecular axis and produces the so-called Λ-doubling: each $E(J, K)$ levels is split into two components, positive ($\varepsilon = +1$) and negative ($\varepsilon = -1$) (or λ-e and λ-f components, respectively). For the NO molecule the Λ-doubling is small, so that for the line-broadening coefficient calculation it is not taken into account and the energy levels are given by Eq. (4.6.3). In the OH molecule, each Λ-doubling component is additionally split by the magnetic hyperfine interaction due to the hydrogen atom (nuclear spin $I = 1/2$) into two components described by the projections of the total angular momentum $\vec{F} = \vec{J} + \vec{I}$ on the molecular axis: $^2\Pi_{1/2}$ (F_1) and $^2\Pi_{1/2}$ (F_2). Since the role of the hyperfine structure and higher-order terms has been found to be negligible, the energies of OH Λ-doublet components are usually calculated as [198]

$$E(J,K) = F(J,K) \pm W(J,K)/2, \qquad (4.6.5)$$

where

$$F(J,K) = B_0[J(J+1) - K^2],$$
$$W(J,1/2) = a(J+1/2), \tag{4.6.6}$$
$$W(J,3/2) = b(J^2 - 1/4)(J+3/2)$$

with $a = 0.316\,\text{cm}^{-1}$, $b = 0.011\,\text{cm}^{-1}$ and B_0 denoting the rotational constant for the fundamental vibrational state.

The transitions between the rotational levels of a polar diatomic molecule are governed by the usual selection rules $\Delta J = 0, \pm 1$. If the Λ-doubling is taken into account (OH case), one more selection rule for the parity of the rotational state appears: $\varepsilon = \pm 1 \rightarrow \varepsilon = \mp 1$; that is, only transitions between pairs of lower or upper levels of Λ-doublets can occur. The selection rule referring to the hyperfine structure is $\Delta F = 0, \pm 1$. The rotational part of the wave function (used for the calculation of the second-order contribution $S_2(b)$) is written as a linear combination of symmetric-top rotational wave functions $|JK\rangle$ and $|J\,{-K}\rangle$:

$$|JK\varepsilon\rangle = \left(|JK\rangle + \varepsilon|J-K\rangle\right)/\sqrt{2} \tag{4.6.7}$$

(here $K = 0, \ldots, J$). For the anisotropy rank of the active molecule $l = 1$ the parity selection rule is $\varepsilon = \pm 1 \rightarrow \varepsilon = \mp 1$ whereas for $l = 2$ it is $\varepsilon = \pm 1 \rightarrow \varepsilon = \pm 1$. The notation of vibrotational lines is the same as for linear molecules: $P(J)$ for the P-branch with $\Delta J = -1$, $m = -J$ and $R(J)$ for the R-branch with $\Delta J = 1$, $m = J+1$. Molecular and spectroscopic parameters of some diatoms are listed in Table 4.46.

Table 4.46. Molecular and spectroscopic parameters of some diatoms [2, 3]*.

Parameter	CO	NO	OH	HF	HCl
ν, cm^{-1}	2169.813	1904.20	3737.76	4139	2991
B_0, cm^{-1}	1.9313	1.7202	18.87	20.96	10.44
μ, D	0.1098	0.15	1.72	1.827	1.108
Q, DÅ	−2.0	−2.5; −1.8	−1.8	2.6	3.8
Ω, DÅ^2	2.56			4.5	4.5
Φ, DÅ^3	−3.4				
α, Å^3	1.95	1.7		0.83	2.58
γ_1	0.16				

* The polarizability anisotropy is defined by $\gamma_1 = (\alpha_{\parallel} - \alpha_{\perp})/(\alpha_{\parallel} + 2\alpha_{\perp})$ with $\alpha_{\parallel}$ and $\alpha_{\perp}$ standing for the parallel and the perpendicular components of the polarizability tensor.

4.6.2. Influence of vibration-rotation coupling on the line shifts

In the framework of semi-classical approaches, the leading term of the line shift corresponding to the radiative transition $i \rightarrow f$ is determined by the difference $\langle i\,|V_{iso}|\,i\rangle - \langle f\,|V_{iso}|\,f\rangle$ of the isotropic potential matrix elements. These elements are calculated without taking into account the spin–orbit interaction, so that the basis of wave functions can be chosen in the form of Eq. (4.6.4). The interaction potential is defined by the dipole, quadrupole and higher-order moments of the interacting molecules, by the molecular polarizabilities (given in Table 4.46) and so on. These physical quantities can be expanded into a series of normal coordinates q of the interacting molecules. For a diatomic molecule colliding with an atom the isotropic potential can therefore be written as [199]

$$V_{iso}(r,q) = V_{iso}(r) + \frac{\partial V_{iso}}{\partial q} q + \cdots, \tag{4.6.8}$$

where $V_{iso}(r)$ stands for the equilibrium molecular configuration and the derivative $\partial V_{iso}(r)/\partial q$ can be expressed through the molecular parameters and their derivatives (like the H_2O case considered in Sec. 3.1). In the linear approximation on q, one obtains [199]:

$$\langle \mathrm{v}_f J_f | V_{iso} | \mathrm{v}_f J_f \rangle - \langle \mathrm{v}_i J_i | V_{iso} | \mathrm{v}_i J_i \rangle = \partial V_{iso} / \partial q \Delta q(\mathrm{v}_f J_f | \mathrm{v}_i J_i), \tag{4.6.9}$$

where

$$\Delta q(\mathrm{v}_f J_f | \mathrm{v}_i J_i) = \langle \mathrm{v}_f J_f | q | \mathrm{v}_f J_f \rangle - \langle \mathrm{v}_i J_i | q | \mathrm{v}_i J_i \rangle. \tag{4.6.10}$$

The matrix elements of the normal coordinate q are calculated in the basis of wave functions given by Eq. (4.6.4), accounting for both vibrational anharmonicity and vibrotational coupling. In the lowest order of the perturbation theory, they read

$$\langle \mathrm{v}J | q | \mathrm{v}J \rangle = -\frac{3}{2} a_1 \gamma^{1/2} \left(\mathrm{v} + \frac{1}{2} \right) + \gamma^{3/2} J(J+1). \tag{4.6.11}$$

If the vibrational dephasing is assumed to be small, the line-shift cross-section is given by

$$\sigma_s = \int_0^\infty \mathrm{v} F(\mathrm{v}) d\mathrm{v} \int_0^\infty 2\pi b db S_1'(b), \tag{4.6.12}$$

where

$$S_1'(b) = \hbar^{-1} \int_{-\infty}^{\infty} \left(\langle \mathrm{v}_f J_f | V_{iso} | \mathrm{v}_f J_f \rangle - \langle \mathrm{v}_i J_i | V_{iso} | \mathrm{v}_i J_i \rangle \right) dt. \tag{4.6.13}$$

Putting Eqs (4.6.9)–(4.6.11) in Eq. (4.6.12) gives

$$\sigma_s = \Delta q(\mathrm{v}_f J_f | \mathrm{v}_i J_i) \sigma^{(1)}, \tag{4.6.14}$$

$$\sigma^{(1)} = \int_0^\infty \mathrm{v}F(\mathrm{v})d\mathrm{v}\int_0^\infty 2\pi b\,db\hbar^{-1}\int_{-\infty}^{\infty}\frac{\partial V_{iso}(r(t))}{\partial q}dt\,. \tag{4.6.15}$$

For the R-branch transitions $J \to J + 1$ ($m = J + 1$)

$$\sigma_s = [2\gamma^{3/2}m - \frac{3}{2}a_1\gamma^{1/2}(\mathrm{v}_f - \mathrm{v}_i)]\sigma^{(1)} \tag{4.6.16}$$

determines the asymptotic line-shift cross-section. For the rotational transitions ($\mathrm{v}_f = \mathrm{v}_i$) the line centre shifts depend on the vibration-rotation coupling:

$$\sigma_s = (2\gamma^{3/2}m)\sigma^{(1)}. \tag{4.6.17}$$

The quantity $\sigma^{(1)}$ can either be computed directly (like the H_2O case) or extracted from fitting to experimental data. In Ref. [199], $\sigma^{(1)} = -(118 \pm 15)\,\text{Å}^2$ was extracted for pure rotational HF-Ar transitions and $\sigma^{(1)} = -90\,\text{Å}^2$ was calculated for the 0–1 HF–Ar fundamental band from the known diagonal matrix elements of the isotropic potential using the relation easily deduced from Eq. (4.6.8):

$$\frac{\partial V_{iso}}{\partial q} = \frac{2\gamma^{-1/2}}{3a_1}\left(\langle \mathrm{v}=0|V_{iso}|\mathrm{v}=0\rangle - \langle \mathrm{v}=1|V_{iso}|\mathrm{v}=1\rangle\right). \tag{4.6.18}$$

The agreement between these two values is quite satisfactory. For the large values of vibrational quantum numbers the approximation (4.6.8) is not valid and the matrix elements in Eq. (4.6.9) have to be calculated exactly.

4.6.3. CO molecule

In the terrestrial atmosphere, CO is formed by the oxidation of methane, other paraffins and organic components. The life-time of atmospheric CO is relatively short (2–3 months) and its sources move constantly. As a result, CO concentration varies significantly in time and with altitude. The data on CO line-broadening and shifting coefficients are included in the spectroscopic database HITRAN. Numerous studies were made for the line shape parameters of CO in mixtures with various atmospheric gases (see Refs [200–217].

The fundamental band line shapes of CO–N_2 were experimentally studied in Refs [200–202]. In Ref. [202], the line shapes were analysed with Lorentzian, Voigt and speed-dependent HC_V profiles at 301 and 348 K. The experimental line-broadening coefficients were approximated by the analytical formulae

$$\gamma(log) = \gamma_0|m|^b, \qquad b = b_0 + b_1\ln m + b_2(\ln m)^2 + \cdots \tag{4.6.19}$$

with all parameters given in Table 4.47. For the temperature exponents of Eq. (3.2.1) a typical value $n = 0.7$ was found.

Table 4.47. Parameters for the $\gamma(log)$ model obtained with Lorentzian (L), Voigt (V) and speed-dependent HC_V profiles for the fundamental band of CO–N_2 (γ_0 is in 10^{-3} cm^{-1} atm^{-1} and T is in K) [202]. Reproduced with permission.

Profile	T	γ_0	b_0	b_1	b_2	b_3	b_4	b_5	b_6	b_7
L	348	72.24121	1.03171	–4.98187	8.94091	–8.42233	4.49026	–1.36280	0.21958	–0.01460
V	348	71.85475	2.04538	–9.62587	17.55630	–16.8520	9.21772	–2.88998	0.48380	–0.03355
HC_V	348	73.22221	1.15896	–5.52787	9.82646	–9.17495	4.85606	–1.46559	0.23523	–0.0156
L	301	80.22492	0.66623	–3.37680	6.08592	–5.76035	3.06696	–0.92324	0.14669	–0.00958
V	301	79.91998	0.52804	–2.76951	5.02829	–4.79972	2.57021	–0.77591	0.12336	–0.00805
HC_V	301	81.42971	0.71831	–3.75500	6.89091	–6.57449	3.51423	–1.06035	0.16874	–0.01103

The 2–0 band of CO–N_2 was studied in Refs [203, 204]. Predoi-Cross *et al.* [204] obtained line-broadening, line-shifting and asymmetry coefficients Y_m. For theoretical computations by the semi-classical RB formalism the interaction potential was composed of electrostatic and atom–atom contributions (atom–atom parameters are given in Appendix B). The electrostatic part was taken in the form

$$V^{(elec)}(r) = 4\varepsilon\left[\left(\frac{\sigma}{r}\right)^{12} - \left(\frac{\sigma}{r}\right)^{6}\right] - 24\varepsilon\frac{d}{\sigma}\left[\left(\frac{\sigma}{r}\right)^{7} - 2\left(\frac{\sigma}{r}\right)^{13}\right]P_1(\cos\theta) + 4\varepsilon\gamma_1\left[\left(\frac{\sigma}{r}\right)^{12} - \left(\frac{\sigma}{r}\right)^{6}\right]P_2(\cos\theta), \quad (4.6.20)$$

where, as usual, θ is the angle between the CO molecular axis and the intermolecular axis, P_1 and P_2 are Legendre polynomials, γ_1 is the polarizability anisotropy and d is the distance between the mass and charge centres.

Recently, the line shapes of purely rotational transitions $J = 3 \leftarrow 2$ and $J = 5 \leftarrow 4$ of CO broadened by N_2 (as well as by O_2, CO_2 and noble gases) were analysed at various temperatures and with various profile models [205]. A study of the N_2 and O_2-broadening coefficients of the line $J = 3 \leftarrow 2$ at 297 K is presented in Ref. [206] (see also references cited therein).

The spectroscopic parameters of 21 $R(J)$ lines from the (3-0) band of self-perturbed CO (6354.1791–6406.70160 cm^{-1}) were reported in [207]. Despite the clearly observed line narrowing, the Voigt profile was used for line shape analysis. An empirical polynomial formula for the room-temperature self-broadening coefficient estimation for this band was proposed:

$$\gamma_{CO-CO}(poly) = \sum_{n=0}^{4} b_n |m|^n, \tag{4.6.21}$$

where $b_0 = (9.392 \pm 0.036) \cdot 10^{-2}$ cm^{-1}, $b_1 = (-6.86 \pm 0.22) \cdot 10^{-3}$ cm^{-1}, $b_2 = (6.46 \pm 0.40) \cdot 10^{-4}$ cm^{-1}, $b_3 = (-2.98 \pm 0.27) \cdot 10^{-5}$ cm^{-1} and $b_4 = (4.94 \pm 0.036) \cdot 10^{-7}$ cm^{-1}. Line shifts in the (3–0) band of CO–CO were also measured in Ref. [208]. Experimental line-broadening and line-shifting coefficients in the (1–0) and (2–0) bands [209, 210] as well as the broadening coefficient of the rotational line $J = 1 \leftarrow 0$ [211] of self-perturbed CO were also obtained.

The case of hydrogen-broadening was considered in Refs [201, 209–210, 212–213]. The perturbation by CO_2 was analysed in Refs [201, 214]. CO–Ar and CO–He line broadening at 301 K was studied in Ref. [215] (Table 4.48). Low-temperature line widths and shifts of CO in He can be found in Ref. [216].

Table 4.48. Broadening coefficients γ(10^{-3} cm^{-1} atm^{-1}, within two standard deviations) for the lines of the fundamental CO band He, Ar at T = 301 K [215]. Reproduced with permission.

m	He		Ar	
	P-branch	*R*- branch	*P*- branch	*R*- branch
1	48.77(9)	48.84(6)	69.45(15)	69.58(17)
2		47.17(8)	64.16(7)	64.09(10)
3	46.45(7)	46.53(6)	59.53(7)	59.55(9)
4		46.09(7)		55.46(12)
5	46.08(5)	46.10(5)	52.42(11)	51.94(11)
6	46.10(5)	45.88(5)	49.91(8)	49.63(10)
7	46.24(6)	46.05(6)	48.14(11)	47.89(11)
8	46.34(5)	46.22(6)	46.85(8)	46.66(12)
9	46.38(6)	46.28(6)	45.83(6)	45.56(14)
10	46.45(13)	46.35(6)	45.20(21)	45.05(10)
11	46.69(4)	46.67(6)	44.48(6)	44.20(15)
12	46.70(4)	46.58(6)	43.95(7)	43.85(12)
13	46.66(5)	46.66(5)	43.44(5)	43.40(8)
14	46.77(7)	46.75(6)	43.06(5)	42.91(5)
15	46.65(8)	46.68(5)	42.70(7)	42.48(10)
16	46.47(6)	46.50(5)	42.28(9)	41.94(10)
17	46.49(8)	46.34(13)	41.84(10)	41.49(13)
18	46.32(6)	46.28(7)	41.02(9)	40.87(10)
19	46.20(8)	46.22(11)	40.45(6)	40.42(6)
20	46.05(13)	45.91(7)	39.78(10)	39.70(6)
21		45.71(11)		39.00(8)
23			37.35(5)	
25			35.66(6)	
27			33.63(9)	
25			33.79(10)	

Systematic computations of CO line broadening by CO_2, H_2O, N_2 and O_2 in the 300–2400 K temperature range were made by Hartmann *et al.* [217] using the RB formalism. At 300 K agreement with experimental data was better than 10%. The temperature exponents were also deduced by a least-squares fitting procedure. Table 4.49 gives these line-broadening parameters. As with the case of CO_2, the temperature-dependent mean half-widths were fitted to Eq. (4.5.4), leading to the parameters collected in Table 4.50. At very high temperatures (above 1500 K) when the line broadening is practically independent on m these mean values provide quite a good approximation.

Table 4.49. CO line-broadening parameters (in 10^{-3} cm^{-1} atm^{-1}) and temperature exponents n (valid for $T > T$(min)) [217]. Reproduced with permission.

$\|m\|$	T(min)	CO–CO_2		CO–H_2O		CO–N_2		CO–O_2	
		γ(300)	n	γ(300)	n	γ(300)	n	γ(300)	n
1	200	112	0.76	109	0.76	76.3	0.76	65.9	0.73
5	200	98.2	0.71	120	0.81	68.3	0.70	55.7	0.61
9	200	82.8	0.63	127	0.80	64.4	0.68	51.6	0.60
13	220	65.6	0.54	127	0.78	61.4	0.66	50.1	0.60
17	290	55.7	0.50	121	0.73	56.4	0.61	48.2	0.57
21	360	50.5	0.50	118	0.71	50.8	0.56	46.5	0.56
25	430	44.8	0.48	109	0.67	47.4	0.53	45.1	0.56
29	500	41.7	0.48	99.5	0.63	43.1	0.49	43.9	0.56
33	570	38.7	0.46	90.4	0.59	38.7	0.45	42.5	0.56
37	640	36.1	0.44	83.5	0.57	34.8	0.40	41.0	0.53
41	700	34.7	0.44	81.7	0.58	33.2	0.38	39.3	0.51
45	780	32.9	0.43	74.3	0.55	32.4	0.40	36.9	0.48
49	850	30.8	0.41	67.8	0.53	29.7	0.37	36.3	0.48
53	910	28.3	0.38	62.6	0.51	27.5	0.35	35.7	0.48
57	980	26.0	0.35	56.3	0.48	25.6	0.33	34.4	0.47
61	1050	23.8	0.32	50.2	0.45	23.8	0.31	33.9	0.47
65	1120	21.8	0.29	44.3	0.41	22.0	0.29	32.8	0.46
69	1190	19.5	0.25	38.5	0.37	20.2	0.26	31.7	0.45
73	1260	17.2	0.20	32.9	0.32	18.7	0.23	30.6	0.44
77	1330	15.4	0.16	27.6	0.26	17.1	0.20	29.5	0.43

Table 4.50. Parameters of Eq. (4.5.4) for CO molecule [217]. Reproduced with permission.

System	$\bar{\gamma}$ (300), 10^{-3} cm^{-1} atm^{-1}	$\bar{N}$ (300)	α
CO–H_2O	122.0	0.774	0.013
CO–CO_2	85.2	0.829	0.040
CO–N_2	65.6	0.760	0.030
CO–O_2	54.1	0.684	0.029

The line-broadening coefficients obtained in Ref. [217] enable a test of the extrapolation properties of the analytical models for the coefficients γ. For linear molecules, the polynomial representations (4.5.2)–(4.5.3) and (4.5.9), the Pade approximant (4.5.5) or the form (4.6.19) can be tested. Moreover, the analytical $\gamma(sur)$ model of Eq. (3.3.4) proposed for asymmetric tops can be applied: for -linear molecules $K_1 = K_2 = 0$ and $x_4 = 0$ so that Eq. (3.3.4) reads

$$\gamma(sur) = x_1 + x_2 / \mathrm{ch}^2 x_3 , \tag{4.6.22}$$

where

$$x_k = x_{k0} + x_{k1}|m| + x_{k2}|m|^2 + \cdots . \tag{4.6.23}$$

Equation (4.6.22) gives a correct $|m|$-dependence if there is a predominant contribution in the intermolecular potential (for example, dipole–dipole interaction between polar molecules). If it is necessary to account for different types of interaction, this equation can be extended by including additional terms:

$$\gamma(sur) = x_1 + x_2 / \mathrm{ch}^2 x_3 + x_4 / \mathrm{ch}^2 x_5 + \cdots . \tag{4.6.24}$$

To study the extrapolation properties of the abovementioned models, their parameters were first obtained by fitting to the RB data [217] for $|m| \le 49$ with almost identical quality. Then they were used in model calculations for $49 \le |m| \le 77$. The results of an extrapolation for CO–N_2 are shown in Fig. 4.18.

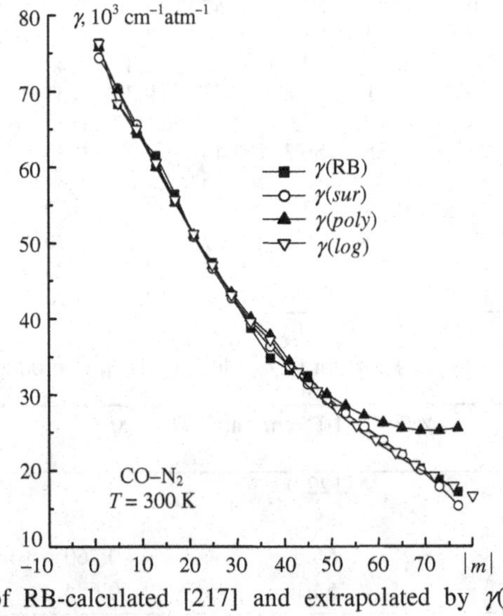

Fig. 4.18. Comparison of RB-calculated [217] and extrapolated by $\gamma(poly)$, $\gamma(log)$ and $\gamma(sur)$ models CO–N_2 line-broadening coefficients.

Since the models $\gamma(log)$ (4.6.19) and $\gamma(sur)$ (4.6.22) demonstrate a good agreement with the theoretical values of Ref. [217], they can be recommended for the line-broadening coefficient computations for high $|m|$-values.

An analogous test was realised for CO line broadening by Ar and He (Fig. 4.19).The parameters of $\gamma(log)$ and $\gamma(sur)$ of Eq. (4.6.24) were determined from fitting to the experimental data of Ref. [215] with $|m| \leq 20$. As can be observed from the figure, the calculated γ-values agree with the experimental data within 3%. The models $\gamma(log)$ and $\gamma(sur)$ give approximately the same broadening coefficients for $20 < |m| \leq 30$. However, for $\gamma(sur)$ the error is maximal for $|m| = 1$, and it is better to exclude the corresponding γ-value from the least-squares fitting.

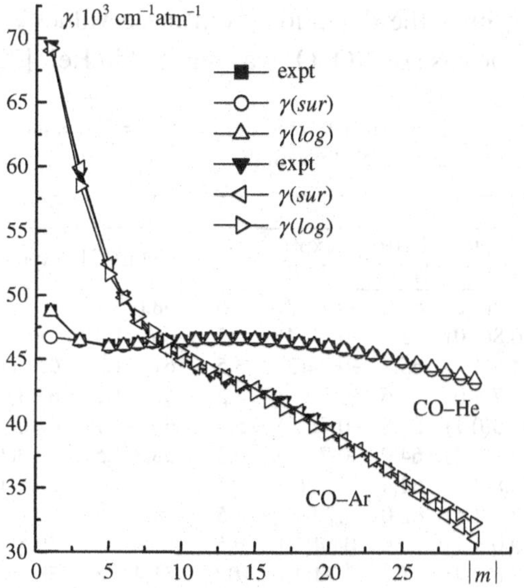

Fig. 4.19. Experimental [215] and calculated with $\gamma(log)$ and $\gamma(sur)$ models CO–He and CO–Ar line-broadening coefficients.

4.6.4. NO molecule

Experimental studies of infrared absorption spectra of NO in mixtures with O_2 and N_2 at various temperatures were realised in Refs [218–226]. The broadening of the fundamental band lines by rare gases was studied in Refs [220, 227]. The self-broadening coefficients of NO [219] have the same order of magnitude as the coefficients for the broadening by O_2 and N_2, so that, for small concentrations of NO in the atmosphere, they do not influence the NO broadening by air.

An analysis of experimental and computed by the semi-classical RB and RBE approaches $NO–N_2$ line-broadening coefficients is given in Ref. [228] (the room-temperature values are presented in Tables 4.51, 4.52). A significant difference between the RB and RBE line widths was stated for high J-values, with the RBE approach giving values closer to the experimental data. It was also found that the line broadening of the lines with $K = 3/2$ is better described by the RBE approach than that of the lines with $K = 1/2$. This result is due to the fact that for $K = 3/2$ the molecule is closer to the linear case ($K= 0$) then for $K = 1/2$ (the conditions of the least moment of inertia are easier satisfied for $K = 3/2$. $NO–N_2$ temperature exponents n for the temperature interval 231–296 K ($T_0 = 296$ K) were obtained by Ballard *et al.* [219] (they are also given in Tables 4.51, 4.52). The mean value of n is 0.708±0.059 for the $^2\Pi_{1/2}$ sub-band and 0.699±0.067 for the $^2\Pi_{3/2}$ sub-band. For individual lines the deviation from these values can reach 15%). An analogous study for the case of $NO–O_2$ was realised in Ref. [229] (Table 4.53).

Table 4.51. Broadening coefficients (in 10^{-3} cm^{-1} atm^{-1}) at $T_0 = 296$ K and temperature exponents for $NO–N_2$ lines in the $^2\Pi_{1/2}$ sub-band.

m	γ expt [222]	γ expt. [219]	γ RBE [228]	n expt [219]	m	γ expt [222]	γ expt[219]	γ RBE [228]	n expt [219]
−1.5	66.8(2.0)	72.2(4.2)	67.8	−0.76(2)	1.5	68.6(3.4)		65.8	
−2.5	58.9(3.0)	66.8(2.0)	65.9	−0.75(4)	2.5	66.7(1.0)		65.1	
−3.5	63.8(1.2)	64.5(1.7)	65.1	−0.74(3)	3.5	61.3(1.2)	65.1(1.2)	64.9	−0.72(1)
−4.5	60.4(1.0)	63.7(1.0)	64.6	−0.70(4)	4.5	60.4(1.2)	63.3(1.0)	64.4	−0.72(2)
−5.5	59.6(1.1)	62.3(0.7)	63.8	−0.72(5)	5.5	60.8(1.7)	61.6(1.0)	63.6	−0.75(4)
−6.5	56.9(0.9)	61.9(1.2)	64.0	−0.70(8)	6.5	58.8(1.5)	60.9(1.7)	62.8	−0.69(3)
−7.5	56.2(2.1)	59.8(1.2)	62.4	−0.77(9)	7.5	58.1(1.1)	60.7(1.0)	62.1	−0.69(3)
−8.5	55.5(0.6)	60.1(1.5)	62.0	−0.70(12)	8.5	55(1.3)	60.2(0.8)	61.5	−0.71(3)
−9.5	56.2(1.5)	60.0(1.7)	61.6	−0.64(10)	9.5	52.6(2.1)	59.5(0.5)	61.2	−0.69(3)
−10.5	56.7(1.8)	58.6(2.4)	61.4	−0.74(13)	10.5	51.4(2.8)	58.9(0.7)	60.9	−0.67(3)
−11.5	55.8(1.2)	59.6(5.0)	61.2	−0.68(15)	11.5	54.9(1.5)	58.1(0.8)	60.6	−0.70(3)
−12.5	55.2(0.6)	57.2(2.9)	61.1	−0.71(7)	12.5	55.5(1.0)	58.1(0.9)	60.3	−0.64(2)
−13.5	54.0(0.6)	55.3(4.2)	60.9	−0.78(10)	13.5	49.6(2.5)	57.5(0.5)	60.0	−0.69(4)
−14.5	57.4(2.6)		60.4		14.5	54.8(1.0)	56.8(1.0)	59.5	−0.62(3)
−15.5	53.1(1.8)		59.9		15.5	52.0(1.2)	56.6(0.9)	59.0	−0.64(2)
−16.5	52.4(1.4)		59.4		16.5	54.6(2.9)	55.3(1.9)	58.3	−0.65(2)
−17.5	56.3(1.9)		59.0		17.5	52.8(2.0)	55.5(1.4)	57.6	−0.60(5)
−18.5	51.3(2.0)		58.4		18.5	53.0(1.4)	54.4(1.5)	56.8	−0.65(3)
−19.5	46.3(1.0)		57.5		19.5	52.7(1.5)	52.6(1.6)	55.9	−0.78(6)
					20.5	52.0(1.7)	54.8(3.9)	55.0	−0.55(9)
					21.5	50.3(3.0)	55.5(1.4)	54.1	

Table 4.52. Broadening coefficients (in 10^{-3} cm^{-1} atm^{-1}) at T_0 = 296 K and temperature exponents for NO–N_2 lines in the $^2\Pi_{3/2}$ sub-band.

m	γ			n	m	γ			n
	expt [222]	expt. [219]	RBE [228]	expt [219]		expt [222]	expt. [219]	RBE [228]	expt [219]
−2.5	67.5(1.5)	75.4(4.5)	67.8	−0.71(11)					
−3.5	66.1(2.4)	66.4(2.8)	67.0	−0.66(8)	3.5	64.6(3.2)	67.0(3.3)	65.9	−0.80(10)
−4.5	64.3(3.0)	67.4(2.0)	66.1	−0.65(4)	4.5	64.9(1.8)	66.2(1.2)	65.3	−0.71(8)
−5.5	62.0(1.6)	63.5(1.6)	65.1	−0.69(9)	5.5	63.0(0.5)	65.8(1.5)	64.5	−0.74(6)
−6.5	61.2(0.6)	64.3(1.5)	64.1	−0.67(9)	6.5	62.3(0.7)	64.3(1.2)	63.6	−0.67(4)
−7.5	60.1(2.3)	63.8(1.7)	63.3	−0.67(3)	7.5	61.8(1.3)	63.8(1.1)	62.8	−0.68(3)
−8.5	59.2(1.8)	61.5(3.3)	62.7	−0.72(6)	8.5	59.1(1.5)	63.3(0.9)	62.1	−0.66(5)
−9.5	60.6(0.3)	63.0(3.3)	62.2	−0.53(8)	9.5	57.8(0.9)	62.0(1.1)	61.6	−0.77(8)
−10.5	57.6(0.9)	62.3(4.0)	61.8	−0.62(17)	10.5	57.0(2.4)	60.6(1.0)	61.2	−0.70(5)
−11.5	57.9(3.4)	58.5(5.4)	61.6	−0.97(14)	11.5	59.3(2.5)	60.4(1.0)	60.9	−0.71(4)
−12.5	58.7(3.8)	58.5(4.5)	61.4	−0.83(10)	12.5	56.0(0.2)	61.4(0.8)	60.6	−0.65(6)
−13.5	56.0(1.0)		61.1		13.5	57.4(1.8)	59.2(1.3)	60.2	−0.69(6)
−14.5	55.5(0.8)		60.6		14.5	58.0(1.2)	57.7(2.0)	59.7	−0.65(6)
−15.5	60.9(2.5)		60.1		15.5	54.8(1.2)	58.8(1.8)	59.1	−0.74(10)
−16.5	56.9(0.4)		59.6		16.5	53.8(2.1)	57.2(2.9)	58.4	−0.61(3)
−17.5					17.5	54.8(0.7)	59.2(2.6)	57.7	−0.68(4)
−18.5	41.8(2.9)		58.5		18.5	54.8(1.9)	55.1(3.5)	56.9	−0.71(17)
−19.5					19.5	47.0(2.7)	58.2(3.8)	56.1	−0.72(4)
−20.5	46.8(7.4)		56.7		20.5	57.8(1.6)	58.4(8.8)	55.2	−0.85(8)
−21.5	51.0(11.0)		56.1		21.5	56.7(2.4)		54.3	

Table 4.53. NO–O_2 line-broadening coefficients (in 10^{-3} cm^{-1} atm^{-1}) at 299 K.

m	$^2\Pi_{1/2}$ (*f*) expt [223]	$^2\Pi_{1/2}$ (*e*) expt [223]	$^2\Pi_{1/2}$ RBE [229]	$^2\Pi_{3/2}$ expt [223]	$^2\Pi_{3/2}$ RBE [229]	m	$^2\Pi_{1/2}$ (*f*) expt [223]	$^2\Pi_{1/2}$ (*e*) expt [223]	$^2\Pi_{1/2}$ RBE [229]	$^2\Pi_{3/2}$ expt [223]	$^2\Pi_{3/2}$ RBE [229]
−1.5	59.67(78)	59.59(87)	49.54			1.5	60.93(40)	60.42(07)	50.23		
−2.5			49.43	59.21(5)	51.30	2.5	57.71(26)	56.98(25)	49.67	58.45(73)	51.55
−3.5	54.59(33)	55.64(32)	49.39	57.97(40)	51.07	3.5	54.65(40)	54.34(67)	49.51	57.41(10)	51.33
−4.5	52.10(45)	52.60(34)	49.29	55.07(33)	50.87	4.5	53.16(35)	51.69(47)	49.37	55.11(25)	51.04
−5.5	50.79(58)	51.63(50)	49.12	54.20(23)	50.63	5.5	51.38(63)	51.01(64)	48.19	54.06(26)	50.75
−6.5	48.90(59)	49.38(36)	48.92	52.70(37)	50.37	6.5	49.95(78)	48.96(63)	48.97	52.92(22)	50.45
−7.5	48.43(27)	48.33(60)	48.70	51.88(33)	50.10	7.5	50.00(30)	47.86(57)	48.74	51.46(35)	50.17
−8.5	48.27(39)	47.75(64)	48.48	50.44(21)	49.84	8.5	49.18(60)	47.44(41)	48.50	50.70(30)	49.89
−9.5	47.38(47)	47.45(62)	48.24	50.85(22)	49.58	9.5	48.03(77)	46.38(54)	48.27	50.35(27)	49.62
−10.5	46.78(59)	46.78(44)	48.01	49.56(20)	49.32	10.5	47.72(89)	45.62(64)	47.03		49.35
−11.5	47.31(36)	47.13(54)	47.75	49.31(22)	49.05	11.5	47.50(67)	46.10(17)	47.77	49.00(30)	49.07
−12.5	47.40(47)	46.17(62)	47.47		48.76	12.5	47.37(65)	45.41(61)	47.48	48.91(28)	48.77
−13.5	46.15(50)	45.07(53)	47.10	48.15(30)	48.38	13.5			47.10	48.66(8)	48.38
−14.5	45.59(11)	44.83(10)	46.64	47.22(37)	47.92	14.5	47.30(40)	43.80(54)	46.65	47.73(10)	47.93
−15.5	46.21(46)	43.43(68)	46.18	46.25(68)	47.46	15.5	48.07(10)	43.88(03)	45.18	47.09(48)	47.46
−16.5	46.74(30)	43.53(34)	45.69	46.98(26)	46.96	16.5	46.65(33)	43.89(09)	45.70	45.65(40)	46.98
−17.5	45.16(61)	44.15(15)	45.26		46.53	17.5	47.12(34)	42.46(40)	44.26	46.12(29)	46.53
−18.5	44.59(74)	42.53(73)	44.86	48.05(1.50)	45.13	18.5	45.63(46)	42.93(45)	44.87	45.80(13)	46.13
−19.5	42.45(83)	41.15(93)	44.46		45.72	19.5	42.99(67)	42.38(23)	44.46	44.73(38)	45.73
−20.5	42.25(36)	40.35(67)	44.03	43.67(16)	45.29	20.5	44.21(53)	39.50(47)	44.05	43.62(85)	45.31
−21.5	43.72(25)	41.13(94)	43.61	45.25(1.80)	44.86	21.5	45.57(62)	40.02(91)	43.59	43.49(96)	44.85

The broadening coefficients γ are generally 17% smaller in the case of NO–O_2 than in the case of NO–N_2. For the fundamental band at 299 K these coefficients (in cm^{-1} atm^{-1}) can be estimated by the formula [223]

$$\gamma(poly) = A + Bm + Cm^2 + Dm^3 \tag{4.6.25}$$

with *A*, *B*, *C* and *D* given in Table 4.54 (NO–O_2 at 299 K and NO–N_2 at 296 K).

Table 4.54. Coefficients of Eq. (4.6.25) for the line-broadening coefficients of the fundamental NO band [223]. Reproduced with permission.

System	γ	*A*	*B*, 10^{-3}	*C*, 10^{-4}	*D*, 10^{-6}
NO–O_2	$\gamma_{3/2}$	0.06463	−2.62	1.47	−3.27
	$\gamma_{1/2}(f)$	0.06515	−3.81	2.71	−6.65
	$\gamma_{1/2}(e)$	0.06433	−3.43	2.14	−4.97
NO–N_2	$\gamma_{3/2}$	0.06954	−1.70	0.91	−2.38
	$\gamma_{1/2}$	0.06910	−2.34	1.36	−3.15
NO–air	$\gamma_{3/2}$	0.06850	−1.89	1.03	−2.57
	$\gamma_{1/2}$	0.06818	−2.61	1.58	−3.71

The narrowing coefficients of the 551.53 kHz NO line broadened by various gases are discussed in Sec. 1.1 in the framework of the soft-collision model.

4.6.5. OH molecule

The OH radical is of particular importance for chemical processes occurring in the terrestrial atmosphere. In the troposphere, it oxidizes many molecules (including CO) and it is the main reagent for the molecules of the type NO_x. In the stratosphere, it is involved in chemical reactions with ozone and contributes to infrared absorption. On the mesosphere level, OH is formed in excited vibrational states ($v \leq 9$) by the reaction H+O_3 and relaxes to the ground state during the night. Emission spectra of OH were recorded in the mesosphere for the rotational numbers $N \approx 33$ [230] (the total angular momentum of the molecule $\vec{J} = \vec{N} + \vec{L} + \vec{S}$, where $\vec{L}$ is the electronic orbital momentum and $\vec{S}$ is the spin).

Experimental determination of OH absorption line profiles was realised for broadening by N_2, O_2, NO_2, He, Ar and Kr in the near IR region, by H_2O and NO_2 in the microwave region and by N_2 and rare gases in the UV region [230–236]. In Ref. [230], the broadening coefficients of OH lines by N_2, O_2, He and Ar pressure were extracted for the *P*-branch of the fundamental (1-0) band using the Voigt profile model (Table 4.55). For Ar-broadening the profile of Rautian and Sobel'man was equally tested, yielding γ and β_0 coefficients (Table 4.56).

Table 4.55. Broadening coefficients (in MHz Torr^{-1}) extracted using the Voigt profile for the P-branch of the band v = 1 ← 0 of OH [230]. Reproduced with permission.

OH-Ar			OH-He			$OH-O_2$			$OH-N_2$		
F	N	γ	F	N	γ	F	N	γ	F	N	γ
1	1	3.37(16)	1	1	1.60(15)	1	1	4.57(16)	1	1	8.25(40)
1	2	2.61(15)	1	2	1.72(14)	1	2	3.88(31)	1	2	7.95(64)
1	3	2.24(13)	1	3	1.88(20)	1	3	3.53(28)	1	3	5.78(36)
1	4	1.47(18)	1	4	1.40(12)	1	4	2.72(44)	1	4	4.81(41)
2	1	2.89(25)	2	1	2.31(18)	2	1	5.02(38)	2	1	8.61(24)
2	3	2.58(12)	2	3	1.58(19)	2	3	3.29(17)	2	3	5.37(35)
2	4	1.26(19)	2	4	2.16(12)	2	4	2.67(30)	2	4	4.29(50)

Table 4.56. Ar-broadening and narrowing coefficients (in MHz Torr^{-1}) obtained with the Rautian–Sobel'man profile for the P-branch of the OH band v = 1 ← 0 [230]. Reproduced with permission.

F	N	γ	β_0
1	1	3.69(18)	1.02(20)
1	2	3.08(17)	1.44(20)
1	3	2.77(15)	2.01(20)
1	4	2.33(21)	2.21(53)
2	1	3.52(26)	1.98(52)
2	3	2.94(14)	1.38(35)
2	4	2.05(22)	1.94(48)

Theoretical RBE computations of rotational and vibrotational line-broadening coefficients for the $OH–N_2$, $OH–O_2$ and OH–Ar systems were made in Ref. [237] for temperatures of 194, 200, 296 and 300 K (Tables 4.57, 4.58); the interaction potential parameters for these systems are given in Appendix B).

Table 4.57. RBE-calculated $OH–N_2$ line-broadening coefficients (in 10^{-3} cm^{-1} atm^{-1}) [237].

J	$^2\Pi_{1/2}$ (200 K)	$^2\Pi_{3/2}$ (200 K)	$^2\Pi_{1/2}$ (300 K)	$^2\Pi_{3/2}$ (300 K)
1/2	147.41		108.81	
3/2	131.26	145.70	100.85	109.42
5/2	90.99	125.98	80.48	96.41
7/2	46.70	96.04	51.50	75.49
9/2	26.47	70.55	31.90	54.38
11/2	19.63	55.01	22.83	40.87

Table 4.58. RBE [237] and experimental [231, 232] $OH–O_2$ coefficients γ(in 10^{-3} cm^{-1} atm^{-1}).

J	T = 296 K			T = 194 K		
	$^2\Pi_{1/2}$ RBE	$^2\Pi_{3/2}$ RBE	$^2\Pi_{3/2}$ expt	$^2\Pi_{1/2}$ RBE	$^2\Pi_{3/2}$ RBE	$^2\Pi_{3/2}$ expt
1/2	45.29			80.92		
3/2	35.51	43.76	66.9(4.4) [232]	70.73	79.07	84.9(3.0) [232]
5/2	23.67	31.69	48.0(12) [231]	59.22	65.70	67.3(3.1) [231]
7/2	14.56	20.05		52.55	53.81	
9/2	10.72	13.45		50.55	49.75	
11/2	9.67	10.98		49.90	48.75	

4.6.6. HF and HCl molecules

Like OH, the line-broadening coefficients of these molecules are required for the monitoring of the upper atmosphere where they are formed as a product of photochemical reactions of gases containing Cl and F with ozone. Despite the difficulty of laboratory measurements, HF and HCl line shapes were recorded for the (1–0) and (2–0) vibrational bands [238–241] as well as for the pure rotational $J = 1 \leftarrow 0$ transitions [242–244] in mixtures with N_2, O_2, Ne, Ar and Xe.

In Ref. [238], HF and HCl line shapes broadened by N_2 and air under conditions like in the upper-atmosphere ($P \leq 1$ atm, $T = 295$ and 202 K) were recorded for the fundamental band (1–0). The collisional parameters were extracted from the experimental line shapes using the soft-collision model of Galatry and the hard-collision model of Rautian and Sobel'man. The line-broadening and line-shifting coefficients were found to be practically the same for both models whereas the narrowing coefficient β_0 appeared 10–50% smaller for the last model. All the coefficients obtained in Ref. [238] using the Galatry profile are given in Table 4.59 (HF) and Table 4.60 (HCl). For the rotational quantum numbers $J \geq 6$, for which the shift is of the same order of magnitude as the line broadening, the experimental line shapes were asymmetric. Since this asymmetry was independent of the shift sign and disappeared with increasing pressure, it was not ascribed to the line interference but rather to a statistical correlation between velocity- and state-changing collisions.

From a theoretical point of view the ratio γ_{air}/γ_{N2} should satisfy the approximate relation

$$\gamma_{air} / \gamma_{N_2} \approx 0.8 + 0.2[Q_{O_2} / Q_{N_2}]^{\eta}, \qquad (4.6.26)$$

where Q_{O2} and Q_{N2} are the O_2 and N_2 quadrupole moments respectively. The exponent η depends on the dominant contribution in the interaction potential: $\eta = 2/3$ for the dipole–quadrupole interactions and $\eta = 0.5$ for the quadrupole-quadrupole interactions; that is, γ_{air}/γ_{N2} equal to 0.88 and 0.90, respectively. According to Tables 4.59, 4.60, the measured γ coefficients demonstrate J-dependent deviations from these values.

According to a cinematic model based on the differential cross-section calculation for hard-sphere molecules, the temperature dependence of γ should be given by $T^{-1/2}$; so one should have $(295/202)^{1/2} \approx 1.21$ for $T = 295$ K and $T = 202$ K [238]. The γ-values obtained for these temperatures show a systematic deviation from this rule. Moreover, the temperature dependences of HCl and HF-broadening coefficients are strictly opposite.

Table 4.59. N_2- and air-broadening, narrowing and shifting coefficients (in 10^{-3} cm^{-1} atm^{-1}) in the v = 1 ← 0 band of HF [238]. Reproduced with permission.

	γ				β				δ	
Line	295 K		202 K		295 K		202 K		295 K	
	N_2	air	N_2	air	N_2	air	N_2	air	N_2	air
R(8)	12.4(4)	12.0(2)			21.3	18.6			−29.2	−28.8
R(7)	14.2(3)	13.9(2)	18.7	19.2	20.8	18.4	10.6	18.1	−29.1	−28.5
R(6)	18.4(2)	16.8(2)	25.1	24.5	21.8	18.4	20.0	20.1	−29.0	−27.9
R(5)	24.8(3)	23.0(4)	30.8	29.4	21.3	21.2	15.8	16.7	−28.0	−27.0
R(4)	37.7(9)	34.2(9)	45.7	40.2	25.0	23.9	21.3	19.2	−24.6	−24.0
R(3)	57.3(2)	50.5(3)	64.2	62.1	28.4	25.4	15.2	29.1	−16.7	−16.7
R(2)	81.8(1.1)	70.5(1.3)	93.6	85.0	32.4	32.2	22.7	29.9	−5.0	−6.3
R(1)	101.1(1.1)	89.5(1.9)	122.3	108.4	51.7	49.8	28.9	20.5	−0.7	−2.3
R(0)	112.6(1.1)	105.0(1.5)	141.5	129.2	64.2	70.8	40.0	36.6	3.3	1.9
P(1)	110.2(4.2)	102.8(1.2)	138.1	128.8	55.8	53.8	22.7	37.4	−9.2	−8.5
P(2)	102.3(1.1)	89.7(3)	122.3	111.7	54.9	39.1	21.6	26.1	−6.1	−7.3
P(3)	83.3(7)	72.7(1.0)	96.2	86.0	39.2	31.7	23.7	31.0	−7.4	−9.8
P(4)	56.5(1.2)	50.2(2)	62.7	60.6	32.9	26.8	27.4	39.4	−6.0	−9.2
P(5)	34.3(6)	31.0(6)	36.9	33.7	26.9	25.9	21.6	16.6	−8.4	−10.9
P(6)	21.6(7)	20.4(6)	24.1	23.4	22.9	25.1	20.8	23.1	−13.1	−15.3
P(7)	14.5(3)	14.5(3)	17.7	16.9	26.5	23.5	17.5	16.7	−18.0	−19.2
P(8)	11.5(3)	11.6(2)			23.7	23.0			−21.0	−21.6
P(9)	9.7(2)	10.3(3)			23.8	25.0			−22.9	−23.1

Table 4.60. N_2- and air-broadening, narrowing and shifting coefficients (in 10^{-3} cm^{-1} atm^{-1}) in the v = 1 ← 0 band of $H^{35}Cl$ [238]. Reproduced with permission.

	γ				β				δ	
Line	295 K		202 K		295 K		202 K		295 K	
	N_2	air	N_2	air	N_2	air	N_2	air	N_2	air
R(10)	12.9(2)	13.0(4)	17.2	15.7	18.7	19.8	32.4	23.9	−12.8	−12.2
R(9)	16.1(3)	15.7(3)	18.2	17.9	20.6	17.2	20.3	18.6	−13.0	−12.7
R(8)	20.4(2)	19.9(3)	23.2	22.3	21.1	19.1	22.7	19.6	−12.9	−12.9
R(7)	26.8(2)	25.6(3)	28.7	28.6	18.8	19.7	13.2	22.1	−13.2	−12.6
R(6)	35.9(4)	33.0(2)	38.4	38.4	20.9	20.9	16.6	23.5	−12.5	−12.2
R(5)	47.3(5)	43.0(4)	54.0	49.1	20.8	20.3	16.6	19.9	−10.9	−10.6
R(4)	59.4(3)	53.7(2)	71.5	66.9	22.7	27.2	10.1	17.2	−7.7	−8.0
R(3)	71.3(6)	62.4(1)	90.0	81.8	25.9	18.1	18.5	24.9	−5.1	−5.6
R(2)	79.5(4)	71.8(4)	105.9	94.4	24.2	23.8	23.4	10.6	−3.8	−4.9
R(1)	88.8(1.0)	79.3(2)	114.5	105.6	29.6	24.1	31.6	32.3	−3.9	−4.1
R(0)	95.9(1.0)	89.4(1.0)	120.1	118.2	52.7	37.8	16.7	35.7	0.3	1.0
P(1)	95.8(1.0)	89.2(5)	120.8	113.9	34.6	15.6	22.2	31.0	−6.1	−7.0
P(2)	88.6(9)	79.9(1.0)	115.9	103.8	26.6	26.6	30.8	26.0	−3.4	−4.3
P(3)	81.0(5)	72.4(6)	106.0	94.9	19.3	20.4	22.1	29.0	−5.2	−5.9
P(4)	72.3(2)	64.4(5)	92.2	81.1	20.5	24.4	26.2	12.5	−6.6	−7.9
P(5)	60.9(5)	54.1(3)	73.2	64.5	20.9	20.3	28.9	14.4	−6.8	−7.7
P(6)	47.7(6)	43.9(4)	54.4	50.1	16.3	21.1	21.5	19.2	−6.5	−7.5
P(7)	36.6(2)	33.7(2)	39.5	37.1	18.7	20.9	25.8	21.9	−6.3	−7.4
P(8)	27.6(7)	26.4(7)	28.3	27.7	21.4	22.2	19.3	17.0	−6.8	−7.9
P(9)	20.9(4)	20.2(3)	21.7	21.3	23.5	20.8	16.6	17.3	−7.9	−8.4
P(10)	15.6(4)	16.0(3)	16.5	16.9	17.6	19.9	18.5	16.7	−8.4	−9.0

Room-temperature N_2- and O_2-broadening coefficients for the lowest rotational transition $J = 1 \leftarrow 0$ (626 GHz) of $H^{35}Cl$ were measured in Ref. [242] using the Voigt profile: $\gamma_{N2} = 3.46(6)$ MHz Torr^{-1} and $\gamma_{O2} = 2.70(5)$ MHz Torr^{-1}. Drouin [243] studied the temperature dependence of the N_2 and O_2-broadening and shifting coefficients of this transition in the interval 195–353 K using the Lorentzian profile; both line width and shift were supposed to follow the power law (3.2.1) with $T_0 = 296$ K. For 296 K the values $\gamma_{N2} = 3.639(7)$ MHz Torr^{-1} and $\gamma_{O2} = 2.595(5)$ MHz Torr^{-1} were obtained, with the corresponding temperature exponents 0.71(3) and 0.79(1), whereas for the line-shifting coefficients $\delta_{N2} = 0.111(3)$ MHz Torr^{-1} and $\delta_{O2} = 0.280(1)$ MHz Torr^{-1} were measured, with the temperature exponents equal to 0.36(19) and 0.50(3), respectively. These coefficients and temperature exponents were also estimated in the case of broadening by air: $\gamma_{air} = 3.420(7)$ MHz Torr^{-1} with $n = 0.73(3)$ and $\delta_{air} = 0.146(3)$ MHz Torr^{-1} with $n = 0.4047$. N_2-, O_2- and Ar-broadened widths and shifts of the same line for $H^{35}Cl$ and $H^{37}Cl$ were extracted from the experimental line shapes with the Galatry profile in Ref. [244].

Experimental values of HCl–Ar γ and δ coefficients in the (1–0) and (2–0) bands for the pressures 10–50 atm were obtained in Ref. [240] (Table 4.61). The absorption coefficient was fitted to the model line shape

$$K(\omega) = n_{HCl} \sum_{i=35,37} \sum_{m} A^i_m F_{G\otimes Rm}(\omega), \tag{4.6.27}$$

where n_{HCl} is the number density of the active molecules, A^i_m is the intensity of the m-th line and $F_{G\otimes Rm}(\omega)$ is the asymmetric Voigt profile obtained by convolution of Gaussian (Doppler effect) and Rosenkranz (collisional broadening) profiles [245]:

$$F_{RZ\,m}(\omega) = \frac{1}{\pi}\frac{\omega}{\omega^i_m}\exp\left[-\beta\frac{\hbar}{2}(\omega-\omega^i_m)\right]\frac{\gamma^i_m + Y^i_m(\omega-\omega^i_m-\delta^i_m)}{(\gamma^i_m)^2+(\omega-\omega^i_m-\delta^i_m)^2}. \tag{4.6.28}$$

In Eq. (4.6.28), all the parameters (except the asymmetry coefficients Y^i_m) were assumed to be identical for two isotopic modifications ^{35}Cl and ^{37}Cl. It was not possible to extract the Y^i_m-values because of their small magnitude, but this fact did not influence the precision in the determination of the line parameters. Since for small $|m|$-values the line intensity A^i_m decreased with increasing Ar pressure, for line shape fitting these line intensities were approximated by

$$A_m(P_{Ar}) = A_m^{HITRAN}[1-\tau_m P_{Ar}] \tag{4.6.29}$$

where A_m^{HITRAN} are the HITRAN database intensities and τ_m are fitting parameters.

Table 4.61. Experimental and theoretical HCl–Ar line-broadening and shifting coefficients (in 10^{-3} cm^{-1} atm^{-1}) with experimental uncertainties of $2\cdot10^{-3}$ cm^{-1} atm^{-1} for γ and 10^{-3} cm^{-1} atm^{-1} for δ [240] (CC results are from Ref. [246]). Reproduced with permission.

	Band 1–0						Band 2–0					
m	γ			δ			γ			δ		
	expt	NG	CC	expt	NG	CC	expt	NG	CC	expt	NG	CC
−15	5.7	3.8		−13.6	−14.6							
−14	6.8	4.4		−13.1	−14.2							
−13	7.3	5.2		−13.6	−14.4							
−12	7.7	6.2		−13.6	−14.7							
−11	8.8	7.5		−13.7	−15.0							
−10	11.2	9.2		−13.8	−15.3							
−9	13.7	11.4		−14.2	−15.7							
−8	16.6	14.2	16.1	−14.4	−16.0	−15.5						
−7	20.4	17.6	19.5	−14.6	−16.2	−15.8	25.9	25.0	24.6	−31.5	−30.3	−27.4
−6	24.8	21.5	23.2	−14.6	−16.3	−16.0	28.0	29.0	28.5	−26.9	−29.9	−27.6
−5	29.4	25.6	27.1	−15.0	−16.1	−16.1	33.6	33.0	32.4	−27.8	−29.0	−27.3
−4	34.0	29.5	30.8	−14.9	−15.4	−15.7	36.4	36.2	36.1	−25.8	−27.4	−26.5
−3	39.0	32.6	34.3	−13.0	−13.7	−13.7	40.2	38.4	38.8	−22.6	−25.1	−23.8
−2	44.0	37.9	39.3	−11.2	−11.8	−11.9	43.9	42.8	43.0	−19.5	−21.9	−21.0
−1	71.0	67.5	67.7	−15.0	−14.0	−15.4	67.6	69.3	69.4	−12.0	−14.7	−15.8
1	58.0	60.9	60.0	11.5	9.5	11.0	53.5	58.6	57.9	5.5	3.6	5.7
2	42.4	38.2	39.4	−7.0	−10.9	−10.3	41.8	43.0	42.9	−16.1	−20.7	−19.2
3	36.8	32.2	33.6	−10.0	−12.9	−12.3	37.2	37.7	37.5	−20.9	−24.6	−22.9
4	32.0	28.4	29.4	−11.2	−13.2	−12.3	33.3	34.2	33.5	−23.0	−25.8	−23.5
5	28.2	24.4	25.9	−11.4	−13.8	−13.1	30.8	30.6	30.2	−24.3	−27.0	−24.5
6	24.2	20.4	22.3	−13.0	−14.8	−13.9	28.2	26.9	26.8	−25.7	−28.0	−25.4
7	20.0	16.7	18.8	−14.2	−14.8	−14.5	25.5	23.2	23.4	−26.7	−28.6	−26.1
8	16.6	13.5	15.6	−14.6	−15.0	−14.9	23.8	19.9	20.3	−27.2	−29.8	−26.4
9	13.3	10.9		−14.9	−15.0		21.5	16.5		−28.2	−30.3	
10	11.1	8.8		−15.0	−14.8							
11	9.8	7.2		−14.9	−14.7							
12	8.0	5.9		−15.3	−14.5							
13	7.9	5.0		−15.0	−14.3							
14	7.2	4.3		−14.8	−14.1							
15	6.8	3.7		−15.5	−13.3							

As mentioned above, laboratory studies are not always feasible in a large range of atmospheric temperatures and pressures, so that theoretical results are also important. Owing to the large values of their rotational constants and significantly spaced energy levels, the HCl and HF molecules represent two examples for which consecutive quantum calculations of γ and δ coefficients were realised. In Ref. [240], the NG method was employed and compared to the CC results [246] (see Sec. 2.8 for more details on these methods). In both methods, the same interaction potential of Eq. (2.5.39) with $0 \le l \le 8$ was used. For the line shift calculations the vibrotational interaction was included via Eqs (4.6.8)–(4.6.10). CC- [246] and NG-calculated [240] γ- and δ-values for HCl–Ar are given in Table 4.61. The HF–Ar case is considered in Ref. [247].

The experimental γ values for HCl–Ar [240] can be used to obtain the parameters of the models $\gamma(log)$ (Eq. (4.6.19)) and $\gamma(sur)$ (Eq. (4.6.22)). As for the case of CO–N_2, these parameters were deduced from an analysis of experimental data with $|m| < 8$ and $|m| < 12$ for the R-branch of the (1–0) band and further used to compute the γ-values for $|m| < 20$. These computed values are compared to the experimental data [240] (available for $|m| \leq 15$) in Fig. 4.20. According to this figure, the $\gamma(sur)$ model reproduces the experimental values if the model parameters are deduced from the measurements with $|m| < 12$. If the experimental data used for parameters fitting are limited by $|m| < 8$, the $\gamma(sur)$-values are smaller or higher than the experimental coefficients, in function of the number of parameters. The model $\gamma(log)$ shows more stable extrapolation results.

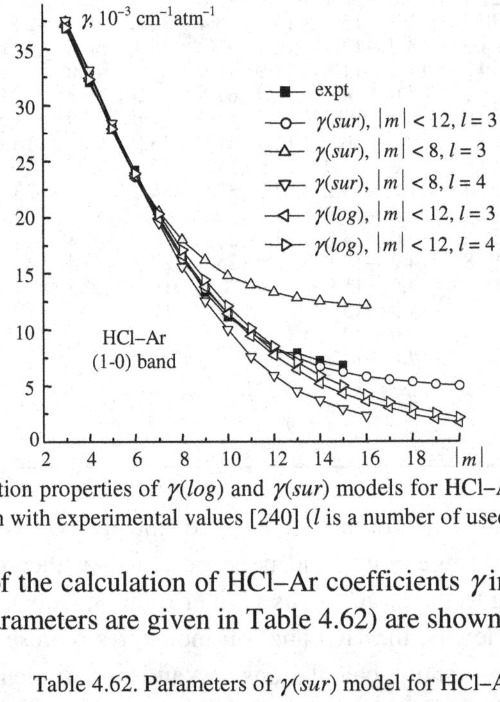

Fig. 4.20. Extrapolation properties of $\gamma(log)$ and $\gamma(sur)$ models for HCl–Ar line-broadening coefficients in comparison with experimental values [240] (l is a number of used parameters).

The results of the calculation of HCl–Ar coefficients γ in the (1–0) and (2–0) bands ($\gamma(sur)$ parameters are given in Table 4.62) are shown in Figs 4.21–4.22.

Table 4.62. Parameters of $\gamma(sur)$ model for HCl–Ar.

	(1–0) band, $\|m\| \neq 1$		(2–0) band, $\|m\| \neq 1$	
	R-branch, $\|m\| \leq 15$	P-branch, $\|m\| \leq 15$	R-branch, $\|m\| \leq 9$	P-branch, $\|m\| \leq 7$
x_{10}	$0.5070(42)\cdot 10^{-2}$	$0.4029(32)\cdot 10^{-2}$	$0.5070\ 10^{-2}$	$0.4029\cdot 10^{-2}$
x_{20}	$0.4022(53)\cdot 10^{-1}$	$0.4350(59)\cdot 10^{-1}$	$0.3534(42)\cdot 10^{-1}$	$0.4138(22)\cdot 10^{-1}$
x_{31}	0.1567(181)	0.1542(112)	0.1043(262)	0.1234(163)
N	14	14	8	6
σ	$1.3\cdot 10^{-4}$	$1.9\cdot 10^{-4}$	$2.8\cdot 10^{-4}$	$2.6\cdot 10^{-4}$

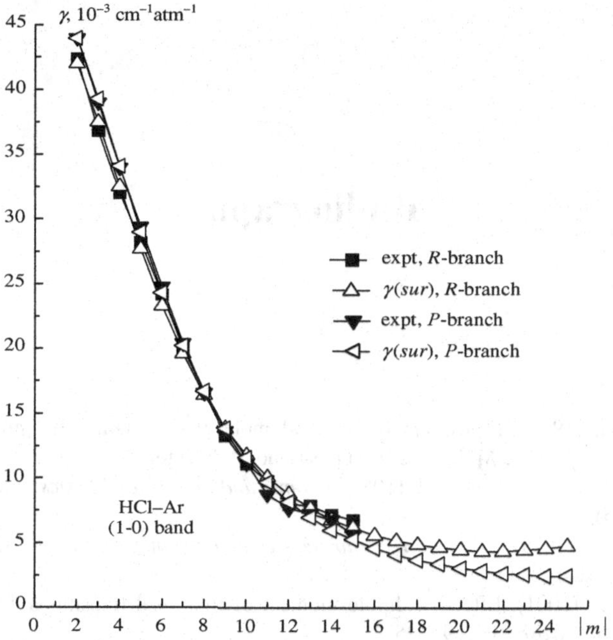

Fig. 4.21. Experimental [240] and calculated by the $\gamma(sur)$ model HCl–Ar line-broadening coefficients in the (1–0) band.

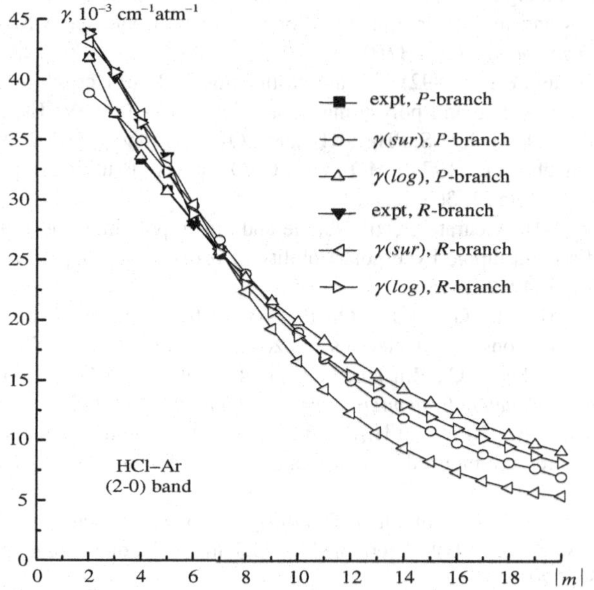

Fig. 4.22. Experimental [240] and calculated by the $\gamma(sur)$ model HCl–Ar line-broadening coefficients in the (2–0) band.

Bibliography

1. Herzberg G. (1945). Molecular spectra and molecular structure. II. Infrared and Raman Spectra of Polyatomic Molecules. Van Nostrand, Princenton.
2. Radzik A.A. and Smirnov B.M. (1980). *Atomic and Molecular Physics*. Atomizdat, Moscow (in Russian).
3. Krasnov K.S. (1979). *Molecular parameters of non-organic substances*. Chemistry, Leningrad (in Russian).
4. Starikov V.I. (2001). Effective dipole moment of rotational transitions of X_2Y–type molecules. *Opt Spectrosc* 91: 189–195.
5. Martin R.L., Davidson E.R. and Eggers Jr. F. (1979). Ab initio theory of the polarizabilit and polarizability derivatives in H_2S. *Chem Phys* 38: 341–348.
6. Rosenberg S., Young R.H. and Schaefer III H.F. (1970). Ground state self-consistent-field wave functions and molecular properties for the iso-electronic series SiH_4, PH_3, H_2S, and HCl. *J Am Chem Soc* 92: 3243–3250.
7. Jain A. and Baluja K.L. (1992). Total (elastic plus inelastic) cross sections for electron scattering from diatomic and polyatomic molecules at 10–5000 eV: H_2, Li_2, HF, CH_4, N_2, CO, C_2H_2, HCN, O_2, HCl, H_2S, PH_3, SiH_4, and CO_2. *Phys Rev A* 45: 202–218.
8. Roos B. and Siegbahn P. (1971). MO-SCF-LCAO studies of sulfur compounds. 1. H_2S and SO_2. *Theor Chim Acta* 21: 368–&.
9. Maroulis G. (1994). Accurate electric dipole and quadrupole moment, dipole polarizability, and first and second dipole hyperpolarizability of ozone from coupled cluster calculations. *J Chem Phys* 101: 4949–4955.
10. Makris C. and Maroulis G. (2005). On the systematic optimization of new basis sets of Gaussian-type functions for calculations on ozone. *Comput Lett* 1: 53–57.
11. Flaud J.M., Camy-Peyret C., Rinsland C.P., Smith M.A.H. and Devi V.M. (1990). *Atlas of ozone spectral parameters from microwave to medium infrared*. Academic Press, Boston.
12. Kissel A., Sumpf B., Kronfeldt H.-D., Tikhomirov B.A. and Ponomarev Yu.N. (2002). Molecular-gas-pressure-induced line shift and line-broadening in the ν_2-band of H_2S. *J Mol Spectrosc* 216: 1–10.
13. Meusel I., Sumpf B. and Kronfeldt H.-D. (1997). Nitrogen broadening of absorption lines in the ν_1, ν_2 and ν_3 bands of H_2S determined by applying an IR diode laser spectrometer. *J Mol Spectrosc* 185: 370–373.
14. Waschull J., Kuhnemann F. and Sumpf B. (1994). Self-, air, and helium broadening in the ν_2 band of H_2S. *J Mol Spectrosc* 165: 150–158.

15. Sumpf B., Meusel I. and Kronfeldt H.-D. (1996). Self- and air-broadening in the ν_1 and ν_3 bands of H_2S. *J Mol Spectrosc* 177: 143–145.
16. Sumpf B. (1997). Experimental investigation of the self-broadening coefficients in the ν_1+ν_3 band of SO_2 and $2\nu_2$ band of H_2S. *J Mol Spectrosc* 181: 160–167.
17. Sumpf B., Meusel I. and Kronfeldt H.-D. (1997). Noble gas broadening in fundamental bands of H_2S. *J Mol Spectrosc* 184: 51–55.
18. Kissel A., Kronfeldt H.-D., Sumpf B., Ponomarev Yu.N., Ptashnik I.V. and Tikhomirov B.A. (1999). Investigation of the line profiles in the ν_2 band of H_2S. *Spectrochim Acta Part A* 55: 2007–2013.
19. Tejwani G.D. and Yeung E.S. (1977). Pressure-broadened linewidths of hydrogen sulfide. *J Quant Spectrosc Radiat Trans* 17: 323–326.
20. Starikov V.I. and Protasevich A.E. (2006). Broadening of absorption lines of the ν_2 band of the H_2S molecule by the pressure of atmospheric gases. *Opt Spectrosc* 101: 523–531.
21. Mason E.A. and Monchick L. (1962). Heat Conductivity of Polyatomic and Polar Gases. *J Chem Phys* 36: 1622–&.
22. Sumpf B., Schone M., Fleischmann O., Heiner Y. and Kronfeldt H.-D. (1997). Quantum number and temperature dependence of foreign gas-broadening coefficients in the ν_1 and ν_3 bands of SO_2: collisions with H_2, He, Ne, Ar, Kr and Xe. *J Mol Spectrosc* 183: 61–71.
23. Pine A.S., Suenram R.D., Brown E.R. and McIntosh K.A. (1996). A terahertz photomixing spectrometer: application to SO_2 self-broadening. *J Mol Spectrosc* 175: 37–47.
24. Prakash V. (1970). Widths of rotational lines of an asymmetric-top molecule SO_2. II. Broadening by a quadrapolar gas. *J Chem Phys* 52: 4674–4677.
25. Hung G., Duan C. and Liu Y. (2002). Nitrogen and self-broadening for far-infrared rotational transitions of SO_2: 702.29336- and 694.49401-GHz lines. *J Mol Spectrosc* 216: 168–169.
26. Sumpf B., Fleischmann O. and Kronfeldt H.-D. (1966). Self-, air-, and nitrogen-broadening in the ν_1 band of SO_2. *J Mol Spectrosc* 176: 127–132.
27. Gamache R.R., Arie E., Boursier C. and Hartmann J.-M. (1998). Pressure-broadening and pressure-shifting of spectral lines of ozone. *Spectrochim Acta Part A* 54: 35–63.
28. Larsen R.W., Nicolaisen F.M. and Sorensen G.O. (2001). Determination of self- air- and oxygen-broadening coefficients of pure rotational absorption lines of ozone and of their temperature dependence. *J Mol Spectrosc* 210: 259–270.
29. Priem D., Colmont J.-M., Rohart F., Wlodarczak G. and Gamache R.R. (2000). Relaxation and lineshape of the 500.4-GHz line of ozone perturbed by N_2 and O_2. *J Mol Spectrosc* 204: 204–215.
30. Yamada M.M. and Amano T. (2005). Pressure broadening of submillimeter-wave lines of O_3. *J Quant Spectrosc Radiat Trans* 95: 221–230.
31. Drouin B.J., Fischer J. and Gamache R.R. (2004). Temperature dependent pressure-induced line shape of O_3 rotational transitions in air. *J Quant Spectrosc Radiat Trans* 83: 63–81.
32. Colmont J.-M., Bakri B., Rohart F., Wlodarczak G., Demaison J., Cazzoli G., Dore L. and Puzzarini C. (2005). Intercomparison between ozone-broadening parameters retrieved from millimeter-wave measurements by using different techniques. *J Mol Spectrosc* 231: 171–187.
33. Drouin B.J. and Gamache R.R. (2008). Temperature dependent air broadened linewidths of ozone rotational transitions. *J Mol Spectrosc* 251: 194–202.
34. Devi V.M., Benner D.C., Smith M.A.H. and Rinsland C.P. (1997). Air-broadening and shift coefficients of O_3 lines in the ν_2 band and their temperature dependence. *J Mol Spectrosc* 182: 221–238.

35. Smith M.A.H., Devi V.M., Benner D.C. and Rinsland C.P. (1997). Temperature dependence of air-broadening and shift coefficients of O_3 lines in the ν_1 band. *J Mol Spectrosc* 182: 239–259.
36. Bouazza S., Barbe A., Plateaux J.J., Rosenmann L., Hartmann J.-M., Camy-Peyret C., Flaud J.M. and Gamache R.R. (1993). Measurements and calculations of room-temperature ozone line-broadening by N_2 and O_2 in the $\nu_1 + \nu_2$ band. *J Mol Spectrosc* 157: 271–289.
37. Neshyba S.P. and Gamache R.R. (1993). Improved line-broadening coefficients for asymmetric rotor molecules with application to ozone lines broadened by nitrogen. *J Quant Spectrosc Radiat Trans* 50: 443–453.
38. Hartmann J.-M., Camy-Peyret C., Flaud J.M., Bonamy J. and Robert D. (1988). New accurate calculations of ozone line-broadening by O_2 and N_2. *J Quant Spectrosc Radiat Trans* 40: 489–495.
39. Barbe A., Regalia L., Plateaux J.J., Von Der Heyden P. and Thomas X. (1996). Temperature dependence of N_2 and O_2 broadening coefficients of ozone. *J Mol Spectrosc* 180: 175–182.
40. Antony B.K., Gamache P.R., Szembek C.D., Niles D.L. and Gamache R.R. (2006). Modified complex Robert–Bonamy formalism calculations for strong to weak interacting systems. *Mol Phys* 104: 2791–2799.
41. Buldyreva J. and Lavrentieva N. (2009). Nitrogen and oxygen broadening of ozone infrared lines in the 5μm region: theoretical predictions by semiclassical and semiempirical methods. *Mol Phys* 107: 1527–1536.
42. http://ozone.iao.ru/
43. Smith M.A.H., Rinsland C.P., Devi V.M., Benner D.C. and Thakur K.B. (1988). Measurements of air-broadened and nitrogen-broadened half-widths and shifts of ozone lines near 9 μm. *J Opt Soc Am B* 5: 585–592.
44. Smith M.A.H., Rinsland C.P., Devi V.M. and Prochaska E.S. (1994). Measurements of pressure broadening and shifts of O_3 lines in the 3-μm region. *J Mol Spectrosc* 164: 239–259; Erratum. *J Mol Spectrosc* 165: 596.
45. Barbe A., Bouazza S. and Plateaux J.J. (1991). Pressure shifts of O_3 broadened by N_2 and O_2. *Appl Opt* 30: 2431–2436.
46. Barbe A., Plateaux J.J. and Bouazza S. (1992). High-resolution infrared spectra of ozone: nitrogen and oxygen broadening coefficients and pressure shifts – analysis of isotopic species. *Proc Tenth All-Union Symp High Res Mol Spectrosc SPIE*, 1811: 83–102.
47. Sokabe N., Hammerich M., Pedersen T., Olafsson A. and Henningsen J. (1992). Photoacoustic spectroscopy of O_3 with a 450-MHz tunable waveguide CO_2 laser. *J Mol Spectrosc* 152: 420–433.
48. Gamache R.R. and Rothman L.S. (1985). Theoretical N_2-broadened halfwidths of $^{16}O_3$. *Appl Opt* 24: 1651–1656.
49. Gamache R.R. and Davies R.W. (1985). Theoretical N_2-, O_2-, and air-broadened halfwidths of $^{16}O_3$ calculated by quantum Fourier transform theory with realistic collision dynamics. *J Mol Spectrosc* 109: 283–299.
50. Lavrentieva N., Osipova A. and Buldyreva J. (2009). Calculations of ozone line shifting induced by N_2 and O_2 pressure. *Mol Phys* 107: 2045–2051.
51. Sumpf B., Pustogov V.V. and Kronfeldt H.D. (1996). Noble gas induced pressure broadening in the ν_3, $\nu_2 + \nu_3 - \nu_2$ and $2\nu_2$ bands of NO_2. *J Mol Spectrosc* 180: 150–153.
52. Dana V., Mandin J.Y., Allout M.Y., Perrin A., Regalia L., Barbe A., Plateaux J.J. and Thomas X. (1997). Broadening parameters of NO_2 lines in the 3.4 μm spectral region. *J Quant Spectrosc Radiat Trans* 57: 445–457.

53. Pustogov V.V., Kuhnemann F., Sumpf B., Heiner Y. and Herrmann K.P. (1994). Pressure broadening of NO_2 by NO_2, N_2, He, Ar, and Kr studied with TDLAS. *J Mol Spectrosc* 167: 288–299.
54. Smith M.A.H., Rinsland C.P., Fridovich B. and Rao K.N. (1985). Intensities and Collision Broadening Parameters from Infrared Spectra. *Mol Spectrosc: Mod Res* 3: 111–228, Academic Press, New York.
55. Smith M.A.H., Rinsland C.P., Devi V.M., Rothman L.S., Fridovich B. and Rao K.N. (1992). Intensities and collision-broadening parameters from infrared spectra: an update. In: Rao K.N. and Weber A. (eds.). *Spectroscopy of the Earth's atmosphere and interstellar medium*, pp. 153–260, Academic Press, New York.
56. Tejwani G.D.T. (1972). Calculation of pressure-broadened linewidths of SO_2 and NO_2. *J Chem Phys* 57: 4676–&.
57. Tejwani G.D.T. and Yeung E.S. (1975). Pressure-broadened linewidths of nitrogen dioxide. *J Chem Phys* 63: 4562–4564.
58. Abeles F.B. and Heggestad H.E. (1973). Ethylene: an urban air pollutant. *J Air Pollut Control Assoc* 23: 517–521.
59. Sinha P., Hobbs P.V., Yokelson R.J., Bertschi I.T., Blake D.R., Simpson I.J., Gao S., Kirschletter T.W. and Novakov T. (2003). Emissions of trace gases and particles from savanna fires in South Africa. *J Geophys Res – Atmos* 108: 8487–&.
60. Kunde V., Aiken A.C., Hanel R., Jennings D.E., Kunde V.G. and Samuelson R.C. (1981). C_4H_2, HC_3N and C_2N_2 in Titan's Atmosphere. *Nature* (London) 292: 686–&.
61. Kostiuk T., Romani P., Espenak F., Livengood T.A. and Goldstein J.J. (1993). Temperature and abundances in the Jovian auroral stratosphere 2. Ethylene as a probe of the microbar region. *J Geophys Res – Planets* 98: 18823–18830.
62. Encrenaz T., Combles M., Zeau Y., Vapillon L. and Berezne J. (1975). A tentative identification of C_2H_4 in the spectrum of Saturn. *Astron Astrophys* 42: 355–356.
63. Brannon J.F. and Varanasi P. (1992). Tunable diode laser measurements on the 951.7393 cm^{-1} line of $^{12}C_2H_4$ at planetary atmospheric temperatures. *J Quant Spectrosc Radiat Trans* 47: 237–242; *J Quant Spectrosc Radiat Trans* 49: 695–696.
64. Blanquet G., Walrand J. and Bouanich J.-P. (2000). Diode-laser measurements of N_2-broadening coefficients in the ν_7 band of C_2H_4. *J Mol Spectrosc* 201: 56–61.
65. Bouanich J.-P., Blanquet G., Walrand J. and Lepère M. (2003). H_2-broadening in the ν_7 band of ethylene by diode-laser spectroscopy. *J Mol Spectrosc* 218: 22–27.
66. Blanquet G., Bouanich J.-P., Walrand J. and Lepère M. (2003). Self-broadening coefficients in the ν_7 band of ethylene at room and low temperatures. *J Mol Spectrosc* 222: 284–290.
67. Bouanich J.-P., Blanquet G., Walrand J. and Lepère M. (2004). Hydrogen-broadening coefficients in the ν_7 band of ethylene at low temperatures. *J Mol Spectrosc* 227: 172–179.
68. Blanquet G., Bouanich J.-P., Walrand J. and Lepère M. (2005). Diode-laser measurements and calculations of N_2-broadening coefficients in the ν_7 band of ethylene. *J Mol Spectrosc* 229: 198–206.
69. Morozhenko V., Kostiuk T., Blass W.E., Hewagama T. and Livengood T.A. (2002). Self- and nitrogen broadening of the ν_{10} $20_{11,10} \leftarrow 19_{10,9}$ ethylene transition at 927.01879 cm^{-1}. *J Quant Spectrosc Radiat Trans* 72: 193–198.
70. Buldyreva J. and Nguyen L. (2008). Extension of the exact trajectory model to the case of asymmetric tops and its application to infrared nitrogen-perturbed linewidths of ethylene. *Phys Rev A* 77: 042720.

71. Nguyen L., Blanquet G., Buldyreva J. and Lepère M. (2008). Measurements and calculations of ethylene line-broadening by argon in the ν_7 band at room temperature. *Mol Phys* 106: 873–880.
72. Nguyen L., Blanquet G., Populaire J.-C. and Lepère M. (2009). Argon-broadening coefficients in the ν_7 band of ethylene at low temperatures. *J Quant Spectrosc Radiat Trans* 110: 367–375.
73. Nguyen L., Blanquet G., Dhyne M. and Lepère M. (2009). Ne- and Kr-broadening coefficients in the ν_7 band of C_2H_4 studied by diode-laser spectroscopy. *J Mol Spectrosc* 254: 94–98.
74. Starikov V.I. and Tyuterev Vl.G. (1997). Intramolecular ro-vibrational interactions and theoretical methods in the spectroscopy of non-rigid molecules. Spektr, Tomsk (in Russian).
75. Baldacchini G., Buffa G., Amato F.D., Pelagalli F. and Tarrini O. (1996). Variation in the sign of the pressure-induced line shifts in the ν_2 band of ammonia with temperature. *J Quant Spectrosc Radiat Trans* 55: 741–743.
76. Baldacchini G., Amato F.D., De Rosa M., Buffa G. and Tarrini O. (1996). Temperature dependence of self-shift of ammonia transitions in the ν_2 band. *J Quant Spectrosc Radiat Trans* 55: 745–753.
77. Nouri S., Orphal J., Aroui H. and Hartmann J.-M. (2004). Temperature dependence of pressure broadening of NH_3 perturbed by H_2 and N_2. *J Mol Spectrosc* 227: 60–66.
78. Bouanich J.-P., Aroui H., Nouri S. and Picard-Bersellini A. (2001). H_2-broadening coefficients in the ν_4 band of NH_3. *J Mol Spectrosc* 206: 104–110.
79. Hadded S., Aroui H., Orphal J., Bouanich J.-P. and Hartmann J.-M. (2001). Line broadening and mixing in NH_3 inversion doublets perturbed by NH_3, He, Ar and H_2. *J Mol Spectrosc* 210: 275–283.
80. Krupnov A. (1996). Experimental nonadditivity of pressure lineshifts in the system of transitions of the ammonia molecule. *J Mol Spectrosc* 176: 124–126.
81. Brown L.R. and Margolis J.S. (1996). Empirical line parameters of NH_3 from 4791 to 5294 cm^{-1}. *J Quant Spectrosc Radiat Trans* 56: 283–294.
82. Fabian M., Ito F. and Yamada K.M.T. (1995). N_2, O_2, and air broadening of NH_3 in ν_2 band measured by FTIR spectroscopy. *J Mol Spectrosc* 173: 591–602.
83. Dhib M., Echargui M.A., Aroui H., Orphal J. and Hartmann J.-M. (2005). Line shift and mixing in the ν_4 and $2\nu_2$ band of NH_3 perturbed by H_2 and Ar. *J Mol Spectrosc* 233: 138–148.
84. Dhib M., Bouanich J.-P. Aroui H. and Broquier M. (2000). Collisional broadening coefficients in the ν_4 band of NH_3 perturbed by He and Ar. *J Mol Spectrosc* 202: 83–88.
85. Aroui H., Nouri S. and Bouanich J.-P. (2003). NH_3 self-broadening coefficients in the ν_2 and ν_4 bands and line intensities in the ν_2 band. *J Mol Spectrosc* 220: 248–258.
86. Dhib M., Echargui M.A., Aroui H. and Orphal J. (2006). Shifting and line mixing parameters in the ν_4 band of NH_3 perturbed by CO_2 and He: Experimental results and theoretical calculations. *J Mol Spectrosc* 238: 168–177.
87. Dhib M., Ibrahim N., Chelin P., Echargui M.A., Aroui H. and Orphal J. (2007). Diode-laser measurements of O_2, N_2 and air-pressure broadening and shifting of NH_3 in the 10 μm spectral region. *J Mol Spectrosc* 242: 83–89.
88. Dhib M., Aroui H. and Orphal J. (2007). Experimental and theoretical study of line shift and mixing in the ν_4 band of NH_3 perturbed by N_2. *J Quant Spectrosc Radiat Trans* 107: 372–384.

89. Aroui H., Chelin P., Echargui M.A., Fellows C.E. and Orphal J. (2008). Line shifts in the ν_2, $2\nu_2$ and ν_4 bands of NH_3 perturbed by O_2. *J Mol Spectrosc* 252: 129–135.
90. Aroui H., Laribi H., Orphal J. and Chelin P. (2009). Self-broadening, self-shift and self-mixing in the ν_2, $2\nu_2$ and ν_4 bands of NH_3. *J Quant Spectrosc Radiat Trans* 110: 2037–2059.
91. Nouri S., Ben Mabrouk K., Chelin P., Aroui H. and Orphal J. (2009). Pressure-induced line mixing in the ν_4 band of NH_3 perturbed by O_2 and self-perturbed. *J Mol Spectrosc* 258: 75–87.
92. Aroui H., Chevalier M., Broquier M., Picard-Bersellini A., Gherissi S. and Legaysommaire N. (1995). Line mixing parameters in the ν_4 rovibrational band of NH_3 perturbed by N_2. *J Mol Spectrosc* 169: 502–510.
93. Aroui H., Picard-Bersellini A., Chevalier M., Broquier M. and Gherissi S. (1996). Pressure-broadening and cross-relaxation rates of rotation-inversion transitions in the ν_4 band of NH_3 perturbed by CO_2. *J Mol Spectrosc* 176: 162–168.
94. Aroui H., Broquier M., Picard-Bersellini A., Bouanich J.-P., Chevalier M. and Gherissi S. (1998). Absorption intensities, pressure-broadening and line mixing parameters of some lines of NH_3 in the ν_4 band. *J Quant Spectrosc Radiat Trans* 60: 1011–1023.
95. Dhib M., Bouanich J.-P., Aroui H. and Picard-Bersellini A. (2001). Analysis of N_2, O_2, CO_2 and air broadening of infrared spectral lines in the ν_4 band of NH_3. *J Quant Spectrosc Radiat Trans* 68: 163–178.
96. Nemtchinov V., Sung K. and Varanasi P. (2004). Measurements of line intensities and half-widths in the 10-μm bands of $^{14}NH_3$. *J Quant Spectrosc Radiat Trans* 83: 243–265.
97. Lerot C., Blanquet G., Bouanich J.-P., Walrand J. and Lepère M. (2006). Hydrogen-broadening coefficients in the ν_5 band of CH_3F at room and low temperatures. *J Mol Spectrosc* 238: 224–233.
98. Bouanich J.-P., Blanquet G., Populaire J.C. and Walrand J. (2001). N_2 broadening for methyl chloride at low temperature by diode-laser spectroscopy. *J Mol Spectrosc* 208: 72–78.
99. Predoi-Cross A., Hambrook K., Brawley-Tremblay M., Bouanich J.-P. and Smith M.A.H. (2006). Measurements and theoretical calculations of N_2-broadening and N_2-shift coefficients in the ν_2 band of CH_3D. *J Mol Spectrosc* 235: 35–53.
100. Leavitt R.P. (1980). Pressure broadening and shifting in microwave and infrared spectra of molecules of arbitrary symmetry: An irreducible tensor approach. *J Chem Phys* 73: 5432–5450.
101. Starikov V.I. (2009). Analytical representation of half-width for molecular ro-vibrational lines in the case of dipole–dipole and dipole–quadrupole interactions. *Mol Phys* 107: 2227–2236.
102. Bouanich J.-P., Walrand J. and Blanquet G. (2005). N_2-broadening coefficients in the ν_2 and ν_4 bands of PH_3. *J Mol Spectrosc* 232: 40–46.
103. Bouanich J.-P. and Blanquet G. (2007). N_2-broadening coefficients in the ν_2 and ν_4 bands of PH_3 at low temperature. *J Mol Spectrosc* 241: 186–191.
104. Salem J., Aroui H., Bouanich J.-P., Walrand J. and Blanquet G. (2004). Collisional broadening and line intensities in the ν_2 and ν_4 bands of PH_3. *J Mol Spectrosc* 225: 174–181.
105. Bouanich J.-P., Salem J., Aroui H., Walrand J. and Blanquet G. (2004). H_2-broadening coefficients in the ν_2 and ν_4 bands of PH_3. *J Quant Spectrosc Radiat Trans* 84: 195–205.
106. Salem J., Bouanich J.-P., Walrand J., Aroui H. and Blanquet G. (2004). Hydrogen line broadening in the ν_2 and ν_4 bands of phosphine at low temperature. *J Mol Spectrosc* 228: 23–30.

107. Levy A., Lacome N. and Tarrago G. (1994). Temperature dependence of collision-broadened lines of phosphine. *J Mol Spectrosc* 166: 20–31.
108. Salem J., Bouanich J.-P., Walrand J., Aroui H. and Blanquet G. (2005). Helium- and argon-broadening coefficients of phosphine lines in the ν_2 and ν_4 bands. *J Mol Spectrosc* 232: 247–254.
109. Blanquet G., Walrand J. and Bouanich J.-P. (1993). Diode-laser measurements of N_2-broadening coefficients in the ν_3 band of $CH_3{}^{35}Cl$. *J Mol Spectrosc* 160: 253–257.
110. Bouanich J.-P., Blanquet G. and Walrand J. (1993). Theoretical O_2- and N_2-broadening coefficients of CH_3Cl spectral lines. *J Mol Spectrosc* 161: 416–426.
111. Lepère M., Blanquet G., Walrand J. and Bouanich J.-P. (1997). Diode-laser measurements of N_2-broadening coefficients in the ν_2 and ν_5 band of $^{12}CH_3F$. *J Mol Spectrosc* 181: 345–351.
112. Lepère M., Blanquet G., Walrand J. and Bouanich J.-P. (1998). Analysis of O_2-broadening of $^{12}CH_3F$ lines in the 6.8 nm spectral region. *J Mol Spectrosc* 192: 17–24.
113. Lance B., Lepère M., Blanquet G., Walrand J. and Bouanich J.-P. (1996). Self-broadening and linestrength in the ν_2 and ν_5 band of $^{12}CH_3F$. *J Mol Spectrosc* 180: 100–109.
114. Rohart F., Ellendt A., Kaghat F. and Mäder H. (1997). Self and polar foreign gas line broadening and frequency shifting of CH_3F: effect of the speed dependence observed by millimeter wave coherent transients. *J Mol Spectrosc* 185: 222–233.
115. Lerot C., Blanquet G., Bouanich J.-P., Walrand J. and Lepère M. (2005). Self-broadening coefficients in the ν_2 and ν_5 bands of $^{12}CH_3F$ at 183 and 298 K. *J Mol Spectrosc* 230: 153–160.
116. Guerin D., Nischan M., Clark D., Dunjko V. and Mantz A.W. (1994). Low-pressure measurements of self and air broadening coefficients in the ν_2 and ν_5 bands of $^{12}CH_3$ F. *J Mol Spectrosc* 166: 130–136.
117. Grigoriev I.M., Bouanich J.-P., Blanquet G., Walrand J. and Lepère M. (1997). Diode-laser measurements of He-broadening coefficients in the ν_6 band of $^{12}CH_3F$. *J Mol Spectrosc* 186: 48–53.
118. Lepère M., Blanquet G., Walrand J. and Bouanich J.-P. (1998). Collisional-broadening coefficients in the ν_6 band of $^{12}CH_3F$ perturbed by Ar. *J Mol Spectrosc* 192: 231–234.
119. Bussery-Honvault B., Moszynski R. and Boissoles J. (2005). Ab initio potential energy surface and pressure broadening coefficients for the He–CH_3F complex. *J Mol Spectrosc* 232: 73–79.
120. Predoi-Cross A., Hambrook K., Brawley-Tremblay S., Bouanich J.-P., Devi V.M. and Smith M.A.H. (2006). Room-temperature broadening and pressure-shift coefficients in the ν_2 band of CH_3D–O_2: Measurements and semi-classical calculations. *J Mol Spectrosc* 236: 75–90.
121. Jacquiez K., Blanquet G., Walrand J. and Bouanich J.-P. (1996). Diode-laser measurements of O_2 broadening coefficients in the ν_3 band of CH_3D. *J Mol Spectrosc* 175: 386–389.
122. Lance B., Ponsar S., Walrand J., Lepère M., Blanquet G. and Bouanich J.-P. (1998). Correlated and noncorrelated line shape models under small lineshift condition: analysis of self-perturbed CH_3D. *J Mol Spectrosc* 189: 124–134.
123. Lerot C., Walrand J., Blanquet G., Bouanich J.-P. and Lepère M. (2003). Diode-laser measurements and calculations of H_2 broadening coefficients in the ν_3 band of CH_3D. *J Mol Spectrosc* 217: 79–86.
124. Lerot C., Walrand J., Blanquet G., Bouanich J.-P. and Lepère M. (2003). H_2-broadening coefficients in the ν_3 band of CH_3D at low temperatures. *J Mol Spectrosc* 219: 329–334.

125. Fejard L., Gabard T. and Champion J.P. (2003). Calculated line broadening coefficients in the ν_2 band of CH_3D perturbed by helium. *J Mol Spectrosc* 219: 88–97.
126. Gabard T. (1998). Calculated helium-broadened line parameters in the ν_4 band of $^{13}CH_4$. *J Quant Spectrosc Radiat Trans* 59: 287–302.
127. Tipping R.H., Brown A., Ma Q., Hartmann J.-M., Boulet C. and Liévin J. (2001). Collision-induced absorption in the ν_2 fundamental band of CH_4. I. Determination of the quadrupole transition moment. *J Chem Phys* 115: 8852–8857.
128. Giulinskii B.I., Perevalov V.I. and Tyuterev V.G. (1987). *The method of irreducible tensor operators in the theory of molecular spectra*. Nauka, Novosibirsk (in Russian).
129. Champion J.-P., Loëte M. and Pierre G. (1992). Spherical top spectra. In: Rao K.N. and Weber A. (eds.). *Spectroscopy of the Earth's atmosphere and interstellar medium*, pp. 339–422, Academic Press, New York.
130. Pine A.S. (1992). Self-broadening, N_2-broadening, O_2-broadening, H_2-broadening, Ar-broadening, and He-broadening in the ν_3 band *Q* branch of CH_4. *J Chem Phys* 97: 773–785.
131. Pine A.S. (1997). N_2 and Ar broadening and line mixing in the P and R branches of the ν_3 band of CH_4. *J Quant Spectrosc Radiat Trans* 57: 157–176.
132. Ballard J. and Johnston W.B. (1986). Self-broadened widths and absolute strengths of $^{12}CH_4$ lines in the 1310–1370 region. *J Quant Spectrosc Radiat Trans* 36: 365–371.
133. Gharavi M. and Buckley S.G. (2005). Diode laser absorption spectroscopy measurement of linestrengths and pressure broadening coefficients of the methane $2\nu_3$ band at elevated temperatures. *J Mol Spectrosc* 229: 78–88.
134. Benner D.C., Devi V.M., Smith M.A.H. and Rinsland C.P. (1993). Air-, N_2-, and O_2-broadening and shift coefficients in the ν_3 spectral region of $^{12}CH_4$. *J Quant Spectrosc Radiat Trans* 50: 65–89.
135. Millot G., Lavorel B. and Steinfeld J.I. (1992). Collisional broadening of rotational lines in the stimulated Raman pentad *Q*-branch of $^{12}CH_4$. *J Quant Spectrosc Radiat Trans* 47: 81–90.
136. Gabard T. (1996). PhD thesis. University of Bourgogne, Dijon, France.
137. Lepère M. (2006). Self-broadening coefficients in the ν_4 band of CH_4 by diode-laser spectroscopy. *J Mol Spectrosc* 238: 193–199.
138. Gabard T. (1997). Argon-broadened line parameters in the ν_3 band of $^{12}CH_4$. *J Quant Spectrosc Radiat Trans* 57: 117–196.
139. Hartmann J.M., Brodbeck C., Flaud P.-M., Tipping P.H., Brown A., Ma Q. and Liévin J. (2002). Collision-induced absorption in the ν_2 fundamental band of CH_4. II. Dependence on the perturber gas. *J Chem Phys* 116: 123–127.
140. Dufour G., Hurtmans D., Henry A., Valentin A. and Lepère M. (2003). Line profile study from diode laser spectroscopy in the $^{12}CH_4$ $2\nu_3$ band perturbed by N_2, O_2, Ar, and He. *J Mol Spectrosc* 221: 80–92.
141. Gabard T., Grigoriev I.M., Grigorovich N.M. and Tonkov M.V. (2004). Helium and argon line broadening in the ν_2 band of CH_4. *J Mol Spectrosc* 225: 123–131.
142. Predoi-Cross A., Brown L.R., Devi V.M., Brawley-Tremblay M. and Benner D.C. (2005). Multispectrum analysis of $^{12}CH_4$ from 4100 to 4635 cm^{-1}: 1. Self-broadening coefficients (widths and shifts). *J Mol Spectrosc* 232: 231–246.
143. Predoi-Cross A., Brawley-Tremblay M., Brown L.R., Devi V.M. and Benner D.C. (2006). Multispectrum analysis of $^{12}CH_4$ from 4100 to 4635 cm^{-1}: II. Air-broadening coefficients (widths and shifts). *J Mol Spectrosc* 236: 201–215.

144. Tejwani G.D.T., Varanasi P. and Fox K. (1975). Collision-broadened linewidths of tetrahedral molecules—II. Computations for CH_4 lines broadened by N_2, O_2, He, Ne, and Ar. *J Quant Spectrosc Radiat Trans* 15: 243–254.
145. Tejwani G.D.T. and Fox K. (1974). Calculated linewidths for $^{12}CH_4$ broadened by N_2 and O_2. *J Chem Phys* 60: 2021–2026.
146. Lyulin O.M., Nikitin A.V., Perevalov V.I., Morino I., Yokota T., Kumazawa R. and Watanabe T. (2009). Measurements of N_2- and O_2-broadening and shifting parameters of the methane spectral lines in the 5550–6236 cm^{-1} region. *J Quant Spectrosc Radiat Trans* 110: 654–668.
147. Smith M.A.H., Benner D.C., Predoi-Cross A. and Devi V.M. (2009). Multispectrum analysis of $^{12}CH_4$ in the ν_4 band: I.: Air-broadened half widths, pressure-induced shifts, temperature dependences and line mixing. *J Quant Spectrosc Radiat Trans* 110: 639–653.
148. Menard-Bourcin F., Menard J. and Boursier C. (2007). Temperature dependence of rotational relaxation of methane in the $2\nu_3$ vibrational state by self- and nitrogen-collisions and comparison with line broadening measurements. *J Mol Spectrosc* 242: 55–63.
149. Mondelain D., Payan S., Deng W., Camy-Peyret C., Hurtmans D. and Mantz A.W. (2007). Measurement of the temperature dependence of line mixing and pressure broadening parameters between 296 and 90 K in the ν_3 band of $^{12}CH_4$ and their influence on atmospheric methane retrievals. *J Mol Spectrosc* 244: 130–137.
150. Antony B.K., Niles D.L., Wroblewski S.B., Humphrey C.M., Gabard T. and Gamache R.R. (2008). N_2-, O_2- and air-broadened half-widths and line shifts for transitions in the ν_3 band of methane in the 2726–3200 cm^{-1} spectral region. *J Mol Spectrosc*, 251: 268–281.
151. Rosenmann L., Hartmann J.M., Perrin M.Y. and Taine J. (1988). Accurate calculated tabulations of IR and Raman CO_2 line broadening by CO_2, H_2O, N_2, O_2 in the 300–2400 K temperature range. *Appl Opt* 27: 3902–3907.
152. Hartmann J.M. and L'Haridon F.L. (1995). Simple modeling of line-mixing effects in IR bands. I. Linear molecules: Application to CO_2. *J Chem Phys* 103: 6467–6478.
153. Thibault F., Boissoles J., Le Doucen R., Menoux V. and Boulet C. (1994). Line mixing effects in the 00^03–00^00 band of CO_2 in helium. I. Experiment. *J Chem Phys* 100: 210–214.
154. Bulanin M.O., Dokuchaev A.B., Tonkov M.V. and Filippov N.N. (1984). Influence of line interference on the vibration-rotation band shapes. *J Quant Spectrosc Radiat Trans* 31: 521–543.
155. Rosenmann L., Perrin M.Y. and Taine J. (1988). Collisional broadening of CO_2 IR lines. I. Diode laser measurements for CO_2–N_2 mixtures in the 295–815 K temperature range. *J Chem Phys* 88: 2995–2998.
156. Rosenmann L., Hartmann J.M., Perrin M.Y. and Taine J. (1988). Collisional broadening of CO_2 IR lines. II. Calculations. *J Chem Phys* 88: 2999–3006.
157. Srivastava R.P. and Zaidi H.R. (1977). Self-broadened widths of rotational Raman and IR lines in CO_2. *Can J Phys* 55: 549–553.
158. Hall R.J. and Stufflebeam J.H. (1984). Quantitative CARS spectroscopy of CO_2 and N_2O. *Appl Opt* 33: 4319–4327.
159. Bykov A.D., Lavrentieva N.N. and Sinitsa L.N. (2000). Calculation of CO_2 broadening and shift coefficients for high-temperature databases. *Atmos Oceanic Opt* 13: 1015–1019.
160. Tashkun S.A., Perevalov V.I., Teffo J.-L., Bykov A.D. and Lavrentieva N.N. (2003). CDSD-1000, the high-temperature carbon dioxide spectroscopic databank. *J Quant Spectrosc Radiat Trans* 82: 165–196.

161. Devi V.M., Benner D.C., Smith M.A.H. and Rinsland C.P. (2003). Nitrogen broadening and shift coefficients in the 4.2–4.5-μm bands of CO_2. *J Quant Spectrosc Radiat Trans* 76: 289–307.
162. Hikida T., Yamada K.M.T., Fukabori M., Aoki T. and Watanabe T. (2005). Intensities and self-broadening coefficients of the CO_2 ro-vibrational transitions measured by a near-IR diode laser spectrometer. *J Mol Spectrosc* 232: 202–212.
163. Hikida T. and Yamada K.M.T. (2006). N_2- and O_2-broadening of CO_2 for the $(3001)_{III} \leftarrow (0000)$ band at 6231 cm^{-1}. *J Mol Spectrosc* 239: 154–159.
164. Toth R.A., Brown L.R., Miller C.E., Devi V.M. and Benner D.C. (2006). Self-broadened widths and shifts of $^{12}C^{16}O_2$: 4750–7000 cm^{-1}. *J Mol Spectrosc* 239: 243–271.
165. Devi V.M., Benner D.C., Brown L.R., Miller C.E. and Toth R.A. (2007). Line mixing and speed dependence in CO_2 at 6248 cm^{-1}: Positions, intensities, and air- and self-broadening derived with constrained multispectrum analysis. *J Mol Spectrosc* 242: 90–117.
166. Li J.S., Liu K., Zhang W.J., Chen W.D. and Gao, X.M. (2008). Pressure-induced line broadening for the $(30012) \leftarrow (00001)$ band of CO_2 measured with tunable diode laser photoacoustic spectrometer. *J Quant Spectrosc Radiat Trans* 109: 1575–1585.
167. Li J.S., Liu K., Zhang W.J., Chen W.D. and Gao, X.M. (2008). Self-, N_2- and O_2-broadening coefficients for the $^{12}C^{16}O_2$ transitions near-IR measured by a diode laser photoacoustic spectrometer. *J Mol Spectrosc* 252: 9–16.
168. Toth R.A., Brown L.R., Miller C.E., Devi V.M. and Benner D.C. (2008). Spectroscopic database of CO_2 line parameters: 4300–7000 cm^{-1}. *J Quant Spectrosc Radiat Trans* 109: 906–921.
169. ftp://ftp.iao.ru/pub/CDSD-1000
170. Yamada K.M.T., Watanabe T. and Fukabori M. (2002). J-dependence of the N_2- and O_2-pressure broadening coefficients observed for NNO $\nu_1 + \nu_3$ band as expressed by rational functions. *J Mol Spectrosc* 216: 170–171.
171. Rohart F., Colmont J.-M., Wlodarczak G. and Bouanich J.-P. (2003). N_2- and O_2-broadening coefficients and profiles for millimeter lines of $^{14}N_2O$. *J Mol Spectrosc* 222: 159–171.
172. Nguyen L., Buldyreva J., Colmont J.-M., Rohart F., Wlodarczak G. and Alekseev E.A. (2006). Detailed profile analysis of milllimeter 502 and 602 Ghz N_2O–$N_2(O_2)$ lines at room temperature for collisional linewidth determination. *Mol Phys* 104: 2701–2710.
173. Schmidt C., Populaire J.C., Walrand J., Blanquet G. and Bouanich J.-P. (1993). Diode-laser measurements of N_2-broadening coefficients in the ν_2 band of HCN at low temperature. *J Mol Spectrosc* 158: 423–432.
174. Rohart F., Nguyen L., Buldyreva J., Colmont J.-M. and Wlodarczak G. (2007). Lineshapes of the 172 and 602 GHz rotational transitions of $HC^{15}N$. *J Mol Spectrosc* 246: 213–227.
175. Yang C., Buldyreva J., Gordon I.E., Rohart F., Cuisset A., Mouret G., Bocquet R. and Hindle F. (2008). Oxygen, nitrogen and air broadening of HCN spectral lines at terahertz frequencies. *J Quant Spectrosc Radiat Trans* 109: 2857–2868.
176. Landrain V., Blanquet G., Lepère M., Walrand J. and Bouanich J.-P. (1997). Diode-laser measurements of H_2-broadening coefficients in the ν_2 band of HCN. *J Mol Spectrosc* 182: 184–188.
177. Devi V.M., Benner D.C., Smith M.A.H., Rinsland C.P., Sharpe S.W. and Sams R.L. (2004). A multispectrum analysis of the $2\nu_2$ spectral region of $H^{12}C^{14}N$: Intensities, broadening and pressure-shift coefficients. *J Quant Spectrosc Radiat Trans* 87: 339–366.

178. Devi V.M., Benner D.C., Smith M.A.H., Rinsland C.P., Predoi-Cross A., Sharpe S.W., Sams R.L., Boulet C. and Bouanich J.-P. (2005). A multispectrum analysis of the ν_2 band of $H^{12}C^{14}N$: Part I. Intensities, broadening, and shift coefficients. *J Mol Spectrosc* 231: 66–84.
179. Valipour H. and Zimmermann D. (2001). Investigation of J-dependence of line shift. Line broadening and line narrowing coefficients in the $\nu_1 + 3\nu_3$ absorption band of acetylene. *J Chem Phys* 114: 3535–3545.
180. Arteaga S.W., Bejger C.M., Gerecke J.L., Hardwick J.L., Martin Z.T., Mayo J., McIlhattan E.A., Moreau J.-M.F., Pilkenton M.J., Polston M.J., Robertson B.T. and Wolf E.N. (2007). Line broadening and shift coefficients of acetylene at 1550 nm. *J Mol Spectrosc* 243: 253–266.
181. Fissiaux L., Dhyne M. and Lepère M. (2009). Diode-laser spectroscopy: Pressure dependence of N_2-broadening coefficients of lines in the $\nu_4 + \nu_5$ band of C_2H_2. *J Mol Spectrosc* 254: 10–15.
182. Dhyne M., Fissiaux L., Populaire J.-C. and Lepère M. (2009). Temperature dependence of the N_2-broadening coefficients of acetylene. *J Quant Spectrosc Radiat Trans* 110: 358–366.
183. Dhyne M., Joubert P., Populaire J.-C. and Lepère M. (2010). Collisional broadening and shift coefficients of lines in the $\nu_4 + \nu_5$ band of C_2H_2 diluted in N_2 from low to room temperatures. *J Quant Spectrosc Radiat Trans* 111: 973–989.
184. Lepère M., Blanquet G., Walrand J., Bouanich J.-P., Herman M. and Vander Auwera J. (2007). Self-broadening coefficients and absolute line intensities in the ν_4+ν_5 band of acetylene. *J Mol Spectrosc* 242: 25–30.
185. Lambot D., Olivier A., Blanquet G., Walrand J. and Bouanich J.-P. (1991). Diode-laser measurements of collisional line broadening in the ν_5 band of C_2H_2. *J Quant Spectrosc Radiat Trans* 45: 145–155.
186. Jacquemart D. (2002). PhD thesis. University Paris VI, Paris, France.
187. Lance B., Blanquet G., Walrand J., Populaire J.-C., Bouanich J.-P. and Robert D. (1999). Inhomogenious lineshape profiles of C_2H_2 perturbed by Xe. *J Mol Spectrosc* 197: 32–45.
188. Bouanich J.-P., Walrand J. and Blanquet G. (2003). Argon-broadening coefficients in the ν_5 band of acetylene at room and low temperatures. *J Mol Spectrosc* 219: 98–104.
189. Buldyreva J., Ivanov S.V. and Nguyen L. (2005). Collisional linebroadening in the atmosphere of light particles: problems and solutions in the framework of semiclassical treatment. *J Raman Spectrosc*, 36: 148–152.
190. Ivanov S.V., Nguyen L. and Buldyreva J. (2005). Comparative analysis of purely classical and semiclassical approaches to collision line broadening of polyatomic molecules: I. C_2H_2–Ar case. *J Mol Spectrosc* 233: 60–67.
191. Nguyen L., Ivanov S., Buzykin O.G. and Buldyreva J. (2006). Comparative analysis of purely classical and semiclassical approaches to collisional line broadening of polyatomic molecules: II. C_2H_2–He case. *J Mol Spectrosc* 239: 101–107.
192. Lambot D., Blanquet G., Walrand J. and Bouanich J.-P. (1991). Diode-laser measurements of H_2-broadening coefficients in the ν_5 band of C_2H_2. *J Mol Spectrosc* 150: 164–172.
193. Bouanich J.-P., Walrand J. and Blanquet G. (2002). Hydrogen-broadening coefficients in the ν_5 band of acetylene at low temperatures. *J Mol Spectrosc* 216: 266–270.
194. Thibault F., Corretja B., Viel A., Bermejo D., Martinez R.Z. and Bussery-Honvault B. (2008). Linewidths of C_2H_2 perturbed by H_2: experiments and calculations from an *ab initio* potential, *Phys Chem* 10: 5419–5428.

195. Bouanich J.-P. and Brodbeck C. (1976). Vibration-rotation matrix elements for diatomic molecules; vibration-rotation interaction functions $F_v^{v'}(m)$ for CO. *J Quant Spectrosc Radiat Trans* 16: 153–163.
196. Dunham I.L. (1932). The energy levels of rotating vibrator. *Phys Rev* 41: 721–731.
197. Makushkin Yu.S. and Tyuterev V.G. (1984). *Perturbation's methods and effective Hamiltonians in molecular spectroscopy*. Nauka, Novosibirsk (in Russian).
198. Herzberg G. (1966). Molecular spectra and molecular structure. I. Spectra of Diatomic Molecules. Van Nostrand, Princeton.
199. Grigoriev I.M., Filippov N.N., Tonkov M.V., Boulet C. and Boissoles J. (2000). Asymptotic behaviour of line shifts in the 0–0 and 0–1 bands of HF in a bath of argon: Influence of vibration–rotation coupling. *J Chem Phys* 113: 2504–2505.
200. Sinclair P.M., Duggan P., Berman R., May A.D. and Drummond J.R. (1997). Line broadening, shifting and mixing in the fundamental band of CO perturbed by N_2 at 301 K. *J Mol Spectrosc* 181: 41–47.
201. Sumpf B., Burrows J.P., Kissel A., Kronfeldt H.D., Kurtz O., Meusel I., Orphal J. and Voigt S. (1998). Line shift investigations for different isotopomers of carbon monoxide. *J Mol Spectrosc* 190: 226–231.
202. Predoi-Cross A., Luo C., Sinclair P.M., Drummond J.R. and May A.D. (1999). Line broadening and temperature exponent of the fundamental band in CO–N_2 mixtures. *J Mol Spectrosc* 198: 291–303.
203. Voigt S., Dreher S., Orphal J. and Burrows J.P. (1996). N_2 broadening in the $^{13}C^{16}O$ 2–0 band around 4167 cm^{-1}. *J Mol Spectrosc* 180: 359–364.
204. Predoi-Cross A., Bouanich J.-P., Benner D.C., May A.D. and Drummond J.R. (2000). Broadening, shifting and line asymmetries in the 2–0 band of CO and CO–N_2: experimental results and theoretical calculations. *J Chem Phys* 113: 158–168.
205. Colmont J.-M., Nguyen L., Rohart F. and Wlodarczak G. (2007). Lineshape analysis of the $J = 3 \leftarrow 2$ and $J = 5 \leftarrow 4$ rotational transitions of room temperature CO broadened by N_2, O_2, CO_2 and noble gases. *J Mol Spectrosc* 246: 86–97.
206. Koshelev M.A. and Markov V.N. (2009). Broadening of the $J = 3 \leftarrow 2$ spectral line of carbon monoxide by pressure of CO, N_2 and O_2. *J Quant Spectrosc Radiat Trans* 110: 526–527.
207. Henningsen J., Simonsen H., Mogelberg T. and Trudso E. (1999). The $0 \rightarrow 3$ overtone band of CO: precise linestrengths and broadening parameters. *J Mol Spectrosc* 193: 354–362.
208. Picque N. and Guelachvili G. (1997). Absolute wavenumbers and self-induced pressure line-shift coefficients for 3–0 vibration-rotation band of $^{12}C^{16}O$. *J Mol Spectrosc* 185: 244–248.
209. Devi M.V., Benner D.C., Smith M.A.H., Rinsland C.P. and Mantz A.W. (2002). Determination of self- and H_2-broadening and shift coefficients in the 2–0 band of $^{12}C^{16}O$ using a multispectrum fitting procedure. *J Quant Spectrosc Radiat Trans* 75: 455–471.
210. Régalia-Jarlot L., Thomas X., Von der Heyden P. and Barbe A. (2005). Pressure-broadened line widths and pressure-induced line shifts coefficients of the (1–0) and (2–0) bands of $^{12}C^{16}O$. *J Quant Spectrosc Radiat Trans* 91: 121–131.
211. Mäder H., Guarnieri A., Doose J., Nissen N., Markov V.N., Shtanyuk A.M., Andrianov A.F., Shapin V.N. and Krupnov A.F. (1996). Comparative studies of $J' \leftarrow J = 1 \leftarrow 0$ CO line parameters in frequency and time domains. *J Mol Spectrosc* 180: 183–187.
212. Sung K. and Varanasi P. (2004). Hydrogen-broadened half-widths and hydrogen-induced line shifts of $^{12}C^{16}O$ relevant to the Jovian atmospheric spectra. *J Quant Spectrosc Radiat Trans* 85: 165–182.

213. Bouanich J.-P. and Predoi-Cross A. (2005). Theoretical calculations for line-broadening and pressure-shifting in the fundamental and first two overtone bands of CO–H_2. *J Mol Struct* 742: 183–190.
214. Sung K. and Varanasi P. (2005). CO_2-broadened half-widths and CO_2-induced line shifts of $^{12}C^{16}O$ relevant to the atmospheric spectra of Venus and Mars. *J Quant Spectrosc Radiat Trans* 91: 319–332.
215. Sinclair P.M., Duggan P., Berman R., Drummond J.R. and May A.D. (1998). Line broadening in the fundamental band of CO in CO–He and CO–Ar mixtures. *J Mol Spectrosc* 191: 258- 264.
216. Thachuk M., Chuaqui C.E. and Le Roy R.J. (1996). Linewidths and shifts of very low-temperature CO in He: a challenge for theory or experiment. *J Chem Phys* 105: 4005–4014.
217. Hartmann J.M., Rosenmann L., Perrin M.Y. and Taine J. (1988). Accurate calculated tabulations of CO line broadening by H_2O, N_2, O_2, and CO_2 in the 200–3000 K temperature range. *Appl Opt* 27: 3063–3065.
218. Houdeau J.P., Boulet C., Bonamy J., Khayar A. and Guelachvili G. (1983). Air broadened NO line widths in a temperature range of atmospheric interest. *J Chem Phys* 79: 1634–1640.
219. Ballard J., Johnston W.B., Kerridge B.J. and Remedios J.J. (1988). Experimental spectral line parameters in the 1–0 band of nitric oxide. *J Mol Spectrosc* 127: 70–82.
220. Chang A.Y., Di Rosa M.D. and Hanson R.K. (1992). Temperature dependence of collisional broadening and shift in the NO $A \leftarrow X$ (0, 0) band in the presence of argon and nitrogen. *J Quant Spectrosc Radiat Trans* 47: 375–390.
221. Di Rosa M.D. and Hanson R.K. (1994). Collision broadening and shift of NO $\gamma(0, 0)$ absorption lines by H_2O, O_2, and NO at 295 K. *J Mol Spectrosc* 164: 97–117.
222. Spencer M.N., Chackerian C., Giver L.P. and Brown L.R. (1997). Temperature dependence of nitrogen broadening of the NO fundamental vibrational band. *J Mol Spectrosc* 181: 307–315.
223. Chackerian C., Freedman R.S., Giver L.P. and Brown L.R. (1998). The NO vibrational fundamental band: O_2-broadening coefficients. *J Mol Spectrosc* 192: 215–219.
224. Allout M.-Y., Dana V., Mandin J.-Y., Von der Heyden P., Décatoire D. and Plateaux J.-J. (1999). Oxygen-broadening coefficients of first overtone nitric oxide lines. *J Quant Spectrosc Radiat Trans* 61: 759–765.
225. Mandin J.-Y., Dana V., Régalia L., Thomas X. and Barbe A. (2000). Nitrogen-broadening in the nitric oxide first overtone band. *J Quant Spectrosc Radiat Trans* 66: 93–100.
226. Colmont J.-M., D'Eu J.F., Rohart F., Wlodarczak G. and Buldyreva J. (2001). N_2- and O_2-broadenings and lineshapes of the 551–53-GHz line of ^{14}NO. *J Mol Spectrosc* 208: 197–208.
227. Pope R.S. and Wolf P.J. (2001). Rare gas pressure broadening of the NO fundamental vibrational band. *J Mol Spectrosc* 208: 153–160.
228. Buldyreva J., Benec'h S. and Chrysos M. (2001). Infrared nitrogen-perturbed NO linewidths in a temperature range of atmospheric interest: an extension of the exact trajectory model. *Phys Rev A* 63: 012708.
229. Buldyreva J., Benec'h S. and Chrysos M. (2001). Oxygen-broadened and air-broadened linewidths for the NO infrared absorption bands by means of the exact trajectory model. *Phys Rev A* 63: 032705.
230. Schiffman A. and Nesbitt D.J. (1994). Pressure broadening and collisional narrowing in OH ($v = 1 \leftarrow 0$) rovibrational transitions with Ar, He, O_2, N_2. *J Chem Phys* 160: 2677–2689.

231. Chance K.V., Jennings D.A., Evenson K.M., Valek M.D., Nolt I.G., Radostitz J.V., Park K. (1991). Pressure broadening of the 118.455 cm^{-1} rotational lines of OH by H_2, He, N_2, and O_2. *J Mol Spectrosc* 146: 375–380.
232. Park K., Zink L.R., Evenson K.M., Chance K.V. and Nolt I.G. (1996). Pressure broadening of the 83.869 cm^{-1} rotational lines of OH by N_2, O_2, H_2, and He. *J Quant Spectrosc Radiat Trans* 55: 285–287.
233. Park K., Zink L.R., Chance K.V., Evenson K.M. and Nolt I.G. (1999). Pressure broadening of the 118.455 cm^{-1} rotational lines of OH by N_2, O_2, H_2, and He. *J Quant Spectrosc Radiat Trans* 61: 715–716.
234. Bastard D., Bretenoux A., Charru A. and Picherit F. (1979). Determination of mean collision cross-sections of free-radical OH with foreign gases. *J Quant Spectrosc Radiat Trans* 21: 369–372.
235. Bastard D., Bretenoux A., Charru A. and Picherit F. (1979). Microwave spectroscopy of free OH radicals in the 37 GHz range = determination of mean collision cross-sections with various gases. *J Phys Lett (Paris)* 40: 533–535.
236. Burrows J.P., Cliff D.I., Davies P.B., Harris G.W., Thrush B.A. and Wilkinson J.P.T. (1979). Pressure broadening of the lowest rotational transition of OH studied by laser magnetic-resonance. *Chem Phys Lett* 65: 197–200.
237. Benec'h S., Buldyreva J. and Chrysos M. (2001). Pressure broadening and temperature dependence of microwave and far infrared rotational lines in OH perturbed by N_2, O_2 and Ar. *J Mol Spectrosc* 210: 8–17.
238. Pine A.S. and Looney J.P. (1987). N_2 and air broadening in the fundamental band of HF and HCl. *J Mol Spectrosc* 122: 41–55.
239. Pine A.S. (1999). Asymmetries and correlations in speed-dependent Dicke-narrowed line shapes in argon-broadened HF. *J Quant Spectrosc Radiat Trans* 62: 397–423.
240. Boulet C., Flaud P.-M. and Hartmann J.-M. (2004). Infrared line collisional parameters of HCl in argon, beyond the impact approximation: measurements and classical path calculations. *J Chem Phys* 120: 11053–11061.
241. Hurtmans D., Henry A., Valentin A. and Boulet C. (2009). Narrowing, broadening and shifting parameters for $R(2)$ and $P(14)$ lines in the HCl fundamental band perturbed by N_2 and rare gases from tunable diode laser spectroscopy. *J Mol Spectrosc* 254: 126–136.
242. Zu L., Hamilton P.A., Chance K.V. and Davies P.B. (2003). Pressure broadening of the lowest rotational transition of $H^{35}Cl$ by N_2 and O_2. *J Mol Spectrosc* 220: 107–112.
243. Drouin B.J. (2004). Temperature-dependent pressure-induced line shape of the HCl $J = 1 \leftarrow 0$ rotational transition in nitrogen and oxygen. *J Quant Spectrosc Radiat Trans* 83: 321–331.
244. Morino I. and Yamada K.M.T. (2005). Absorption profiles of HCl for the $J = 1–0$ rotational transition: Foreign-gas effects measured for N_2, O_2, and Ar. *J Mol Spectrosc* 233: 77–85.
245. Rosenkranz P.W. (1975). Shape of the 5 mm oxygen band in the atmosphere. *IEEE Trans. Antennas Propag* 23: 498–506.
246. Roche C.F., Hutson J.M. and Dickinson A.S. (1995). Calculations of line width and shift cross sections for HCl in Ar. *J Quant Spectrosc Radiat Trans* 53: 153–164.
247. Hartmann J.-M. and Boulet C. (2000). Line shape parameters for HF in a bath of argon as a test of classical path model. *J Chem Phys* 113: 9000–9010.

Appendix A

Matrix elements of operators of physical quantities

Line-width and line-shift calculations require the vibration-rotation matrix elements of various physical quantities like dipole moment or polarizability. To evaluate these matrix elements it is convenient to use the irreducible tensor formalism, which considers the operator of any physical quantity as an l-th rank tensor T^l with the components

$$T_m^l = \sum_{n=-l}^{l} D_{nm}^l A_{ln} \tag{A.1}$$

($-l \leq m \leq l$) in the space-fixed coordinate frame; A_{ln} are its components in the molecular frame, and D_{nm}^l are the rotational matrices defined in Ref. [1]. For ele.ctrostatic interactions the components A_{lk} are usually noted q_{lk}.

For $l = 1$, we have

$$T_0^1 \equiv M_0^1 = \mu D_{00}^1 \tag{A.2}$$

in the case of a linear AB molecule or a symmetric top, and

$$T_0^1 \equiv M_0^1 = \frac{\mu}{\sqrt{2}}(D_{-10}^1 - D_{10}^1) \tag{A.3}$$

in the case of an asymmetric X_2Y molecule with the dipole moment μ along the molecular x-axis ($\mu_x \neq 0$, $\mu_y = \mu_z = 0$ in Eq. (2.5.2) and $M_X = M_Y = 0$, $M_Z \neq 0$ in Eq. (2.5.3)).

For $l = 2$, we have

$$T_m^2 \equiv Q_m^2 = \sum_{n=-2}^{2} D_{nm}^2 q_{2n} \text{ ,} \tag{A.4}$$

where q_{2n} are the spherical components of the quadrupole moment in the molecular frame. In the principle symmetry, axes of the molecule are chosen to be the molecular frame axes, we have $q_{2-1} = q_{21} = 0$, $q_{2-2} = q_{22}$ and

$$Q_0^2 = q_{20} D_{00}^2 + q_{22}(D_{-20}^2 + D_{20}^2)\,. \tag{A.5}$$

For $l = 3$, only one component

$$T_0^3 = q_{3-1}(D_{-10}^1 - D_{10}^1) \tag{A.6}$$

is usually used with $q_{3-1} = \Omega/\sqrt{2}$ (Ω is the octupole moment).

The evaluation of matrix elements for electrostatic interactions therefore needs the matrix elements of the operators M_0^1 and Q_0^2. For the operator T_0^3 the matrix elements are calculated in the same way as for the operator M_0^1 but μ is replaced by Ω. Generally, the matrix elements of the components T_m^l are defined by Eq. (2.5.25) and expressed in terms of the reduced matrix elements of Eq. (2.5.30). As a result, the interruption function $S(b)$ depends on the matrix elements

$$A_l(J\tau; J'\tau') = \sum_k A_{lk} \langle J\tau \| D_{k\circ}^l \| J'\tau' \rangle \tag{A.7}$$

(the index τ distinguishes the $2J+1$ states with the total angular moment $\vec{J}$).

The matrix element of Eq. (A.7) is given by Eq. (2.6.11) for linear molecules:

$$A_l^{(n)}(J; J') = A_{l0}^{(n)}(2J'+1)^{1/2} C_{J'0l0}^{J0} \tag{A.8}$$

(the parameters $A_{l0}^{(n)}$ are defined in Sec. 2.6) and has the form

$$A_l^{(n)}(JK; J'K') = A_{l0}^{(n)}(2J'+1)^{1/2} C_{J'K'l0}^{JK} \delta_{KK'} \tag{A.9}$$

for symmetric tops.

In general, the wave functions of the asymmetric top $|J\tau\rangle$ can be expressed as a linear combination of the rotational wave functions of symmetric top $|JK\rangle$, but it is more convenient to write them in the symmetrized symmetric-top basis $|J\kappa\gamma\rangle$:

$$|J\tau\rangle = \sum_{\kappa=0}^{J} C_\kappa^{(J\tau)} |J\kappa\gamma\rangle\,, \tag{A.10}$$

where $\gamma = \pm 1$ determines the symmetry of the rotational wave functions and the coefficients $C_\kappa^{(J\tau)}$ are obtained by diagonalization of the rotational Hamiltonian. These symmetrized rotational functions are related to the usual ones via

$$\begin{aligned} |J\kappa\gamma\rangle &= (|JK\rangle + \gamma|J-K\rangle)/2^{1/2} \quad (K \neq 0), \\ |J\kappa=0\rangle &= |JK=0\rangle. \end{aligned} \tag{A.11}$$

These functions can be divided into four different rotational sub-bases: E^+ (even integer κ, $\gamma = 1$), E^- (even integer $\kappa \geq 2$, $\gamma = -1$), O^+ (odd integer κ, $\gamma = 1$) and O^-

(odd integer κ, $\gamma = -1$) [2, 3]. The products of the rotational wave functions (A.11) with the vibrational wave functions Ψ_V of the symmetry type Γ_V produce the vibrotational functions $|VJ\kappa\gamma\rangle$ of the overall symmetry Γ. For a given J-value, these symmetries are given in the fourth column of Table A1 [4]. The last four columns of this table show the dimension of the rotational basis n_R, the parity of the rotational (pseudo-)quantum numbers K_a, K_c and the dimension n_H of the Hamiltonian sub-matrix (n_V means the total number of the interacting vibrational states and n_B is the number of these states with symmetry B_1).

Table A1. Symmetry types for the symmetrized vibrotational basis of X_2Y molecules [4]. Reproduced with permission.

J	Rotational sub-basis	Γ_v	Γ	Dimension of the rotational basis, n_R	K_a	K_c	Dimension of the Hamilnonian sub-matrix, n_H
even	O^-	B_1	A_1	$J/2$	o	e	$n_V(J/2+1)-n_B$
	E^+	A_1		$J/2+1$	e	e	
	E^-	B_1	A_2	$J/2$	e	o	$n_V(J/2)$
	O^+	A_1		$J/2$	o	o	
	E^+	B_1	B_1	$J/2+1$	e	e	$n_V(J/2)+n_B$
	O^-	A_1		$J/2$	o	e	
	O^+	B_1	B_2	$J/2$	o	o	$n_V(J/2)$
	E^-	A_1		$J/2$	e	o	
odd	O^+	B_1	A_1	$(J+1)/2$	o	e	$n_V(J-1)/2+n_B$
	E^-	A_1		$(J-1)/2$	e	e	
	E^+	B_1	A_2	$(J+1)/2$	e	o	$n_V(J+1)/2$
	O^-	A_1		$(J+1)/2$	o	o	
	E^-	B_1	B_1	$(J-1)/2$	e	e	$n_V(J+1)/2-n_B$
	O^+	A_1		$(J+1)/2$	o	e	
	O^-	B_1	B_2	$(J+1)/2$	o	o	$n_V(J+1)/2$
	E^+	A_1		$(J+1)/2$	e	o	

In the basis of the wave functions given by Eq. (A.11), the required matrix elements read:

$$\begin{aligned} q_1(J'\tau';J\tau) = {} & \frac{\mu(2J+1)^{1/2}}{2^{1/2}}\{\sum_{\kappa\geq 1} C_{\kappa}^{(J\tau)} C_{\kappa+1}^{(J'\tau')*} C_{J1\kappa 1}^{J'\kappa+1} - \sum_{\kappa>1} C_{\kappa}^{(J\tau)} C_{\kappa-1}^{(J'\tau')*} C_{J1\kappa-1}^{J'\kappa-1}\} \\ & + \mu(2J+1)^{1/2}\{\frac{1}{2} C_0^{(J\tau)} C_1^{(J'\tau')*}[1-\gamma'(-1)^{J+J'+1}] C_{J101}^{J'1} \\ & - \frac{1}{2} C_1^{(J\tau)} C_0^{(J'\tau')*} < [1-\gamma(-1)^{J+J'+1}] C_{J11-1}^{J'0}\} \end{aligned} \tag{A.12}$$

$$
\begin{aligned}
q_2(J'\tau';J\tau) &= q_{20}(2J+1)^{1/2}\sum_{\kappa} C_{\kappa}^{(J\tau)} C_{\kappa}^{(J'\tau')*} C_{J2\kappa 0}^{J'\kappa} + q_{22}(2J+1)^{1/2} \\
&\times\{\sum_{\kappa\ge 2} C_{\kappa}^{(J\tau)} C_{\kappa+2}^{(J'\tau')*} C_{J2\kappa 2}^{J'\kappa+2} + \sum_{\kappa>2} C_{\kappa}^{(J\tau)} C_{\kappa-2}^{(J'\tau')*} C_{J2\kappa-2}^{J'\kappa-2} \\
&+\frac{1}{2^{3/2}} C_1^{(J\tau)} C_1^{(J'\tau')*} [\gamma C_{J2-12}^{J'1} + \gamma' C_{J21-2}^{J'-1}] \\
&\times[C_{K=2}^{(J\tau)} C_{K'=0}^{(J'\tau')^*}[1+\gamma(-1)^{J+J'}] C_{J22-2}^{J'0} \\
&+ C_{K=0}^{(J\tau)} C_{K'=2}^{(J'\tau')^*}[1+\gamma'(-1)^{J+J'}] C_{J202}^{J'2}]\}.
\end{aligned}
\tag{A.13}
$$

The matrix elements $q_3(J'\tau'; J\tau)$ are obtained from Eq. (A.12) replacing μ by Ω.

The matrix elements for the operators corresponding to induction and dispersion interactions are determined by the matrix elements $C_{l_1 l_2}^{(k_1 k_1';k_2 k_2')}(1'2';12)$ of Eq. (2.6.9). Some matrix elements for the dipole, quadrupole and octupole moment operators of the H_2O molecule are given in Tables A2–A4.

Table A2. Matrix elements $q_1(i'; i)$ for the ground state of the H_2O molecule ($\mu = -1.854$ D).

$J'K_a'K_c'$	JK_aK_c	$q_1(i'; i)$	$J'K_a'K_c'$	JK_aK_c	$q_1(i'; i)$
0 0 0	1 1 1	1.854	2 0 2	1 1 1	1.612
1 0 1	1 1 0	–2.27	2 0 2	2 1 1	–2.669
1 0 1	2 1 2	2.271	2 0 2	3 1 3	2.722
1 1 1	0 0 0	–1.854	2 0 2	3 3 1	–0.214
1 1 1	2 0 2	–1.612	2 1 2	1 0 1	–2.271
1 1 1	2 2 0	2.070	2 1 2	2 2 1	–1.694
1 1 0	1 0 1	–2.271	2 1 2	3 0 3	–2.447
1 1 0	2 2 1	2.273	2 1 2	3 2 1	1.783

Table A3. Matrix elements $q_2(i'; i)$ for the ground state of the H_2O molecule ($q_{xx} = -0.13$ Å D, $q_{yy} = -2.5$ Å D, $q_{zz} = 2.63$ Å D).

$J'K_a'K_c'$	JK_aK_c	$q_2(i'; i)$	$J'K_a'K_c'$	JK_aK_c	$q_2(i'; i)$
0 0 0	2 0 2	2.346	1 1 1	3 1 2	2.122
0 0 0	2 2 0	2.214	1 1 1	3 3 1	2.445
1 0 1	1 0 1	–2.881	1 1 0	1 1 0	2.738
1 0 1	2 2 1	–1.935	1 1 0	2 1 2	2.253
1 0 1	3 0 3	2.867	1 1 0	3 1 2	3.251
1 0 1	3 2 1	2.778	1 1 0	3 3 0	2.593
1 1 1	1 1 1	0.142	2 0 2	0 0 0	2.346
1 1 1	2 1 1	4.188	2 0 2	2 0 2	–3.471

Table A4. Matrix elements $q_3(i'; i)$ for the ground state of the H_2O molecule (Ω = 2.0 Å^2 D).

$J'K_a'K_c'$	JK_aK_c	$q_3(i'; i)$	$J'K_a'K_c'$	JK_aK_c	$q_3(i'; i)$
1 0 1	2 1 2	–2.138	2 1 2	1 0 1	2.138
1 1 1	2 0 2	–1.907	2 1 2	2 2 1	–1.464
1 1 1	2 2 0	0.273	2 1 2	3 0 3	–0.343
1 1 0	2 2 1	0.534	2 1 2	3 2 1	–1.746
2 0 2	1 1 1	1.907	2 1 1	2 0 2	1.876
2 0 2	2 1 1	1.876	2 1 1	2 2 0	–1.216
2 0 2	3 1 3	–1.422	2 1 1	3 2 2	–1.581
2 0 2	3 3 1	–0.279	2 2 1	1 1 0	–0.534

Bibliography

1. Landau L.D. and Lifshitz, E.M. (1977). *Quantum Mechanics (Non-relativistic Theory).* Pergamon Press, Oxford.
2. Camy-Peyret C. and Flaud J.-M. (1975). PhD thesis. University Pierre and Marie Curie, Paris (in French).
3. Camy-Peyret C. and Flaud J.-M. (1974). The interacting states (020), (100), and (001) of H_2 ^{16}O. *J Mol Spectrosc* 51: 142–150.
4. Kwan Y.Y. (1976). The interacting states of an asymmetric top molecule XY_2 of the group C_{2v}. Application to five interacting states: (101), (021), (120), (200) and (011) of H_2 ^{16}O. *J Mol Spectrosc* 71: 260–280.

Appendix B

Parameters of intermolecular interaction potentials

For practical calculations of collisional line widths and shifts, the interaction potential energy of two colliding molecules 1 and 2 is often approximated by the sum of atom–atom and electrostatic contributions. The atom–atom potential is defined as a sum of pairwise Lennard–Jones isotropic interactions between the atoms of these molecules [1]:

$$V^{(aa)} = \sum_{i,j} V_{ij}^{(aa)} = \sum_{i,j} \left(\frac{d_{ij}}{r_{1i,2j}^{12}} - \frac{e_{ij}}{r_{1i,2j}^{6}} \right), \tag{B.1}$$

where d_{ij} and e_{ij} are the atomic pair energy parameters and $r_{1i,2j}$ is the distance between the i-th atom of the molecule 1 and the j-th atom of the molecule 2. Since for each atomic pair the Lennard–Jones interaction can be also written as

$$V_{ij}^{(aa)} = 4\varepsilon_{ij}[(\sigma_{ij} / r_{1i,2j})^{12} - (\sigma_{ij} / r_{1i,2j})^{6}], \tag{B.2}$$

instead of d_{ij} and e_{ij} the parameters ε_{ij} and σ_{ij} are often used. For the interaction between different kinds of atoms these parameters ε_{ij} and σ_{ij} are calculated from the parameters ε_{ii}, ε_{jj} and σ_{ii}, σ_{jj} for the identical atoms, according to the conventional combination rules [2]

$$\varepsilon_{ij} = \sqrt{\varepsilon_{ii}\varepsilon_{jj}}, \quad \sigma_{ij} = (\sigma_{ii} + \sigma_{jj})/2. \tag{B.3}$$

The isotropic part of the atom–atom intermolecular potential (B.1)

$$V_{iso}(r) = 4\varepsilon_{12}[(\sigma_{12} / r)^{12} - (\sigma_{12} / r)^{6}] \tag{B.4}$$

governs the relative molecular trajectory and therefore determines the resonance functions. The parameters ε_{12} and σ_{12} for the molecular pair are also estimated by

$$\varepsilon_{12} = \sqrt{\varepsilon_1 \varepsilon_2}, \quad \sigma_{12} = (\sigma_1 + \sigma_2)/2, \tag{B.5}$$

where $\varepsilon_{1(2)}$, $\sigma_{1(2)}$ are the Lennard–Jones parameters for two identical molecules 1 (molecules 2).

The anisotropic part $V_{aniso} = V^{(aa)} - V_{iso}$ defines the interruption functions $S(b)$. To calculate V_{aniso} as a function of the intermolecular distance r, the two-centre expansion [3] should be used for the interatomic distances $r_{1i,2j}$:

$$r_{1i,2j}^{-n} = \sum_{l_1 l_2 l} f_{l_1 l_2 l}^{n}(r_{1i}, r_{2j}, r) \sum_{m_1 m_2 m} C_{l_1 m_1 l_2 m_2}^{lm} Y_{l_1 m_1}(\Omega_{1i}) Y_{l_2 m_2}(\Omega_{2j}) Y_{lm}(\Omega)^* , \quad \text{(B.6)}$$

$$f_{l_1 l_2 l}^{n}(r_{1i}, r_{2j}, r) \equiv \frac{(-1)^{l_2}}{r^n} \left[\frac{(4\pi)^3 (2l_1+1)(2l_2+1)}{2l+1} \right]^{1/2} C_{l_1 0 l_2 0}^{l0}$$

$$\times \sum_{p,q} \left(\frac{r_{1i}}{r} \right)^p \left(\frac{r_{2j}}{r} \right)^q \frac{(n+p+q-l-3)!!(n+p+q+l-2)!!}{(n-2)!(p-l_1)!!(p+l_1+1)!!(q-l_2)!!(q+l_2+1)!!} \quad \text{(B.7)}$$
$$\times \{1 + \delta_{n1}(\delta_{pl_1}\delta_{ql_2}\delta_{p+q,l} - 1)\},$$

which needs the distances r_{1i}, r_{2j} of each atom from the molecular centre of mass. In Eq. (B.6), Ω_{1i}, Ω_{2j} and Ω represent the orientations of the $\vec{r}_{1i}$, $\vec{r}_{2j}$ and $\vec{r}$ vectors in the laboratory-fixed frame. Since for polyatomic molecules the atoms i, j may be situated out of the molecular axes described by Ω_1 and Ω_2, the spherical harmonics $Y_{l_1 m_1}(\Omega_{1i})$ and $Y_{l_2 m_2}(\Omega_{2j})$ should be expressed in terms of Ω_1 and Ω_2. For example, we have [4]:

$$Y_{l_1 m_1}(\Omega_{1i}) = \sum_{k_1} D_{m_1 k_1}^{l_1}(\Omega_1)^* Y_{l_1 k_1}(\Omega'_{1i}), \quad \text{(B.8)}$$

where Ω'_{1i} is the orientation of the i-th atom in the molecular frame (for the perturbing molecule the index 1 should be replaced by the index 2).

The atom–atom potential parameters for some molecular pairs are gathered in Table B1. The N_2 and O_2 molecular parameters needed for the calculation of long-range contributions are given in Table B2. Table B3 contains the parameters for rare gases. For the isotropic potential energy of methane interacting with various gases, the Lennard–Jones parameters are collected in Table B4.

Table B1. Lennard–Jones parameters for atom–atom and intermolecular interactions*.

System	d_{ij} (10^{-7} erg Å^{12}) or ε_{ij} (K)	e_{ij} (10^{-10} erg Å^6) or σ_{ij} (Å)	r_{1i}/r_{2j} (Å) β (°)	ε_{12} (K) σ_{12} (Å)
H_2O–H_2O [5]	$d_{OO} = 0.348$ $d_{HH} = 3.26\cdot10^{-3}$ $d_{OH} = 0.0$	$e_{OO} = 0.0$ $e_{HH} = 0.0$ $e_{OH} = 0.075$	$r_{1H} = 0.9197$ $r_{1O} = 0.0656$ 110°58′	$\varepsilon_{12} = 67.4$ $\sigma_{12} = 3.22$
H_2O–N_2 [6]	$d_{ON} = 0.360$ $d_{HN} = 0.035$	$e_{ON} = 0.0$ $e_{HN} = 0.0$	$r_{2N} = 0.55$	$\varepsilon_{12} = 96.8$ $\sigma_{12} = 3.28$
H_2O–N_2 [7]	$\varepsilon_{HN} = 20.46$ $\varepsilon_{NO} = 43.90$	$\sigma_{HN} = 2.990$ $\sigma_{NO} = 3.148$		$\varepsilon_{12} = 109.44$ $\sigma_{12} = 3.51$

H_2O–O_2 [6]	$d_{OO} = 0.215$ $d_{HO} = 0.021$	$e_{OO} = 0.0$ $e_{HO} = 0.144$	$r_{2O} = 0.6037$	$\varepsilon_{12} = 120.0$ $\sigma_{12} = 3.10$
H_2O–Ar [6]	$d_{OAr} = 0.806$ $d_{HAr} = 0.078$	$e_{OAr} = 0.0$ $e_{HAr} = 0.0$		$\varepsilon_{12} = 89.0$ $\sigma_{12} = 3.33$
NO–N_2 [8]	$d_{NN} = 0.2910$ $d_{NO} = 0.1425$	$e_{NN} = 0.2502$ $e_{NO} = 0.1911$	$r_{1N} = 0.614$ $r_{2N} = 0.548$ $r_{2O} = 0.603$	$\varepsilon_{12} = 118.8$ $\sigma_{12} = 3.58$ $\varepsilon_{12} = 104.4$ $\sigma_{12} = 3.58$
OH–N_2 [9]	$d_{HN} = 0.1139$ $d_{HN} = 0.0959$ $d_{ON} = 0.1627$ $d_{ON} = 0.1843$	$e_{HN} = 0.1195$ $e_{HN} = 0.1077$ $e_{ON} = 0.3183$ $e_{ON} = 0.2148$	$r_{1H} = 0.9135$ $r_{1O} = 0.0571$ $r_{2N} = 0.5488$	
OH–O_2 [9]	$d_{HO} = 0.0680$ $d_{OO} = 0.2478$	$e_{HO} = 0.0980$ $e_{OO} = 0.2605$	$r_{2O} = 0.6030$	
OH–Ar [9]	$d_{HAr} = 0.1763$ $d_{OAr} = 0.3390$	$e_{HAr} = 0.1945$ $e_{OAr} = 0.3877$		
H_2O–CO [10]	$\varepsilon_{HC} = 32.18$ $\varepsilon_{HO} = 24.19$ $\varepsilon_{OC} = 40.43$ $\varepsilon_{OO} = 51.73$	$\sigma_{HC} = 2.810$ $\sigma_{HO} = 2.850$ $\sigma_{OC} = 3.285$ $\sigma_{OO} = 3.01$		$\varepsilon_{12} = 164.35$ $\sigma_{12} = 3.60$
N_2O–N_2 [11]	$d_{NN} = 0.476$	$e_{NN} = 0.316$		$\varepsilon_{12} = 141.9$ $\sigma_{12} = 3.782$
N_2O–O_2 [11]	$d_{NO} = 0.255$	$e_{NO} = 0.250$		$\varepsilon_{12} = 157.7$ $\sigma_{12} = 3.656$
N_2O–He [11]	$d_{NHe} = 0.0440$ $d_{OHe} = 0.0222$	$e_{NHe} = 0.0594$ $e_{OHe} = 0.0467$		$\varepsilon_{12} = 47.4$ $\sigma_{12} = 3.227$
N_2O–Ar [11]	$d_{NAr} = 0.879$ $d_{OAr} = 0.468$	$e_{NAr} = 0.575$ $e_{OAr} = 0.452$		$\varepsilon_{12} = 165.2$ $\sigma_{12} = 3.648$
N_2O–Xe [11]	$d_{NXe} = 3.738$ $d_{OXe} = 2.065$	$e_{NXe} = 1.362$ $e_{OXe} = 1.071$		$\varepsilon_{12} = 224.5$ $\sigma_{12} = 3.967$
H_2O–Ar	$d_{OAr} = 0.5008$ $d_{HAr} = 0.1621$	$e_{OAr} = 0.466$ $e_{HAr} = 0.183$		$\varepsilon_{12} = 107.98$ $\sigma_{12} = 3.496$
H_2O–He	$d_{OHe} = 0.0265$ $d_{HHe} = 0.0008$	$e_{OHe} = 0.0058$ $e_{HHe} = 0.0022$		$\varepsilon_{12} = 31.54$ $\sigma_{12} = 3.072$
H_2O–Ne	$d_{ONe} = 0.074$ $d_{HNe} = 0.023$	$e_{ONe} = 0.13$ $e_{HNe} = 0.005$		$\varepsilon_{12} = 58.86$ $\sigma_{12} = 3.168$
H_2O–Kr	$d_{OKr} = 0.858$ $d_{HKr} = 0.281$	$e_{OKr} = 0.66$ $e_{HKr} = 0.26$		$\varepsilon_{12} = 129.01$ $\sigma_{12} = 3.594$
H_2O–Xe	$d_{OXe} = 2.34$ $d_{HXe} = 0.78$	$e_{OXe} = 1.17$ $e_{HXe} = 0.46$		$\varepsilon_{12} = 146.66$ $\sigma_{12} = 3.844$
CH_4–Ar [12]	$d_{CAr} = 1.0866$ $d_{HAr} = 0.075$	$e_{CAr} = 0.1924$ $e_{HAr} = 0.22$		
HCN–HCN [13]	$\varepsilon_{HH} = 11.25$ $\varepsilon_{NN} = 37.21$ $\varepsilon_{CC} = 52.44$	$\sigma_{HH} = 2.869$ $\sigma_{NN} = 3.292$ $\sigma_{CC} = 3.728$		$\varepsilon_{12} = 81.31$ $\sigma_{12} = 3.918$
CO–CO [14]	$\varepsilon_{CC} = 143$ $\varepsilon_{OO} = 198$ $\varepsilon_{CO} = 168$ $d_{CC} = 0.82$ $d_{OO} = 0.25$ $d_{OC} = 0.46$	$\sigma_{CC} = 3.56$ $\sigma_{OO} = 3.14$ $\sigma_{CO} = 3.35$ $e_{CC} = 0.40$ $e_{OO} = 0.26$ $e_{OC} = 0.33$	$r_{1C} = 0.64$ $r_{1O} = 0.48$	

CO–CO [15]	$\varepsilon_{OO} = 51.8037$ $\varepsilon_{CC} = 31.5550$	$\sigma_{OO} = 3.0058$ $\sigma_{CC} = 3.56379$	$r_{1C} = 0.6461$ $r_{1O} = 0.4847$	
CO_2–CO_2 [15]	$\varepsilon_{OO} = 51.8037$ $\varepsilon_{CC} = 92.0400$	$\sigma_{OO} = 3.0058$ $\sigma_{CC} = 2.93738$	$r_{1C} = 1.162$	
N_2–N_2 [15] O_2–O_2 [15] H_2–H_2 [15]	$\varepsilon_{NN} = 37.2072$ $\varepsilon_{OO} = 51.8037$ $\varepsilon_{HH} = 11.254$	$\sigma_{NN} = 3.29155$ $\sigma_{OO} = 3.00578$ $\sigma_{HH} = 2.68259$	$r_{1N} = 0.550$ $r_{1O} = 0.605$ $r_{1H} = 0.3754$	
O_3–N_2 [16]	$d_{ON} = 0.392$	$e_{ON} = 0.319$	$r_{OO} = 1.278$	$\varepsilon_{12} = 150$ $\sigma_{12} = 3.93$
O_3–O_2 [16]	$d_{OO} = 0.293$	$e_{OO} = 0.276$	156.8°	$\varepsilon_{12} = 174$ $\sigma_{12} = 3.77$

*For the H_2O molecule r_{1H}, r_{1O} denote the distances of the H and O atoms from the centre of mass G; β is the angle $\angle$HGH.

Table B2. Molecular parameters for N_2, O_2 and H_2 ($\mu = 0$).

Molecule	ν (cm^{-1})	B (cm^{-1})	u (eV)	Q (DÅ)	α (Å^3)	γ	m (a.m.u.)	Ω (DÅ^2)	Φ (DÅ^3)	σ_{11} (Å)	ε_{11} (K)
N_2	2359	1.9982	15.581	−1.52	1.74	0.176	28.006	0.0	6.0	3.816	92.91
O_2	1580	1.4376	12.071	−0.39	1.57	0.239	31.990	0.0	8.0	3.45	128
H_2	4401	60.68	15.426	−0.65	0.806	0.11	2.016	0.65	0.126	2.944	32.0

Table B3. Parameters of rare gas atoms [2, 17].

Parameter	He	Ne	Ar	Kr	Xe
α (Å^3)	0.205	0.397	1.642	2.48	4.01
u (eV)	24.588	21.565	15.76	14.00	12.130
m (a.m.u.)	4.0	20.18	39.9	83.8	131.30
σ_{ii} (Å)	2.59	2.78	3.504	3.60	4.100
ε_{ii} (K)	6.03	34.9	117.7	171	221

Table B4. Lennard–Jones parameters for isotropic interaction of CH_4 with various perturbers [18].

Parameter	CH_4–He	CH_4–Ar	CH_4–Kr	CH_4–H_2	CH_4–N_2	CH_4–CO_2
σ_{12} (Å)	6.017	6.835	7.005	6.398	7.080	7.222
ε_{12} (K)	38.24	131.3	157.3	72.47	114.4	181.6

For a linear or symmetric-top active molecule colliding with an atom, the model interaction potential of Smith–Giraud–Cooper [19] can also be used:

$$V(r,\theta) = 4\varepsilon\left\{\left[\left(\frac{\sigma}{r}\right)^{12} - \left(\frac{\sigma}{r}\right)^{6}\right] + \left[R_1\left(\frac{\sigma}{r}\right)^{12} - A_1\left(\frac{\sigma}{r}\right)^{7}\right]P_1(\cos\theta) + \left[R_2\left(\frac{\sigma}{r}\right)^{12} - A_2\left(\frac{\sigma}{r}\right)^{6}\right]P_2(\cos\theta)\right\}, \quad \text{(B.9)}$$

where r is the distance between the centre of mass of the active molecule and the perturbing atom, θ is the angle between the molecular symmetry axis and the intermolecular axis, and P_1 and P_2 are the Legendre polynomials. The parameters of this potential for some molecular systems are given in Table B5.

Table B5. Parameters of the Smith–Giraud–Cooper potential [19] for some systems*.

	A_1	R_1	A_2	R_2
HCl-Ar [19]	0.33	0.37	0.14	0.65
NH_3 -He [20]	0.18	0.32	0.20	0.20
NH_3 -Ar [20][a]	0.20	0.50	0.20	0.40
NH_3 -Ar [21][b]	0.20	0.40	0.20	0.30
CH_3 F–He [22]	0.030	0.95	0.006	0.022

* For the HCl–Ar pair ε_{12} = 202 K, σ_{12} = 3.37 Å, for all other systems ε_{12} and σ_{12} are calculated by the combination rules; *a* and *b* – for ν_4 and $2\nu_2$ bands, respectively.

For the CO–N_2 system (Sec. 4.5), the model potential

$$V(r,\theta) = 4\varepsilon\left[\left(\frac{\sigma}{r}\right)^{12} - \left(\frac{\sigma}{r}\right)^{6}\right] + 24\varepsilon\frac{d}{\sigma}\left[2\left(\frac{\sigma}{r}\right)^{13} - \left(\frac{\sigma}{r}\right)^{7}\right]P_1(\cos\theta) + 4\varepsilon\gamma\left[\left(\frac{\sigma}{r}\right)^{12} - \left(\frac{\sigma}{r}\right)^{6}\right]P_2(\cos\theta) \tag{B.10}$$

can be used, where d is the distance between the centre of mass and the centre of charges for the active molecule and γ is its anisotropy.

Bibliography

1. MacRury T.B., Steele W.A. and Berne B.J. (1976). Intermolecular potential models for anisotropic molecules, with applications to N_2, CO_2, and benzene. *J Chem Phys* 64: 1288–1299.
2. Hirschfelder J.O., Curtiss C.F. and Bird R.B. (1964). *Molecular Theory of Gases and Liquids.* Wiley, New York.
3. Yasuda H. and Yamamoto T. (1971). On the two-center expansion of an arbitrary function. *Prog Theor Phys* 45: 1458–1465.
4. Gray C.G. and Gubbins K.E. (1984). *Theory of molecular fluids, Volume 1: Fundamentals.* Clarendon press, Oxford.
5. Labani B., Bonamy J., Robert D. and Hartmann J.-M. (1987). Collisional broadening of rotation-vibration lines for asymmetric-top molecules. III. Self-broadening case; application to H_2O. *J Chem Phys* 87: 2781–2769.
6. Labani B., Bonamy J., Robert D., Hartmann J.-M. and Taine J. (1986). Collisional broadening of rotation-vibration lines for asymmetric top molecules. I. Theoretical model for both distant and close collisions. *J Chem Phys* 84: 4256–4267.

7. Lynch R., Gamache R.R. and Neshyba S.P. (1996). Fully complex implementation of the Robert–Bonamy formalism: half-widths and line shifts of H_2O broadened by N_2. *J Chem Phys* 105: 5711–5721.
8. Buldyreva J., Benec'h S. and Chrysos M. (2001). Infrared nitrogen-perturbed NO line widths in a temperature range of atmospheric interest: An extension of the exact trajectory model. *Phys Rev A* 63: 012708.
9. Benec'h S., Buldyreva J. and Chrysos M. (2001). Pressure broadening and temperature dependence of microwave and infrared rotational lines in OH perturbed by N_2, O_2, and Ar. *J Mol Spectrosc* 210: 8–17.
10. Gamache R.R., Lynch R., Plateaux J.J. and Barbe A. (1997). Half-widths and line shifts of water vapour broadening by CO_2: measurements and complex Robert–Bonamy formalism calculations. *J Quant Spectrosc Radiat Trans* 57: 485–496.
11. Margottin-Maclou M., Rachet F., Henry A. and Valentin A. (1996). Pressure-induced line shifts in the ν_3 band of nitrous oxide perturbed by N_2, O_2, He, Ar and Xe. *J Quant Spectrosc Radiat Trans* 56: 1–16.
12. Gabard T. (1997). Argon-broadened line parameters in the ν_3 band of $^{12}CH_4$. *J Quant Spectrosc Radiat Trans* 57: 177–196.
13. Schmidt C., Populaire J.-C., Walrand J., Blanquet G. and Bouanich J.-P. (1993). Diode-laser measurements of N_2-broadening coefficients in the ν_2 band of HCN at low temperature. *J Mol Spectrosc* 158: 423–432.
14. Robert D. and Bonamy J. (1979). Short range effects in semiclassical molecular line broadening calculations. *J Phys (Paris)* 40: 923–943.
15. Bouanich J.-P. (1992). Site-site Lennard–Jones potential parameters for N_2, O_2, H_2, CO and CO_2. *J Quant Spectrosc Radiat Trans* 47: 243–250.
16. Bouazza S., Barbe A., Plateaux J.J., Rosenmann L., Hartmann J.-M., Camy-Peyret C., Flaud J.M. and Gamache R.R. (1993). Measurements and calculations of room-temperature ozone line-broadening by N_2 and O_2 in the $\nu_1+\nu_3$ band. *J Mol Spectrosc* 157: 271–289.
17. Smith M.A.H., Rinsland C.P., Fridovich B. and Rao K.N. (1985). Intensities and Collision-Broadening Parameters from Infrared Spectra. *Mol Spectrosc: Mod Res* 3: 111–228, Academic Press, New York.
18. Hartmann J.-M., Brodbeck C., Flaud P.M., Tipping R.H., Brown A., Ma Q. and Liévin J. (2002). Collision-induced absorption in the ν_2 fundamental band of CH_4. II. Dependence on the perturber gas. *J Chem Phys* 116: 123–127.
19. Smith E.W., Giraud M. and Cooper J. (1976). A semiclassical theory for spectral line broadening in molecules. *J Chem Phys* 65: 1256–1267.
20. Dhib M., Bouanich J.-P., Aroui H. and Broquier M. (2000). Collisional broadening coefficients in the ν_4 band of NH_3 perturbed by He and Ar. *J Mol Spectrosc* 202: 83–88.
21. Dhib M., Echargui M.A., Aroui H., Orphal J. and Hartmann J.-M. (2005). Line shift and mixing in the ν_4 and $2\nu_2$ band of NH_3 perturbed by H_2 and Ar. *J Mol Spectrosc* 233: 138–148.
22. Grigoriev I.M., Bouanich J.-P., Blanquet G., Walrand J. and Lepère M. (1997). Diode-laser measurements of He-broadening coefficients in the ν_6 band of $^{12}CH_3F$. *J Mol Spectrosc* 186: 48–53.

Appendix C

Relations used in calculation of resonance functions

The second-order contributions ${}^{l_1l_2}S_2$ are defined by the resonance functions containing real and imaginary parts. The real parts are given by

$$M^{\,l}_{pp'} = {}^{(nn')}C^{\,l}_{pp'}\,{}^{l_1l_2}f_p^{\,p'} = \sum_{m=-l}^{l} I^{\,p*}_{lm} I^{\,p'}_{lm}\,, \tag{C.1}$$

where ${}^{l_1l_2}f_p^{\,p'}(k)$ are the resonance functions normalised to unity at $k = 0$ and ${}^{(nn')}C^{\,l}_{pp'}$ are the coefficients appearing in the ${}^{l_1l_2}S_2$ contributions (for electrostatic interactions $l = l_1 + l_2$). The integrals $I^{\,p}_{lm}$ read

$$I^{\,p}_{lm} = \int_{-\infty}^{\infty} dt \frac{e^{i\omega t} C_{lm}(\hat{r})}{r(t)^p} \tag{C.2}$$

and need the spherical harmonics

$$C_{lm}(\hat{r}) \equiv C_{lm}(\theta,\varphi) = \sqrt{4\pi/(2l+1)}\,Y_{lm}(\theta,\varphi) \tag{C.3}$$

with the angles θ, φ describing the orientation of the intermolecular vector $\vec{r}$ in the laboratory frame. These spherical harmonics are given in Table C1 for $l \le 4$.

Table C1. Spherical harmonics $C_{lm}(\theta, \varphi)$ for $l \le 4$.

$C_{00} = 1$	$C_{20} = 2^{-1}(3\cos^2\theta - 1)$
$C_{10} = \cos\theta$	$C_{2\pm1} = \mp (3/2)^{1/2} \sin\theta\cos\theta\, e^{\pm i\varphi}$
$C_{1\pm1} = \mp (2)^{-1/2} \sin\theta\, e^{\pm i\varphi}$	$C_{2\pm2} = 2^{-1} (3/2)^{1/2} \sin^2\theta\, e^{\pm 2i\varphi}$
$C_{30} = 2^{-1}(5\cos^2\theta - 3)\cos\theta$	$C_{40} = (5/8)(7\cos^4\theta - 6\cos^2\theta + 3/5)$
$C_{3\pm1} = \mp (3^{1/2}/4)(5\cos^2\theta - 1)\sin\theta\, e^{\pm i\varphi}$	$C_{4\pm1} = \mp 2^{-1}\sin\theta(7\cos^2\theta - 3)\cos\theta\, e^{\pm i\varphi}$
$C_{3\pm2} = 2^{-1}(15/2)^{1/2} \sin^2\theta\cos\theta\, e^{\pm 2i\varphi}$	$C_{4\pm2} = 8^{-1}\, 10^{1/2} \sin^2\theta(7\cos^2\theta - 1)e^{\pm 2i\varphi}$
$C_{3\pm3} = \mp 4^{-1} 5^{1/2} \sin^3\theta\, e^{\pm 3i\varphi}$	$C_{4\pm3} = \mp 4^{-1} (35)^{1/2} \sin^3\theta\cos\theta\, e^{\pm 3i\varphi}$
	$C_{4\pm4} = 16^{-1}\, 70^{1/2} \sin^4\theta\, e^{\pm 4i\varphi}$

The resonance functions are independent of the choice of collision plane and the definition of the collision angle Ψ. This plane and this angle can therefore be taken as the most convenient for calculation. In any case, the integrals of Eq. (C.2) can be written as

$$I_{lm}^{p} = \sum_{s} a(l,m,s) \int_{-\infty}^{\infty} dt \frac{e^{i\omega t} (\cos\Psi)^m \sin(\Psi)^s}{[r(t)]^p}, \qquad \text{(C.4)}$$

where the coefficients $a(l,m,s)$ should be taken from Table C1 according to the definition of θ and φ. For example, for $\theta = \pi/2 - \Psi$, $\varphi = \pi/2$ (the collision plane is the ZOY plane) we have $a(2,2,0) = a(2,-2,0) = -2^{-1}(3/2)^{1/2}$).

Taking $\cos\Psi = b/r(t)$, $\sin\Psi = vt/r(t)$, $r(t) = (b^2+v^2t^2)^{1/2}$ for the model of straight-line trajectory and introducing the dimensionless variables $x = vt/b$, $k = \omega b/v$, we can reduce the integrals of Eq. (C.4) to the form completely defined by the integrals

$$I_n = \int_{-\infty}^{\infty} \frac{e^{ikx} dx}{(1+x^2)^n}, \qquad \text{(C.5)}$$

which can be expressed in terms of Whittaker functions $W_{\lambda,\mu}(z) = W_{\lambda,-\mu}(z)$:

$$I_n = \int_{-\infty}^{\infty} \frac{e^{ikx} dx}{(1+x^2)^n} = \frac{2\pi k^{n-1}}{2^n \Gamma(n)} W_{0,\frac{1}{2}-n}(2k). \qquad \text{(C.6)}$$

For $\lambda = 0$ these functions reduce to the Bessel functions of the second kind $K_\mu(z)$:

$$W_{0,\mu}(z) = (z/\pi)^{1/2} K_\mu(z/2), \qquad \text{(C.7)}$$

which obey the relations

$$zK_{\mu-1}(z) - zK_{\mu+1}(z) = -2\mu K_\mu(z), \qquad \text{(C.8)}$$

$$K_{\mu+1}(z) = K_{\mu-1}(z) + (2\mu/z)K_\mu(z), \qquad \text{(C.9)}$$

$$K_{1/2}(z) = K_{-1/2}(z) = \sqrt{\pi/(2z)}e^{-z}. \qquad \text{(C.10)}$$

In terms of the Bessel functions, Eq. (C.6) reads

$$I_n = \frac{2\pi k^{n-1}}{2^n \Gamma(n)} \sqrt{\frac{2k}{\pi}} K_{n-\frac{1}{2}}(k) \qquad \text{(C.11)}$$

and for integer n can be written as

$$I_n = \frac{\pi J_n e^{-k}}{2^{n-1}(n-1)!}, \qquad \text{(C.12)}$$

where the polynomials J_n satisfy the recursion relation

$$J_n = k^2 J_{n-2} + (2n-3)J_{n-1}. \qquad \text{(C.13)}$$

The integrals I_n and the polynomials J_n (needed for the resonance functions of straight-line and parabolic trajectory models) are gathered in Table C2.

Table C2. Integrals I_n and polynomials J_n.

$I_1 = \pi J_1 e^{-k}$	$J_1 = 1$
$I_2 = \pi(2^1)^{-1} J_2 e^{-k}$	$J_2 = 1 + k$
$I_3 = \pi(2!2^2)^{-1} J_3 e^{-k}$	$J_3 = k^2 + 3k + 3$
$I_4 = \pi(3!2^3)^{-1} J_4 e^{-k}$	$J_4 = k^3 + 6k^2 + 15k + 15$
$I_5 = \pi(4!2^4)^{-1} J_5 e^{-k}$	$J_5 = k^4 + 10k^3 + 45k^2 + 105k + 105$
$I_6 = \pi(5!2^5)^{-1} J_6 e^{-k}$	$J_6 = k^5 + 15k^4 + 105k^3 + 420k^2 + 945k + 945$
$I_7 = \pi(6!2^6)^{-1} J_7 e^{-k}$	$J_7 = k^6 + 21k^5 + 210k^4 + 1260k^3 + 4725k^2 + 10395k + 10395$
$I_8 = \pi(7!2^7)^{-1} J_8 e^{-k}$	$J_8 = k^7 + 28k^6 + 378k^5 + 3150k^4 + 17325k^3 + 62370k^2 + 135135k + 135135$
$I_9 = \pi(8!2^8)^{-1} J_9 e^{-k}$	$J_9 = k^8 + 36k^7 + 630k^6 + 6930k^5 + 51975k^4 + 270270k^3 + 945945k^2 + 2027025k + 2027025$

The resonance functions for different contributions (proportional to r^{-p}) to the long-range interaction potential of Eq. (2.5.6) are given in Table C3; $(p - l)$ is odd for electrostatic terms and even for induction and dispersion terms.

Table C3. Resonance functions for different contributions to the long-range interaction potential.

l_1	l_2	l	p	${}^{l_1 l_2}f_p^p(k)$
1	1	2	3	${}^{11}f_3^3(k) \equiv f_1(k)$
1	2	3	4	${}^{12}f_4^4(k) \equiv f_2(k)$
1	3	4	5	${}^{13}f_5^5(k) \equiv f_3(k)$
2	1	3	4	${}^{21}f_4^4(k) \equiv f_2(k)$
2	2	4	5	${}^{22}f_5^5(k) \equiv f_3(k)$
0	2	2	6	${}^{02}f_6^6(k) \equiv g_1(k)$
2	2	4	6	${}^{22}f_6^6(k) \equiv g_2(k)$
0	1	1	7	${}^{01}f_7^7(k) \equiv g_3(k)$
2	1	1,3	7	${}^{12}f_7^7(k) \equiv g_5(k)$

Bibliography

1. Gradshteyn I.S. and Ryzhik I.M. (2007). *Table of integrals, series and products.* Elsevier Academic Press.

Appendix D

Second-order contributions from atom–atom potential in the parabolic trajectory model

The second-order contributions to the interruption function can be written as a series over the anisotropy ranks l_1, l_2 of two colliding molecules:

$$S_2 = \sum_{l_1,l_2} ({}^{l_1 l_2}S_{2,i}^{outer} + {}^{l_1 l_2}S_{2,f}^{outer} + {}^{l_1 l_2}S_2^{middle}), \tag{D.1}$$

where

$${}^{l_1 l_2}S_{2,i}^{outer} + {}^{l_1 l_2}S_{2,f}^{outer} = \sum_{p,p'} [{}^{l_1 l_2}S_{p,p'}^{outer}(i) + {}^{l_1 l_2}S_{p,p'}^{outer}(f)], \tag{D.2}$$

$${}^{l_1 l_2}S_{p,p'}^{middle} = \sum_{p,p'} {}^{l_1 l_2}S_{p,p'}^{middle}(i,f). \tag{D.3}$$

The general S_2 expression for the case of two linear colliders was obtained by Robert and Bonamy [1]. The case of an asymmetric X_2Y molecule perturbed by a diatom AB was considered by Labani *et al.* [2]; some corrections to their results were made by Neshyba and Gamache [3]. Labani *et al.* [4] also studied the case of two asymmetric molecules of X_2Y type.

For perturbation by an atom ($l_2 = 0$), only the terms with $l_1 = 1, 2$ are usually retained in the atom–atom interaction potential. For $l_1 = 1, l_2 = 0$, the *middle*-term vanishes (${}^{10}S_2^{middle} = 0$), and only

$${}^{10}S_{p,p'}^{outer}(i) = \frac{{}^{10}a_p^{p'} A_p B_{p'}}{r_c^{p+p'-2} (\hbar v_c')^2 (2J_i + 1)} \sum_{i'} C_{10}(i';i) {}^{10}f_p^{p'}(k) \tag{D.4}$$

(and the analogous term for the final f-state) contribute to the interruption function. If in Eq. (D.4) $p,p' = 7$ or 9, then A_p, $B_{p'}$ correspond to ${}^0E_p^{10}$, ${}^0E_{p'}^{10}$, and if $p,p' = 13$ or 15, then A_p, $B_{p'}$ give ${}^0\Delta_p^{10}$, ${}^0\Delta_{p'}^{10}$ in notations of Refs [2, 3]. The functions $C_{10}(i',i)$ are determined by the reduced matrix elements of the rotational matrices

$$C_{10}(i',i)=\frac{1}{2}\left(\langle i'\|D^{1}_{-1\circ}\|i\rangle-\langle i'\|D^{1}_{1\circ}\|i\rangle\right)^{2}, \tag{D.5}$$

and the coefficients ${}^{10}a_p^{p'}$ read [2]

$$\begin{aligned}
{}^{10}a_7{}^{7} &= 75\pi^2/128,\\
{}^{10}a_7{}^{9} &= 875\pi^2/128,\\
{}^{10}a_7{}^{13} &= -3465\pi^2/2048,\\
{}^{10}a_7{}^{15} &= -165165\pi^2/4096,\\
{}^{10}a_9{}^{9} &= 30625\pi^2/1536,\\
{}^{10}a_9{}^{15} &= -1926925\pi^2/8192,\\
{}^{10}a_9{}^{13} &= -40425\pi^2/4096,\\
{}^{10}a_{13}{}^{13} &= 160083\pi^2/131072,\\
{}^{10}a_{13}{}^{15} &= 7630623\pi^2/131072,\\
{}^{10}a_{15}{}^{15} &= 363726363\pi^2/524288.
\end{aligned} \tag{D.6}$$

For $l_1 = 2$, $l_2 = 0$, we have

$$\begin{aligned}
{}^{20}S^{outer}_{p,p'}(i)+{}^{20}S^{outer}_{p,p'}(f) &= \frac{{}^{20}a_p^{p'}}{(\hbar v_c')^2 r_c^{p'+p-2}(2J_i+1)}\Bigg[\sum_{i'}C_p(i',i)C_{p'}(i',i)\,{}^{20}f_p^{p'}\\
&\quad+\sum_{f'}C_p(f',f)C_{p'}(f',f)\,{}^{20}f_p^{p'}\Bigg],
\end{aligned} \tag{D.7}$$

$$\begin{aligned}
{}^{20}S^{middle}_{p,p'}(i,f) &= \frac{{}^{20}a_p^{p'}}{(\hbar v_c')^2 r_c^{p'+p-2}(2J_i+1)}\,{}^{20}f_p^{p'}(0)\times\\
&\quad\times[C_p(i,i)C_{p'}(f,f)+C_p(f,f)C_{p'}(i,i)]W,
\end{aligned} \tag{D.8}$$

where $W=(-1)^{J_1+J_f}(2J_i+1)^{1/2}W(J_iJ_fJ_iJ_f;12)$ and $W(\ldots)$ is the Racah coefficient [5]. The functions C_p are defined as

$$\begin{aligned}
C_p(i',i) &= -\frac{1}{2}({}^0A_p-2\,{}^2A_p)\langle i'\|D^2_{0\circ}\|i\rangle+\frac{1}{2\sqrt{6}}(3\,{}^0A_p+2\,{}^2A_p)\\
&\quad\times\left[\langle i'\|D^2_{2\circ}\|\;i\rangle+\langle i'\|D^2_{-2\circ}\|i\rangle\right]
\end{aligned} \tag{D.9}$$

with ${}^{|m_1|}A_p = {}^{|m_1|}E_p{}^{20}$ for $p = 8$, 10 and ${}^{|m_1|}A_p = {}^{|m_1|}\Delta_p{}^{20}$ for $p = 14$, 16, respectively. The coefficients ${}^{20}a_p^{p'}$ are given by [2]

$$\begin{aligned}
{}^{20}a_8{}^{8} &= 1935\pi^2/1152,\\
{}^{20}a_8{}^{10} &= 175\pi^2/16,\\
{}^{20}a_8{}^{14} &= -9471\pi^2/1024,
\end{aligned}$$

$$
\begin{aligned}
{}^{20}a_8{}^{16} &= -448305\pi^2/4096,\\
{}^{20}a_{10}{}^{10} &= 9125\pi^2/512,\\
{}^{20}a_{10}{}^{16} &= -731445\pi^2/2048,\\
{}^{20}a_{10}{}^{14} &= -123585\pi^2/4096,\\
{}^{20}a_{14}{}^{14} &= 8377677\pi^2/655360,\\
{}^{20}a_{14}{}^{16} &= 99198099\pi^2/327680,\\
{}^{20}a_{16}{}^{16} &= 4698750771\pi^2/2621440.
\end{aligned}
\tag{D.10}
$$

For the H_2O molecule interacting with a rare gas atom A the quantities ${}^{|m_1|}\mathrm{E}_p{}^{10}$, ${}^{|m_1|}\Delta_p{}^{10}$, ${}^{|m_1|}\mathrm{E}_p{}^{20}$, ${}^{|m_1|}\Delta_p{}^{20}$ are determined as [2, 3]

$$
\begin{aligned}
{}^{0}\mathrm{E}_7{}^{10} &= 2e_{\mathrm{HA}}r_{1\mathrm{H}}\cos(\beta) - e_{\mathrm{OA}}r_{1\mathrm{O}},\\
{}^{0}\Delta_{13}{}^{10} &= 2d_{\mathrm{HA}}\, r_{1\mathrm{H}}\cos(\beta) - d_{\mathrm{OA}}r_{1\mathrm{O}},\\
{}^{0}\mathrm{E}_9{}^{10} &= (2/5)e_{\mathrm{HA}}\,(r_{1\mathrm{H}})^3\cos(\beta) - (1/5)e_{\mathrm{OA}}(r_{1\mathrm{O}})^3,\\
{}^{0}\Delta_{15}{}^{10} &= (2/5)d_{\mathrm{HA}}\,(r_{1\mathrm{H}})^3\cos(\beta) - (1/5)d_{\mathrm{OA}}(r_{1\mathrm{O}})^3,\\
{}^{0}\mathrm{E}_8{}^{20} &= e_{\mathrm{HA}}\,(r_{1\mathrm{H}})^2\,[3\cos^2(\beta) - 1] + e_{\mathrm{OA}}(r_{1\mathrm{O}})^2,\\
{}^{0}\Delta_{14}{}^{20} &= d_{\mathrm{HA}}\,(r_{1\mathrm{H}})^2\,[3\cos^2(\beta) - 1] + d_{\mathrm{OA}}(r_{1\mathrm{O}})^2,\\
{}^{0}\mathrm{E}_{10}{}^{20} &= (1/7)e_{\mathrm{HA}}\,(r_{1\mathrm{H}})^4\,[3\cos^2(\beta) - 1] + (1/7)e_{\mathrm{OA}}(r_{1\mathrm{O}})^4,\\
{}^{0}\Delta_{16}{}^{20} &= (1/7)d_{\mathrm{HA}}(r_{1\mathrm{H}})^4\,[3\cos^2(\beta) - 1] + (1/7)d_{\mathrm{OA}}(r_{1\mathrm{O}})^4,\\
{}^{2}\mathrm{E}_8{}^{20} &= -(3/2)e_{\mathrm{HA}}\,(r_{1\mathrm{H}})^2\sin^2(\beta),\\
{}^{2}\Delta_{14}{}^{20} &= -(3/2)d_{\mathrm{HA}}\,(r_{1\mathrm{H}})^2\sin^2(\beta),\\
{}^{2}\mathrm{E}_{10}{}^{20} &= -(3/14)e_{\mathrm{HA}}\,(r_{1\mathrm{H}})^4\sin^2(\beta),\\
{}^{2}\Delta_{16}{}^{20} &= -(3/14)d_{\mathrm{HA}}\,(r_{1\mathrm{H}})^4\sin^2(\beta).
\end{aligned}
\tag{D.11}
$$

In Eq. (D.10), like Appendix B, e_{HA}, d_{HA} and e_{OA}, d_{OA} are the parameters of the Lennard–Jones potential:

$$
V_{ij}(r_{1i,2j}) = \frac{d_{ij}}{r_{1i,2j}^{12}} - \frac{e_{ij}}{r_{1i,2j}^{6}}
\tag{D.12}
$$

for the pairs i,j = H, A and O, A, whereas $r_{1\mathrm{H}} = 0.9197$ Å, $r_{1\mathrm{O}} = 0.0656$ Å are the distances of the atoms H and O to the centre of mass G of the H_2O molecule; $\beta = 111.5°$ is the angle ∠HGH.

The quantities ${}^{|m_1|}\mathrm{E}_p{}^{l_1l_2}$, ${}^{|m_1|}\Delta_p{}^{l_1l_2}$ and other necessary coefficients for the interactions X_2Y-AB can be found, for example, in Ref. [3]. The corresponding formulae for the X_2Y-X_2Y case are given in Ref. [4].

Bibliography

1. Robert D. and Bonamy J. (1979). Short range force effects in semiclassical molecular line broadening calculations. *J Phys (Paris)* 40: 923–943.
2. Labani B., Bonamy J., Robert D., Hartmann J.-M. and Taine J. (1986). Collisional broadening of rotation-vibration lines for asymmetric top molecules. I. Theoretical model for both distant and close collisions. *J Chem Phys* 84: 4256–4267.
3. Neshyba S.P. and Gamache R.R. (1993). Improved line-broadening coefficients for asymmetric rotor molecules with application to ozone lines broadened by nitrogen. *J Quant Spectrosc Radiat Trans* 50: 443–453.
4. Labani B., Bonamy J., Robert D. and Hartmann J.-M. (1987). Collisional broadening of rotation-vibration lines for asymmetric top molecules. III. Self-broadening case; application to H_2O. *J Chem Phys* 87: 2781–2769.
5. Varshalovich D.A., Moskalev A.N. and Khersonskii V.K (1988). *Quantum theory of angular momentum.* World Scientific, New Jersey.

Appendix E

Resonance functions in the parabolic trajectory model

The resonance functions ${}^{l_1 l_2} f_p^{p'}$ are defined by Eqs (C.1) and (C.2). For the parabolic trajectories, the relations required for the calculation of integrals (C.2) are

$$\sin\Psi = \varphi_1 z r_c / r, \tag{E.1}$$

$$\cos\Psi = (1+\varphi_2 z^2) r_c / r, \tag{E.2}$$

$$r = (r_c^2 + \mathrm{v}_c' t)^{1/2}, \tag{E.3}$$

where the notations $\varphi_1 = \mathrm{v}_c / \mathrm{v}_c'$, $\varphi_2 = (1-\varphi_1^2)/2$, $z = \mathrm{v}_c' t / r_c$ are introduced; r_c and v_c are the distance of the closest approach and the relative velocity of the colliding molecular pair at this distance, whereas v_c' is the apparent velocity [1]. The resonance parameter k is defined as $k = \omega r_c / \mathrm{v}_c'$.

For the isotropic intermolecular potential in the Lennard–Jones form

$$V_{iso}(r) = 4\varepsilon_{12}[(\sigma_{12}/r)^{12} - (\sigma_{12}/r)^6], \tag{E.4}$$

its reduced value $V_{iso}^{\circ}(r) \equiv 2V_{iso}(r)/(m^{\circ}\mathrm{v}^2)$ reads

$$V_{iso}^{\circ}(r) = \lambda[\beta^{12}(r_c/r)^{12} - \beta^6 (r_c/r)^6], \tag{E.5}$$

where $\lambda \equiv 8\varepsilon_{12}/(m^{\circ}\mathrm{v}^2)$ and $\beta \equiv \sigma_{12}/r_c$. For this particular isotropic Lennard–Jones potential we have $\varphi_1 = \{[1-\lambda(\beta^{12}-\beta^6)]/[1+\lambda(5\beta^{12}-2\beta^6)]\}^{1/2}$. (The straight-line trajectory model is obtained with $\lambda = 0$, so that $\varphi_1 = 1$ and $\varphi_2 = 0$.)

If the collision plane is the XOZ plane ($\varphi = 0$) and $\theta = \pi/2 - \Psi$, then for $l = 1$ the integrals of Eq. (C.4) read

$$I_{10}^p = \frac{1}{\mathrm{v}_c' r_c^{p-1}} \frac{ik}{(p-1)} \varphi_1 I_{\frac{p-1}{2}}, \tag{E.6}$$

$$I_{11}^p = -\frac{1}{\sqrt{2}} \frac{1}{\mathrm{v}_c' r_c^{p-1}} [I_{\frac{p+1}{2}} + \varphi_2 (I_{\frac{p-1}{2}} - I_{\frac{p+1}{2}})], \tag{E.7}$$

and the resonance functions ${}^{10}f_p^{p'}$ are obtained from Eq. (C.1) as

$$
{}^{10}f_p^{p'} = N(I_{10}^{p*}I_{10}^{p'} + 2I_{11}^{p*}I_{11}^{p'}), \tag{E.8}
$$

where N is the normalisation coefficient. For $p, p' = 7, 9, 13, 15$ the integrals are given in Table E1 (the coefficient $\pi/(v_c' r_c^{p-1})$ is omitted for brevity).

Table E1. Trajectory integrals for $l = 1$.

$I_{10}^{7} = i\varphi_1 k(6\cdot 2^4)^{-1}J_3 e^{-k}$	$I_{11}^{7} = (2^{1/2}\cdot 3!\cdot 2^4)^{-1}\{J_4 + \varphi_2[6J_3 - J_4]\}e^{-k}$
$I_{10}^{9} = i\varphi_1 k(8\cdot 3!\cdot 2^4)^{-1}J_4 e^{-k}$	$I_{11}^{9} = (2^{1/2}\cdot 4!\cdot 2^5)^{-1}\{J_5 + \varphi_2[8J_4 - J_5]\}e^{-k}$
$I_{10}^{13} = i\varphi_1 k(12\cdot 5!\cdot 2^6)^{-1}J_6 e^{-k}$	$I_{11}^{13} = (2^{1/2}\cdot 6!\cdot 2^7)^{-1}\{J_7 + \varphi_2[12J_6 - J_7]\}e^{-k}$
$I_{10}^{15} = i\varphi_1 k(14\cdot 6!\cdot 2^7)^{-1}J_7 e^{-k}$	$I_{11}^{15} = (2^{1/2}\cdot 7!\cdot 2^8)^{-1}\{J_8 + \varphi_2[14J_7 - J_8]\}e^{-k}$

The corresponding resonance functions read

$$
\begin{aligned}
{}^{10}f_7^{7} &= e^{-2k}\{(k\varphi_1 J_3)^2 + [J_4 + \varphi_2(6J_3 - J_4)]^2\}/[J_4(0)]^2,\\
{}^{10}f_9^{9} &= e^{-2k}\{(k\varphi_1 J_4)^2 + [J_5 + \varphi_2(8J_4 - J_5)]^2\}/[J_5(0)]^2,\\
{}^{10}f_{13}^{13} &= e^{-2k}\{(k\varphi_1 J_6)^2 + [J_7 + \varphi_2(12J_6 - J_7)]^2\}/[J_7(0)]^2,\\
{}^{10}f_{15}^{15} &= e^{-2k}\{(k\varphi_1 J_7)^2 + [J_8 + \varphi_2(14J_7 - J_8)]^2\}/[J_8(0)]^2,\\
{}^{10}f_7^{9} &= e^{-2k}\{(k\varphi_1)^2 J_3 J_4 + [J_4 + \varphi_2(6J_3 - J_4)][J_5 + \varphi_2(8J_4 - J_5)]\}/[J_4(0)\,J_5(0)],\\
{}^{10}f_7^{13} &= e^{-2k}\{(k\varphi_1)^2 J_3 J_6 + [J_4 + \varphi_2(6J_3 - J_4)][J_7 + \varphi_2(12J_6 - J_7)]\}/[J_4(0)J_7(0)],\\
{}^{10}f_7^{15} &= e^{-2k}\{(k\varphi_1)^2 J_3 J_7 + [J_4 + \varphi_2(6J_3 - J_4)][J_8 + \varphi_2(14J_7 - J_8)]\}/[J_4(0)J_8(0)],\\
{}^{10}f_9^{13} &= e^{-2k}\{(k\varphi_1)^2 J_4 J_6 + [J_5 + \varphi_2(8J_4 - J_5)][J_7 + \varphi_2(12J_6 - J_7)]\}/[J_5(0)J_7(0)],\\
{}^{10}f_9^{15} &= e^{-2k}\{(k\varphi_1)^2 J_4 J_7 + [J_5 + \varphi_2(8J_4 - J_5)][J_8 + \varphi_2(14J_7 - J_8)]\}/[J_5(0)J_8(0)],\\
{}^{10}f_{13}^{15} &= e^{-2k}\{(k\varphi_1)^2 J_6 J_7 + [J_7 + \varphi_2(12J_6 - J_7)][J_8 + \varphi_2(14J_7 - J_8)]\}/[J_7(0)J_8(0)],
\end{aligned} \tag{E.9}
$$

where the constants $J_n(0) \equiv J_n(k = 0)$ are calculated from Table C2.

For $l = 2$ and $\varphi = 0$, $\theta = \pi/2 - \Psi$ the integrals of Eq. (C4) can be written as

$$
\begin{aligned}
I_{20}^{p} &= \frac{1}{v_c' r_c^{p-1}}(\varphi_4 I_{\frac{p}{2}} - \frac{3}{2}\varphi_1^2 I_{\frac{p+2}{2}}),\\
I_{21}^{p} &= \sqrt{\frac{3}{2}}\frac{1}{v_c' r_c^{p-1}} ik\varphi_1(\frac{\varphi_3}{p} I_{\frac{p}{2}} + \frac{\varphi_2}{p-2} I_{\frac{p-2}{2}}),\\
I_{22}^{p} &= \frac{1}{2}\sqrt{\frac{3}{2}}\frac{1}{v_c' r_c^{p-1}}(\varphi_1^2 I_{\frac{p+2}{2}} + 2\varphi_2 I_{\frac{p+2}{2}}),
\end{aligned} \tag{E.10}
$$

where $\varphi_3 = (1 + \varphi_1^2)/2$ and $\varphi_4 = (1 - 3\varphi_2)$ (I_n are given in Table C2).

The resonance functions ${}^{20}f_p^{p'}$ defined as

$$ {}^{20}f_p^{p'} = N(I_{20}^{p*}I_{20}^{p'} + 2I_{21}^{p*}I_{21}^{p'} + 2I_{22}^{p*}I_{22}^{p'})/n_p^{p'} \tag{E.11}$$

take the form

$$\begin{aligned}
{}^{20}f_8^{8} &= e^{-2k}[(J_{20}^{8})^2 + 12(k\varphi_1 J_{21}^{8})^2 + 3(J_{22}^{8})^2]/(n_8^{8})^2,\\
{}^{20}f_{10}^{10} &= e^{-2k}[(J_{20}^{10})^2 + 12(k\varphi_1 J_{21}^{10})^2 + 3(J_{22}^{10})^2]/(n_{10}^{10})^2,\\
{}^{20}f_{14}^{14} &= e^{-2k}[(J_{20}^{14})^2 + 12(k\varphi_1 J_{21}^{14})^2 + 3(J_{22}^{14})^2]/(n_{14}^{14})^2,\\
{}^{20}f_{16}^{16} &= e^{-2k}[(J_{20}^{16})^2 + 12(k\varphi_1 J_{21}^{16})^2 + 3(J_{22}^{16})^2]/(n_{16}^{16})^2,\\
{}^{20}f_8^{10} &= e^{-2k}[J_{20}^{8}J_{20}^{10} + 12(k\varphi_1)^2 J_{21}^{8}J_{21}^{10} + 3J_{22}^{8}J_{22}^{10}]/n_8^{10},\\
{}^{20}f_8^{14} &= e^{-2k}[J_{20}^{8}J_{20}^{14} + 12(k\varphi_1)^2 J_{21}^{8}J_{21}^{14} + 3J_{22}^{8}J_{22}^{14}]/n_8^{14},\\
{}^{20}f_{10}^{14} &= e^{-2k}[J_{20}^{10}J_{20}^{14} + 12(k\varphi_1)^2 J_{21}^{10}J_{21}^{14} + 3J_{22}^{10}J_{22}^{14}]/n_{10}^{14},\\
{}^{20}f_{10}^{16} &= e^{-2k}[J_{20}^{10}J_{20}^{16} + 12(k\varphi_1)^2 J_{21}^{10}J_{21}^{16} + 3J_{22}^{10}J_{22}^{16}]/n_{10}^{16},\\
{}^{20}f_{14}^{16} &= e^{-2k}[J_{20}^{14}J_{20}^{16} + 12(k\varphi_1)^2 J_{21}^{14}J_{21}^{16} + 3J_{22}^{14}J_{22}^{16}]/n_{14}^{16},\\
{}^{20}f_8^{16} &= e^{-2k}[J_{20}^{8}J_{20}^{16} + 12(k\varphi_1)^2 J_{21}^{8}J_{21}^{16} + 3J_{22}^{8}J_{22}^{16}]/n_8^{16},
\end{aligned} \tag{E.12}$$

with the coefficients $n_p^{p'}$ determined from the equation

$$ {}^{20}f(k=0,\ \varphi_1=\varphi_3=\varphi_4=1,\ \varphi_2=0)=1 \tag{E.13}$$

(e.g, $n_8^8 = [J_{20}^8(0)]^2 + 3[J_{22}^8(0)]^2$) and the polynomials J_{2n}^p given in Table E2.

Table E2. Polynomials J_{2n}^p for the resonance functions ${}^{20}f_p^{p'}$.

$J_{20}^{8} = 16\varphi_4 J_4 - 3\varphi_1^2 J_5$	$J_{21}^{14} = \varphi_3 J_7 + 14\varphi_2 J_6$
$J_{20}^{10} = 20\varphi_4 J_5 - 3\varphi_1^2 J_6$	$J_{21}^{16} = \varphi_3 J_8 + 16\varphi_2 J_7$
$J_{20}^{14} = 28\varphi_4 J_7 - 3\varphi_1^2 J_8$	$J_{22}^{8} = \varphi_1^2 J_5 + 8\varphi_2 J_4$
$J_{20}^{16} = 32\varphi_4 J_8 - 3\varphi_1^2 J_9$	$J_{22}^{10} = \varphi_1^2 J_6 + 20\varphi_2 J_5$
$J_{21}^{8} = \varphi_3 J_4 + 8\varphi_2 J_3$	$J_{22}^{14} = \varphi_1^2 J_8 + 28\varphi_2 J_7$
$J_{21}^{10} = \varphi_3 J_5 + 10\varphi_2 J_4$	$J_{22}^{16} = \varphi_1^2 J_9 + 32\varphi_2 J_8$

The imaginary parts of resonance functions in the parabolic trajectory approximation were calculated in Ref. [2].

Bibliography

1. Robert D. and Bonamy J. (1979). Short range force effects in semiclassical molecular line broadening calculations. *J Phys (Paris)* 40: 923–943.
2. Lynch R., Gamache R.R. and Neshyba S.P. (1996). Fully complex implementation of the Robert–Bonamy formalism: half widths and line shifts of H_2O broadened by N_2. *J Chem Phys* 105: 5711–5721.

Appendix F

Resonance functions in the exact trajectory model

In the exact trajectory model, the real parts of the resonance functions corresponding to the long-range interactions are also calculated by Eq. (C.1); Eqs (2.10.7) and (2.10.8) are also valid. For the isotropic intermolecular potential in the Lennard–Jones form, Eq. (C.4) reduces to Eq. (2.10.10). The calculation of resonance functions can be made (see Refs [1, 2]) with the formulae of the straight-line trajectory approximation if the correspondence between the physical quantities of these formulae and the integrals of Eq. (2.10.10) is established (see Sec. 2.10). So calculated resonance functions can be further approximated [3] by the model functions $f_n^{model}(k, \lambda, \beta)$, as described in Sec. 2.11. An example of the tabulated resonance functions $g_3(k, \lambda, \beta) \equiv {}^{10}f_7^7$ is given in Table F1.

Table F1. Resonance function $g_3(k, \lambda, \beta) \equiv {}^{10}f_7^7$ in the exact trajectory model.

	$\beta = 0.3$					$\beta = 0.4$				
k	$\lambda = 0.5$	$\lambda = 1.0$	$\lambda = 2.0$	$\lambda = 3.0$	$\lambda = 4.0$	$\lambda = 0.5$	$\lambda = 1.0$	$\lambda = 2.0$	$\lambda = 3.0$	$\lambda = 4.0$
0.0	1.0007	1.0012	1.0021	1.0030	1.0039	1.0028	1.0053	1.0103	1.0154	1.0205
0.1	0.9991	0.9996	1.0005	1.0014	1.0023	1.0012	1.0037	1.0087	1.0137	1.0189
0.2	0.9944	0.9948	0.9957	0.9966	0.9975	0.9964	0.9989	1.0038	1.0089	1.0140
0.3	0.9864	0.9869	0.9878	0.9886	0.9895	0.9885	0.9909	0.9959	1.0009	1.0059
0.4	0.9755	0.9759	0.9768	0.9776	0.9785	0.9775	0.9799	0.9848	0.9897	0.9948
0.5	0.9616	0.9620	0.9628	0.9637	0.9646	0.9635	0.9659	0.9707	0.9756	0.9806
0.6	0.9448	0.9453	0.9461	0.9469	0.9478	0.9468	0.9491	0.9539	0.9587	0.9635
0.7	0.9255	0.9259	0.9267	0.9275	0.9284	0.9274	0.9297	0.9343	0.9390	0.9437
0.8	0.9037	0.9041	0.9049	0.9057	0.9065	0.9055	0.9077	0.9122	0.9168	0.9214
0.9	0.8796	0.8800	0.8808	0.8815	0.8823	0.8814	0.8835	0.8879	0.8924	0.8968
1.0	0.8535	0.8539	0.8547	0.8554	0.8562	0.8552	0.8573	0.8616	0.8658	0.8702
1.1	0.8257	0.8261	0.8268	0.8275	0.8282	0.8273	0.8294	0.8334	0.8375	0.8417
1.2	0.7964	0.7967	0.7974	0.7981	0.7988	0.7979	0.7999	0.8037	0.8077	0.8116
1.3	0.7658	0.7661	0.7668	0.7674	0.7681	0.7673	0.7691	0.7728	0.7765	0.7803
1.4	0.7342	0.7345	0.7351	0.7357	0.7364	0.7356	0.7373	0.7408	0.7443	0.7479
1.5	0.7019	0.7022	0.7028	0.7034	0.7039	0.7032	0.7049	0.7081	0.7114	0.7148
2.0	0.5378	0.5380	0.5384	0.5388	0.5392	0.5387	0.5398	0.5419	0.5441	0.5463
2.5	0.3866	0.3867	0.3869	0.3871	0.3873	0.3871	0.3876	0.3888	0.3900	0.3912

3.0	0.2626	0.2626	0.2627	0.2628	0.2628	0.2627	0.2630	0.2634	0.2638	0.2642
3.5	0.1697	0.1697	0.1697	0.1697	0.1697	0.1697	0.1697	0.1697	0.1696	0.1696
4.0	0.1050	0.1050	0.1050	0.1049	0.1049	0.1049	0.1048	0.1046	0.1043	0.1041
4.5	0.0626	0.0625	0.0625	0.0624	0.0624	0.0624	0.0623	0.0620	0.0617	0.0614
5.0	0.0360	0.0360	0.0359	0.0359	0.0359	0.0359	0.0358	0.0355	0.0352	0.0350
5.5	0.0201	0.0201	0.0201	0.0200	0.0200	0.0201	0.0199	0.0197	0.0195	0.0193
6.0	0.0110	0.0109	0.0109	0.0109	0.0109	0.0109	0.0108	0.0107	0.0105	0.0104
6.5	0.0058	0.0058	0.0058	0.0058	0.0058	0.0058	0.0057	0.0056	0.0055	0.0054
7.0	0.0030	0.0030	0.0030	0.0030	0.0030	0.0030	0.0030	0.0029	0.0028	0.0028
7.5	0.0015	0.0015	0.0015	0.0015	0.0015	0.0015	0.0015	0.0015	0.0014	0.0014

	$\beta = 0.5$					$\beta = 0.6$				
k	$\lambda = 0.5$	$\lambda = 1.0$	$\lambda = 2.0$	$\lambda = 3.0$	$\lambda = 4.0$	$\lambda = 0.5$	$\lambda = 1.0$	$\lambda = 2.0$	$\lambda = 3.0$	$\lambda = 4.0$
0.0	1.0096	1.0190	1.0384	1.0587	1.0798	1.0263	1.0537	1.1134	1.1807	1.2570
0.1	1.0079	1.0174	1.0368	1.0570	1.0781	1.0246	1.0520	1.1116	1.1789	1.2552
0.2	1.0031	1.0125	1.0318	1.0520	1.0730	1.0197	1.0470	1.1064	1.1735	1.2497
0.3	0.9951	1.0045	1.0237	1.0437	1.0645	1.0116	1.0387	1.0978	1.1645	1.2405
0.4	0.9841	0.9933	1.0123	1.0321	1.0528	1.0004	1.0272	1.0857	1.1520	1.2277
0.5	0.9700	0.9791	0.9979	1.0174	1.0378	0.9861	1.0126	1.0704	1.1361	1.2113
0.6	0.9532	0.9621	0.9805	0.9997	1.0198	0.9690	0.9950	1.0520	1.1168	1.1913
0.7	0.9336	0.9423	0.9604	0.9792	0.9989	0.9491	0.9746	1.0305	1.0942	1.1679
0.8	0.9116	0.9201	0.9377	0.9560	0.9752	0.9267	0.9515	1.0061	1.0686	1.1411
0.9	0.8873	0.8955	0.9126	0.9304	0.9491	0.9019	0.9261	0.9792	1.0402	1.1111
1.0	0.8610	0.8689	0.8854	0.9026	0.9206	0.8751	0.8984	0.9499	1.0090	1.0782
1.1	0.8328	0.8405	0.8563	0.8729	0.8902	0.8464	0.8688	0.9184	0.9755	1.0425
1.2	0.8032	0.8105	0.8256	0.8414	0.8581	0.8162	0.8376	0.8851	0.9399	1.0044
1.3	0.7722	0.7792	0.7936	0.8086	0.8244	0.7846	0.8050	0.8502	0.9025	0.9642
1.4	0.7403	0.7469	0.7605	0.7747	0.7897	0.7520	0.7713	0.8140	0.8636	0.9222
1.5	0.7077	0.7138	0.7266	0.7400	0.7540	0.7186	0.7368	0.7770	0.8237	0.8789
2.0	0.5416	0.5457	0.5541	0.5630	0.5723	0.5489	0.5609	0.5875	0.6183	0.6547
2.5	0.3886	0.3908	0.3954	0.4000	0.4049	0.3926	0.3990	0.4130	0.4288	0.4471
3.0	0.2633	0.2641	0.2657	0.2673	0.2690	0.2647	0.2670	0.2717	0.2766	0.2818
3.5	0.1697	0.1696	0.1695	0.1693	0.1690	0.1696	0.1693	0.1686	0.1673	0.1652
4.0	0.1046	0.1042	0.1032	0.1022	0.1011	0.1038	0.1024	0.0993	0.0954	0.0907
4.5	0.0620	0.0615	0.0603	0.0591	0.0578	0.0610	0.0594	0.0558	0.0517	0.0469
5.0	0.0355	0.0350	0.0340	0.0329	0.0318	0.0347	0.0332	0.0301	0.0267	0.0230
5.5	0.0198	0.0194	0.0186	0.0178	0.0169	0.0191	0.0180	0.0157	0.0132	0.0107
6.0	0.0107	0.0104	0.0099	0.0093	0.0087	0.0102	0.0094	0.0079	0.0063	0.0048
6.5	0.0056	0.0055	0.0051	0.0047	0.0044	0.0053	0.0048	0.0038	0.0029	0.0020
7.0	0.0029	0.0028	0.0026	0.0023	0.0021	0.0027	0.0024	0.0018	0.0013	0.0008
7.5	0.0015	0.0014	0.0013	0.0011	0.0010	0.0014	0.0008	0.0006	0.0003	0.0001

	$\beta = 0.7$					$\beta = 0.8$				
k	$\lambda = 0.5$	$\lambda = 1.0$	$\lambda = 2.0$	$\lambda = 3.0$	$\lambda = 4.0$	$\lambda = 0.5$	$\lambda = 1.0$	$\lambda = 2.0$	$\lambda = 3.0$	$\lambda = 4.0$
0.0	1.0548	1.1157	1.2619	1.4514	1.6983	1.0638	1.1348	1.3014	1.4880	1.5193
0.1	1.0531	1.1140	1.2601	1.4499	1.6984	1.0621	1.1331	1.3001	1.4890	1.5336
0.2	1.0481	1.1088	1.2547	1.4454	1.6987	1.0572	1.1281	1.2962	1.4917	1.5752
0.3	1.0398	1.1002	1.2458	1.4377	1.6986	1.0490	1.1199	1.2896	1.4955	1.6408
0.4	1.0284	1.0882	1.2334	1.4267	1.6972	1.0376	1.1084	1.2802	1.4996	1.7251
0.5	1.0138	1.0730	1.2173	1.4121	1.6936	1.0232	1.0937	1.2676	1.5026	1.8213
0.6	0.9962	1.0546	1.1978	1.3938	1.6866	1.0057	1.0758	1.2518	1.5034	1.9218
0.7	0.9758	1.0332	1.1747	1.3714	1.6751	0.9854	1.0549	1.2326	1.5004	2.0187

0.8	0.9528	1.0089	1.1483	1.3449	1.6579	0.9625	1.0311	1.2098	1.4925	2.1045
0.9	0.9273	0.9820	1.1187	1.3141	1.6342	0.9370	1.0045	1.1833	1.4786	2.1725
1.0	0.8997	0.9527	1.0860	1.2792	1.6034	0.9094	0.9754	1.1532	1.4579	2.2175
1.1	0.8701	0.9213	1.0505	1.2401	1.5651	0.8797	0.9440	1.1195	1.4298	2.2357
1.2	0.8389	0.8879	1.0125	1.1972	1.5191	0.8483	0.9105	1.0825	1.3942	2.2253
1.3	0.8062	0.8530	0.9724	1.1508	1.4659	0.8155	0.8753	1.0425	1.3513	2.1861
1.4	0.7725	0.8169	0.9303	1.1012	1.4059	0.7815	0.8386	0.9997	1.3013	2.1194
1.5	0.7379	0.7797	0.8868	1.0490	1.3398	0.7466	0.8007	0.9546	1.2451	2.0280
2.0	0.5618	0.5896	0.6607	0.7679	0.9551	0.5682	0.6051	0.7107	0.9049	1.3460
2.5	0.3995	0.4142	0.4506	0.5017	0.5783	0.4032	0.4231	0.4772	0.5615	0.6686
3.0	0.2672	0.2722	0.2831	0.2943	0.2992	0.2688	0.2759	0.2916	0.3027	0.2592
3.5	0.1694	0.1687	0.1653	0.1565	0.1334	0.1697	0.1691	0.1642	0.1450	0.0828
4.0	0.1024	0.0992	0.0903	0.0760	0.0513	0.1021	0.0983	0.0863	0.0632	0.0237
4.5	0.0594	0.0557	0.0465	0.0339	0.0171	0.0589	0.0545	0.0428	0.0257	0.0069
5.0	0.0332	0.0300	0.0227	0.0140	0.0048	0.0328	0.0291	0.0203	0.0100	0.0023
5.5	0.0179	0.0156	0.0106	0.0054	0.0011	0.0177	0.0150	0.0093	0.0039	0.0009
6.0	0.0094	0.0078	0.0047	0.0019	0.0002	0.0093	0.0075	0.0041	0.0015	0.0004
6.5	0.0048	0.0038	0.0020	0.0006	0.0000	0.0047	0.0037	0.0018	0.0006	0.0002
7.0	0.0024	0.0018	0.0008	0.0002	0.0000	0.0024	0.0018	0.0008	0.0002	0.0001
7.5	0.0012	0.0008	0.0003	0.0001	0.0000	0.0012	0.0008	0.0003	0.0001	0.0000

	$\beta = 0.9$					$\beta = 1.0$				
k	$\lambda = 0.5$	$\lambda = 1.0$	$\lambda = 2.0$	$\lambda = 3.0$	$\lambda = 4.0$	$\lambda = 0.5$	$\lambda = 1.0$	$\lambda = 2.0$	$\lambda = 3.0$	$\lambda = 4.0$
0.0	0.9238	0.8623	0.7690	0.7005	0.6468	0.5701	0.4125	0.2730	0.2069	0.1677
0.1	0.9224	0.8611	0.7681	0.6998	0.6464	0.5693	0.4120	0.2727	0.2067	0.1676
0.2	0.9183	0.8575	0.7653	0.6978	0.6452	0.5670	0.4105	0.2719	0.2062	0.1671
0.3	0.9114	0.8515	0.7606	0.6944	0.6431	0.5632	0.4081	0.2705	0.2052	0.1664
0.4	0.9019	0.8431	0.7541	0.6895	0.6400	0.5580	0.4047	0.2686	0.2039	0.1654
0.5	0.8898	0.8325	0.7458	0.6832	0.6358	0.5514	0.4004	0.2661	0.2022	0.1642
0.6	0.8752	0.8196	0.7356	0.6753	0.6304	0.5434	0.3953	0.2632	0.2002	0.1627
0.7	0.8583	0.8046	0.7237	0.6659	0.6235	0.5342	0.3894	0.2599	0.1979	0.1609
0.8	0.8391	0.7876	0.7100	0.6549	0.6152	0.5239	0.3828	0.2561	0.1953	0.1590
0.9	0.8180	0.7688	0.6946	0.6423	0.6053	0.5126	0.3755	0.2519	0.1925	0.1568
1.0	0.7950	0.7482	0.6777	0.6282	0.5938	0.5004	0.3676	0.2475	0.1894	0.1545
1.1	0.7704	0.7261	0.6593	0.6127	0.5807	0.4873	0.3592	0.2427	0.1862	0.1521
1.2	0.7443	0.7026	0.6397	0.5958	0.5661	0.4736	0.3504	0.2377	0.1827	0.1495
1.3	0.7171	0.6780	0.6188	0.5777	0.5501	0.4592	0.3412	0.2324	0.1791	0.1468
1.4	0.6889	0.6524	0.5970	0.5585	0.5327	0.4445	0.3317	0.2270	0.1754	0.1441
1.5	0.6600	0.6260	0.5744	0.5384	0.5143	0.4293	0.3219	0.2215	0.1717	0.1412
2.0	0.5118	0.4900	0.4556	0.4305	0.4125	0.3516	0.2716	0.1928	0.1520	0.1265
2.5	0.3734	0.3616	0.3413	0.3246	0.3108	0.2772	0.2227	0.1644	0.1324	0.1117
3.0	0.2585	0.2538	0.2441	0.2343	0.2245	0.2120	0.1786	0.1381	0.1139	0.0977
3.5	0.1711	0.1711	0.1686	0.1642	0.1584	0.1584	0.1409	0.1148	0.0973	0.0849
4.0	0.1091	0.1117	0.1137	0.1131	0.1108	0.1162	0.1099	0.0948	0.0827	0.0735
4.5	0.0675	0.0711	0.0754	0.0772	0.0774	0.0840	0.0850	0.0779	0.0701	0.0635
5.0	0.0407	0.0444	0.0495	0.0526	0.0543	0.0601	0.0653	0.0638	0.0593	0.0548
5.5	0.0240	0.0273	0.0324	0.0359	0.0383	0.0427	0.0499	0.0521	0.0500	0.0472
6.0	0.0140	0.0167	0.0211	0.0245	0.0271	0.0301	0.0381	0.0425	0.0422	0.0406
6.5	0.0080	0.0101	0.0138	0.0168	0.0193	0.0212	0.0290	0.0347	0.0356	0.0350
7.0	0.0046	0.0061	0.0090	0.0116	0.0138	0.0148	0.0220	0.0282	0.0300	0.0301
7.5	0.0026	0.0037	0.0059	0.0080	0.0098	0.0103	0.0167			

	$\beta = 1.02$					$\beta = 1.08$		$\beta = 0.0$
k	$\lambda = 0.5$	$\lambda = 1.0$	$\lambda = 2.0$	$\lambda = 3.0$	$\lambda = 4.0$	$\lambda = 0.5$	$\lambda = 1.0$	$\lambda = 0.0$
0.0	0.4957	0.3424	0.217	0.1608	0.1283	0.3066	0.188	1.000
0.1	0.4950	0.3420	0.216	0.1606	0.1282	0.3063	0.188	0.998
0.2	0.4931	0.3408	0.216	0.1602	0.1279	0.3053	0.188	0.993
0.3	0.4900	0.3389	0.215	0.1596	0.1275	0.3037	0.187	0.985
0.4	0.4856	0.3363	0.213	0.1586	0.1268	0.3015	0.186	0.974
0.5	0.4801	0.3330	0.212	0.1575	0.1259	0.2987	0.184	0.961
0.6	0.4735	0.3291	0.209	0.1561	0.1249	0.2953	0.183	0.944
0.7	0.4659	0.3245	0.207	0.1545	0.1238	0.2914	0.181	0.925
0.8	0.4574	0.3193	0.204	0.1527	0.1225	0.2871	0.179	0.903
0.9	0.4480	0.3137	0.201	0.1508	0.1210	0.2823	0.176	0.879
1.0	0.4379	0.3076	0.198	0.1486	0.1195	0.2771	0.174	0.853
1.1	0.4271	0.3012	0.195	0.1464	0.1178	0.2716	0.171	0.825
1.2	0.4157	0.2943	0.191	0.1440	0.1161	0.2658	0.168	0.796
1.3	0.4039	0.2872	0.187	0.1415	0.1143	0.2598	0.165	0.765
1.4	0.3917	0.2799	0.183	0.1390	0.1125	0.2536	0.162	0.734
1.5	0.3792	0.2724	0.179	0.1364	0.1105	0.2472	0.159	0.701
2.0	0.3149	0.2335	0.159	0.1227	0.1006	0.2141	0.142	0.537
2.5	0.2529	0.1953	0.138	0.1089	0.0905	0.1814	0.125	0.386
3.0	0.1978	0.1604	0.118	0.0958	0.0807	0.1510	0.109	0.262
3.5	0.1516	0.1299	0.101	0.0836	0.0716	0.1242	0.094	0.169
4.0	0.1145	0.1043	0.085	0.0727	0.0634	0.1012	0.081	0.105
4.5	0.0854	0.0831	0.072	0.0631	0.0559	0.0819	0.069	0.062
5.0	0.0631	0.0658	0.060	0.0546	0.0493	0.0659	0.059	0.036
5.5	0.0464	0.0520	0.051	0.0472	0.0433	0.0528	0.050	0.020
6.0	0.0339	0.0409	0.042	0.0407	0.0381	0.0422	0.042	0.010
6.5	0.0247	0.0322	0.035	0.0351	0.0335	0.0336	0.036	0.005
7.0	0.0179	0.0252	0.029	0.0303	0.0294	0.0268	0.030	0.003
7.5	0.0130	0.0198	0.025	0.0261	0.0258	0.0213	0.025	0.001

Bibliography

1. Lavrentieva N.N. and Starikov V.I. (2006). Approximation of resonance functions for exact trajectories in the pressure-broadening theory. Real parts. *Mol Phys* 104: 2759–2766.
2. Lavrentieva N.N. and Starikov V.I. (2005). Approximation of resonance functions for real trajectories in the pressure-broadening theory. *Atmos Oceanic Opt* 18: 814–819.
3. Starikov V.I. (2008). Calculation of the self-broadening coefficients of water-vapor absorption lines using exact trajectory model. *Opt Spectrosc* 104: 513–523.

Index